ALGAL FLORA OF SUNDARBANS MANGALS

ALGAL FLORA OF SUNDARBANS MANGALS

Dr. Neera Sen Ph.D
&
Dr. Kumudranjan Naskar Ph.D

2003
Daya Publishing House
Delhi-110 035

ISBN 81-7035–286–X

Published by : **Daya Publishing House**
1123/74, Deva Ram Park
Tri Nagar, Delhi-110 035
Phone : 7103999
Fax : (011) 7199029
e-mail : dayabooks@vsnl.com
website : www.dayabooks.com

Showroom : 4762-63/23, Ansari Road
New Delhi-110002
Ph. : 3245578, 3244987

Laser composed by : P. K. Jain & Co., New Delhi

Printed at : Chawla Offset Printers, Delhi-110 052

Dedicated to our Dearest Parents

Nikhilesh Chandra Sen and Krishna Sen	Late Ram Sadhan Naskar and Late Tulasi Bala Naskar

Acknowledgement

The authors wish to express their sincere and warmest gratitude to the Indian Council of Agricultural Research for providing the research grants for the ongoing National Fellow Scheme, but for which, such an extensive study on the algal flora of Sundarbans would not have been possible. Sincere appreciation is made for the constant encouragement and guidance of Dr. D.N. Guha Bakshi, Ex-Deputy Director, Botanical Survey of India. We are also grateful to Prof. S.C. Santra, Professor, Department of Environmental Science, University of Kalyani, Prof. D.K. Bhattacharya, Professor, Department of Oil Technology, University of Calcutta and Dr. B.K. De, Lecturer, Institute of Science and Technology, Ahmedabad for their personal interest and valuable guidelines in the research work. Sincere acknowledgement must be made of the constant encouragement and logistic support provided by Dr. G.G. Maiti, Reader and Head, Department of Botany, University of Kalyani, Dr. G.N. Chattopadhyay, Soil Scientist, Department of Agriculture, Viswabharati University, Dr. A. Mahapatra, Principal Scientist, Central Institute of Jute and Allied Fibres, Mrs. Sandhya Mahapatra, Ex-Deputy Director, Botanical Survey of India. Dr. A. K. Ghosh, Retd. Principal Scientist, Central Inland Fisheries Research Station. Dr. P. K. Pandit, Retd. Principal scientist, Central Inland Fisheries Research Station, Dr. D. Nath, Principal Scientist, Central Inland Fisheries Research Institute Dr. A. Ghosh, Principal Scientist, Central Inland Fisheries Research Station and Mr. H.C. Karmakar, Principal Scientist, Central Inland Fisheries Research Station.

Special acknowledgement is made of the help rendered by Mr. S. Mukherjee, Conservator of Forest, Department of Forest, Govt. of West Bengal, Mr. A. Sarkar, Lecturer, K.N. College, Berhampore, Mrs. Dolanchampa Sarkar, Senior Soil Surveyor, AISLUS, Govt. of India, Mr. Ashok Ghosh, Research Fellow, Mrs. Manjistha Dasgupta, Research Fellow, Dr. R.N. Mandal, Scientist, Central Institute of Freshwater Aquaculture and Mr. A. Bhunia, Research Fellow. We are also indebted to Mr. R. Das for his yeoman service, Mr. D. Saha, Mr. A. Barui and Mr. J. Patra for their active cooperation. Special mention must be made of Mr. Dinataran Sarkar and Mrs. Usha Rani Sarkar of Masjidbati, Sundarbans for their constant cooperation during our field visits to Sundarbans.

Special and heartfelt gratitude goes out to our family members Mr. Nikhilesh Chandra Sen, Mrs. Krishna Sen, Mr. Tushar Sengupta, Mr. Bijan Sarkar, Mrs. Sipra Naskar, Mrs. Biva Naskar, Mr.Soumya Naskar and Mr. Saikat Naskar, for their constant help, encouragement and support.

Last but not least, sincere gratitude goes out to our publisher Mr. Anil Mittal, Daya Publishing House, New Delhi for their enthusiastic zeal in publication of the book.

Neera Sen
Kumudranjan Naskar

Preface

The mangrove ecosystem in the Indian Sundarbans has great role in the estuarine fish production and sustenance of large number of aquatic, amphibious and terrestrial life forms; this is because an ideal mangrove habitat is known to release more than 6000 tons litter per hectare per annum (on dry weight basis) and this huge volume of organic matter after complete decomposition or mineralisation releases nutrients to the estuarine, in-shore and off-shore water and soil phases. As such, the soil and water in the estuaries and deltatic regions dominated with dense mangrove vegetation are considered as one of the most nutrient rich and productive areas among the tropical forest zones. The Sundarbans, dominated with large number of mangrove associated and major and minor mangal species along with vast number of aquatic and terrestrial mangrove habitat faunal assemblage are categorized as the *World Heritage Site* by IUCN and UNESCO (1984). It is worth mentioning that though the mangrove ecosystem is governed by a large number of biotic components, the algae have a very major role as the primary producer group of organisms of the first order, both in the aquatic, as well as, in the terrestrial habitats. In this mangrove habitat the algal species in the form of planktonic, benthic and periphytic vegetation have been overlooked and find insignificant mention in discussions pertaining to the Sundarbans mangals of India, but their impact and importance is very notable and is a subject of high scientific value for investigation. These algal species in the saline estuarine water or deltaic forest land maintain different entities from their fresh water counterparts and also play the most important role of harnessing the solar energy and fixing inorganic substances of the sea water and silty soil in this newly formed areas, consequently transforming them into available important food for the consumer group of aquatic, as well as, terrestrial fauna of this estuarine-delta sea land interphase zone. For all these reasons it was found imperative to collect and identify the different algal flora of this less explored mangrove habitat of Indian Sundarbans and to find out their direct and indirect use and impact as fish food and development of fishery activities in this amply bestowed fish cradle.

Taking these propositions as priority areas, exploration, identification and estimation of the nutrient value of dominant algal flora of Sundarbans have been carried out and based on these field and laboratory studies the collected information and data have been highlighted in the present book.

Neera Sen
Kumudranjan Naskar

Contents

		Pages
	Acknowledgement	*vii*
	Preface	*ix*
1.	Introduction	1
2.	Geology and Geography of Indian Sundarbans	35
3.	Factors Governing Habitats of Algae	49
4.	Phycological Studies : India & Bangladesh	51
5.	Field and Laboratory Methodologies for the Investigation of Algae	55
6.	Taxonomy and Morphology	71
7.	Ecology	201
8.	Applied Phycology	247
	References	270
	Algal Species Index	313

LIST OF TABLES

		Pages
Table 1	Total Area in the Sundarbans Biosphere Reserve in India	3
Table 2	The World Mangrove Zones	5
Table 3	The World Mangrove Species	7
Table 4	Indian Mangrove Zones/Areas, Species Composition & Important Associated Rivers/Deltas/Estuaries	10
Table 5	Major and Minor Elements of Mangroves & Mangrove Associates of Indian Sundarbans	14
Table 6	The Faunal Habitats in the Indian Sundarbans Mangals	25
Table 7	Distribution of Mangrove Inhabiting Fauna in Relation to their Habitats	26
Table 8	Shell-fish & Fin-fish Species of Economic Importance from the Estuaries and Off-shores of Indian Sundarbans	28
Table 9	List of Extinct & Threatened/ Endangered Fauna in the Indian Sundarbans	30
Table 10	Climate of the Sundarbans in West Bengal	37
Table 11	Population in the Rural Blocks of Mangrove Reclaimed Areas of Indian Sundarbans	39

		Pages
Table 12	Devastating Cyclones, Storms, Famines & Other Natural Calamities Since 1582 to 1991 Recorded form Sundarbans	42
Table 13	Chemical Properties of Estuarine Water in Sundarbans (Zone-I, Eastern Part of the Sundarbans Mangal)	47
Table 14	Chemical Properties of Estuarine Water in Sundarbans (Zone-II, Central Part of the Sundarbans Mangal)	47
Table 15	Chemical Properties of Estuarine Water in Sundarbans (Zone-III, Western Part of the Sundarbans Mangal)	48
Table 16	Habitat of Identified Cyanophyta Members from Indian Sundarbans	107
Table 17	Habitat of Identified Chlorophyta Members from Indian Sundarbans	145
Table 18	Habitat of Identified Chrysophyta & Phaeophyta Members from Indian Sundarbans	180
Table 19	Habitat of Identified Rhodophyta Members from Indian Sundarbans	199
Table 20	Distribution, Abundance & Habitat Ecology of the Studied Algal Species from Indian Sundarbans	203
Table 21	Range of Physico-chemical Parameters & Algal Species Encountered in Zone-I, During Different Seasons	217

		Pages
Table 22	Range of Physico-chemical Parameters & Algal Species Encountered in Zone-II, During Different Seasons	220
Table 23	Range of Physico-chemical Parameters & Algal Species Encountered in Zone-III, During Different Seasons	223
Table 24	SDI, Species Richness, Chlorophyll–a Content, Light Intensity & Primary Productivity of Water Samples from Estuarine Wetlands During Summer	230
Table 25	SDI, Species Richness, Chlorophyll–a Content, Light Intensity & Primary Productivity of Water Samples from Estuarine Wetlands During Monsoon	231
Table 26	SDI, Species Richness, Chlorophyll–a Content, Light Intensity & Primary Productivity of Water Samples from Estuarine Wetlands During Post-Monsoon	232
Table 27	SDI, Species Richness, Chlorophyll–a Content, Light Intensity & Primary Productivity of Water Samples from Estuarine Wetlands During Winter	233
Table 28	List of Zooplankters Collected & Identified from the Indian Sundarbans	236
Tabel 29	Average Physico-chemical Parameters and Net Primary Productivity in Different Stretches of an Estuarine River of Indian Sundarbans During Different Seasons	238

		Pages
Table 30	Bio-Chemistry of Selected Common Macrophytic Algae from Sundarbans	248
Table 31a	Fatty Acid Composition & Total Lipids of the Macrophytic Algal Species	250
Table 31b	Fatty Acid Composition & Total Lipids of the Macrophytic Algal Species	251
Table 31c	Fatty Acid Composition & Total Lipids of the Macrophytic Algal Species	252
Table 32	Food & Feeding Habits of Some Common Estuarine and Brackish Water Shell-fish and Fin-Fish Species	266

MAPS

Map 1	Indian Mangrove Zones	12
Map 2	River System in Indian Sundarbans in West Bengal	36
Map 3	Different Rural Blocks in Indian Sundarbans, West Bengal	41
Map 4	The Eastern, Central & Western Zones of Indian Sundarbans, Demarcated for the Present Study	56
Map 5	Sampling Sites in the Indian Sundarbans	59

FIGURES

		Pages
Figure 1	Abundance of Mangroves in Different Parts of the Indian Sundarbans	18
Figure 2	Compartmental Model of the Food Chains Operating in the Indian Sundarbans Mangals	31
Figure 3	Trophic Levels in the Indian Sundarbans	34

Chapter 1

Introduction

1.1. Sundarbans Mangrove Ecosystem

Sundarbans, the largest single chunk of prograding delta at the mouth of the Ganga-Brahmaputra estuary is a unique ecosystem in more than a single perspective. Besides being the only mangrove tiger-land on the face of this earth and a mangrove area with the highest floral diversity, it is the most intriguing and bewildering virgin land, a lot of which is yet to be explored and unfolded. The fact that IUCN has declared it a *World Heritage Site* substantiates the claim that Sundarbans is indeed, in need of a lot of care and good attention, which would go into the protection and conservation of this precious biodiversity rich ecosystem. There has been a lot of interpretation, as well as, misinterpretation during the past few decades regarding the area coverage of Sundarbans in West Bengal, India. According to Naskar and Guha Bakshi (1987) the Sundarbans mangrove ecosystem in India is spread over an area of 4,266.6 sq. km. in the south-eastern estuarine tracts of the Ganga-Brahmaputra system. In the Indian Sundarbans, six major estuarine rivers, *viz.,* Baratala or Muri Ganga, Saptamukhi, Thakuran or Jamira, Matla, Goasaba and Herobhanga or Harinbhanga meet the Bay of Bengal in their southern mouths and are inter connected with each other through numerous criss-cross creeks, canals and small rivers creating about 110 islands, of which around 56 islands are still part of the Sundarbans mangrove ecosystem and the rest have been cleared, reclaimed and converted to human habitation areas. These tidal estuarine rivers carry the sea-water from the Bay during high tide and inundate the mangrove forests at regular intervals. The river Hugli (Ganga) in the extreme west is the only river carrying freshwater from upstream of lower Ganga delta in Indian Sundarbans, since most of the other estuarine rivers have lost their earlier connections with the river Ganga since 12th century and due to the neo-tectonic movement of the river Ganga towards east (Morgan & Mc Intire, 1959).

With a view to conserve this fragile ecosystem along with its vulnerable biodiversity, a few major projects were initiated in the early 1970's. In 1973, the Sundarbans Development Board was formed with the Head Quarter at Mayukh Bhavan, Salt Lake, Calcutta - 700 091 along with several branches in different parts of Sundarbans with a view to undertake developmental activities particularly agriculture, fisheries and possible socio - economic developmental works in the mangrove reclaimed areas of Indian Sundarbans. In the same year, *i.e.,* 1973, about 2,585 sq. km. of the south-eastern part of Indian Sundarbans was declared as 'Sundarbans Tiger Reserve' and in 1976 in an attempt to save the estuarine crocodile, *Crocodylus porosus* the 'Crocodile Breeding and Rearing Project' was initiated at Bhagbatpur in a 30 ha inter-tidal land under Namkhana Range. Within the 2,585 sq. km. of the 'Sundarbans Tiger Reserve' area about 1,330 sq. km. inter-tidal mangrove forest is demarcated as Core Area which falls under strict management measures and the rest 1,255 sq. km. area is demarcated as *Buffer Zone,* where limited forestry and fishing activities are allowed with prior permission from the Forest Department or Sundarbans Tiger Reserve. In 1976, within the buffer zone an area of 304 sq. km. of Sajnekhali was declared as *Wildlife Sanctuary* and outside the Sundarbans Tiger Reserve an area of 38 sq. km. at Lothian Island and 5.85 sq. km area at Holiday Island were also declared as Wildlife Sanctuaries with the objective to conserve and maintain the habitat of wild deer, wild boar, different species of mangrove habitat flora and fauna. In 1984, only 1,330 sq. km. area of the core area of Sundarbans Tiger Reserve was declared as *National Park* for protecting the interesting and vulnerable flora and fauna of the mangrove areas. In 1989, the Sundarbans was included under the *Man and Biosphere Reserve* programme and the entire 9,630 sq. km. area of Sundarbans was taken under this Sundarbans Biosphere Reserve Project (Table 1).

The genesis and development of this ecosystem has been and is still a factor of interaction among the three major components ruling this ecosystem, *viz.*, waterways, islands and the forests. The dynamic relationship in various facets of these three major components with the human group has been a deciding factor of the fate of these mangals. On one hand Sundarbans has been bestowed with the highest flora, marsh and swamp flora (Naskar & Guha Bakshi, 1987, Naskar, 1993 and Naskar & Mandal, 1999) and important faunal components, *viz.,* the Royal Bengal Tigers, spotted deers, estuarine crocodiles, fishing cats, number of local

Table 1 : Total Area in the Sundarbans Biosphere Reserve in India

Identity of the Geographical Area	*Covered Area*
Total Area under the Sundarbans Biosphere Reserve, West Bengal	9630.0 sq. km.
Sundarbans Mangrove Ecosystem, Ganga delta, India	4266.6 sq. km.
Total Forest Land in Indian Sundarbans (55% of total)	2347.0 sq. km.
Total Water-shed in Indian Sundarbans (45% of total)	1920.0 sq. km.
Sundarbans Tiger Reserve area in India	2585.1 sq. km.
Mangrove Forest Area outside S.T.R.	1681.5 sq. km
Mangrove Reclaimed Human Habitat Zones	5363.8 sq. km.

(Source : Naskar, 2000)

and migratory birds, numerous shell-fish and fin-fish species and on the other it is unfortunately an ecosystem on which depends the life of a huge and evergrowing population. This ecosystem provides food, shelter and livelihood in many direct and indirect ways to the sprawling population of about 40 lakhs (32 lakhs according to the 1991 Census) inhabiting the Sundarbans of West Bengal in particular and providing a number of commercially viable and potentially exportable commodities, *viz.*, timber, honey, wax, seafood etc. to the whole nation in general (Naskar, 1988).

Due to its immense value and potential resources, the Sundarbans have been getting a lot of attention from unwanted and undesirable sectors which have 'sustainable development' and 'wise exploitation for long term benefits' as the last thing on their mind. Sundarbans have to be made bereft of such attention to save this glorious gift and instead a lot of research with a holistic approach is needed to understand this ecosystem with all its potentials for conserving it as it was gifted to us for the coming generations.

The photosynthetic kingdom responsible for the sustenance of this unique ecosystem comprises mainly of the mangrove vegetation and the other photosynthetic organisms in the form of varied algal groups. Though the mangroves are a well known group of plants, the initial and primary role in maintaining the sustainability of the ecosystem is played by the varied groups of algae which must claim their position as subject of immense importance due to their role in paving way for succession of higher flora and also as food for the innumerable faunal components thriving in the marshy and saline areas of mangrove regions. The stricking fact is that, as the Indian Sundarbans coastal regions are impregnated with high silt load along with strong tidal current the growth and regeneration of the sea grass or the coral reef ecosystem are completely absent. Only, one dominant submerged species of flowering plants, *viz. Ruppia maritima* L. (Rupiaceae) commonly grows as the submerged species in the stagnant or lower current influenced brackishwater areas of the Indian Sundarbans. At times, *R. maritima* L. is accompanied by *Enteromorpha* spp., *Chaetomorpha* spp. and *Lola* spp. in the brackishwater bheries (fisheries) of the Sundarbans and in the long run they increase the gross productivity of the fisheries they inhabit. Besides these, on decomposition the *Ruppia* sp. and the algal flora release nutrients to the soil and water phases of these fisheries and are sometimes preferred by the brackishwater fish and shrimp species as food.

1.2. Mangrove Ecosystems of the World

The mangrove ecosystem or the mangals are restricted in the tropical and sub-tropical zones of the world mainly within the latitudes 32°00′ N (Japan) and 33°00′ (S.E. Australia) (Macintosh, 1984); but in Australia, mangroves have been reported from Victoria Coast at Corner Inlet extended over 38°00′ S to 45°00′ S latitudes (*Avicennia marina* Vierh.), this forms the southern limit of the world distribution of mangroves (Tomlinson, 1986). The dominant mangrove zones in the world are found to be distributed in South - East Asia, North - East Australia and South - East Africa between the longitudes 30°00′ E and 165°00′ E (Macintosh, 1984), these form the mangroves of the Old World. In the New World, mangroves are distributed in North America at Louisiana, Pacific Coasts of North - West Mexico, Bermuda Islands and Pacific Coasts of South America (Macintosh, 1984). Some sporadic mangals have also been reported by different workers from the West Coast of Africa and East Coast of South America, *i.e.,* on Atlantic Ocean. The actual area under the world mangrove ecosystem have not been well estimated or are not unanimously accepted by experts all over the world; the main reason for such a disparity being the fact that the definitions of mangroves and mangals are both highly variable among the different experts in this field. The Indo - West Pacific mangroves or the Old World mangroves are estimated to be spread over an area of 1,02,631 sq. km. and the New World mangroves over an area of 71,029 sq. km (Macintosh, 1984; UNESCO, 1984; IUCN, 1984). Naskar and Mandal (1999) have given an elaborate account of the world distribution of mangroves. The World mangroves zones, both Old World and New World have been tabulated (Table 2).

OLD WORLD MANGALS

Table 2 : The World Mangrove Zones

Sl. No.	Country/Zones of Mangroves	Areas under Mangals (Km^2)	Latitudes	Longitudes
1	Indonesia	42,150	6°N - 11°S	95o E - 141°E
2	Australia	11,500	8°S - 38° 45′S	115°E - 155°E
3	Bangladesh	5,800	21°N - 22°31′N	89°10′E - 92°15′E
4	India	6,749	7°N - 23°N	69°E - 89°5′E
5	Thailand	2,873	6°N - 13°13′N	99°E - 103°E
6	Burma	5,200	15°N - 20°N	93°E - 97°E
7	Papua New Guinea	4,116	4°S - 10°S	141°E - 156°E
8	Philippines	2,321	5°N - 18°N	119°30′E - 126°E
9	Sabah	3,503	4°N - 7°N	115°E - 119°E
10	Madagascar Rapublic	3,207	11°S - 24°30′S	43°30′E - 50°20′E
11	Vietnam	2,800	9°N - 21° N	104°E - 108°E
12	Pakistan	2,495	24°10′N -25°37′N	61°38′E - 68°10′E
13	Sarawak	1,728	1°20′S - 5°N	110°E - 115°E
14	Peninsular Malaysia	1,055	1°30′N - 7°N	100°30′E - 104°10′E
15	Mozambique	850	10°27′S - 26°57′S	32°E - 40°30′E
16	Kenya	530	1°S - 4°30′S	39°30′E - 41°E
17	Kampuchea	500	10°10′N - 12°N	103°E - 104°30′E
18	Tanzania	1,155	5°S - 10°15′S	39°E - 41°E
19	New Zealand	198	35°S - 38°S	173°E - 178°E
20	Fiji	197	16°S - 20°S	178°E - 180°E
21	Sri Lanka	32	6°10′N - 9°50′N	80°E - 82°E
22	Singapore	5	1°N - 3°N	102°30′E - 104°10′E
23	Hongkong	65	22°30′N - 23°N	113°30′E - 114°E
24	South Africa	negligible	35°S - 24°S	20°E - 31°E
25	Other Pacific Islands	100	35°N - 35°S	120°E - 120°W
26	Maldives	negligible	4°S - 7°N	73°E - 75°E
27	Mauritius	20	19°59′S - 20°32′S	57°18′E - 57°47′E
28	Seychelles	negligible	4°33′S - 4°47′S	55°22′E - 55°32′E

(Source : Naskar & Mandal, 1999)

Contd...

Tabl 2 : The World Mangrove Zones

	Sl. No.	Country/Zones of Mangroves	Areas under Mangals (Km^2)	Latitudes	Longitudes
NEW WORLD MANGALS	1	Brazil	25,000	0° - 30°S	30°W - 52°W
	2	Cameroon	2,000	2°N - 5°N	8°E - 10°E
	3	Columbia	4,400	1°30′N - 12°N	72°W - 74°W
	4	Ecuador	2,159	1°20′N - 3°S	79°30′W - 81°W
	5	Gabon	2,500	0°30′N - 1°N	6°30′E - 10°E
	6	Mexico	6;600	15°N - 31°N	87°W - 114°W
	7	Nigeria	9,730	5°N - 5°30′N	5°E - 9°E
	8	Panama	4,860	7°N - 9°N	78°W - 82°W
	9	Senegal	5,000	12°N - 17°N	16°W - 17°W
	10	U.S.A.	2,050	35°N - 26°N	60°W - 121°W
	11	Venezuela	6,730	9°N - 12°N	60°W - 72°10′W

(Source : Naskar & Mandal, 1999)

Table 3 : The World Mangrove Species *(Modified After Tomlinson, 1986).*

Sl. No.	*Old World Mangroves*	*Sl. No.*	*New World Mangroves*
*1	*Rhizophora apiculata* Blume	1	*Rhizophora harrisonii* Leechman
*2	*Rhizophora mucronata* Lamk.	2	*Rhizophora mangle* L.
3	*Rhizophora stylosa* Griff.	3	*Rhizophora racemosa* Meyer
4	*Rhizophora samoensis* Salveza	4	*Avicennia germinans* Stearn
5	*Rhizophora X lamarckii* Montr.	5	*Avicennia schaueriana* Stapf. & Lee
6	*Rhizophora X selala* Tomlinson	6	*Avicennia bicolor* Standley
*7	*Bruguiera gymnorhiza* Lamk.	7	*Laguncularia racemosa* Gaertn. f.
*8	*Bruguiera parviflora* Wt. & Arn.	8	*Paelliciera rhizophoreae* Tri & Pin.
*9	*Bruguiera sexangula* Poir		
*10	*Bruguiera cylindrica* Blume		
11	*Bruguiera hainesii* Rogers		
12	*Bruguiera exaristata* Ding Hou		
*13	*Ceriops tagal* Robinson		
*14	*Ceriops decandra* Ding Hou		
*15	*Kandelia candel* Druce		
*16	*Avicennia alba* Blume		
*17	*Avicennia officinalis* L.		
*18	*Avicennia marina* Vierh.		
19	*Avicennia lanata* Ridley		
20	*Avicennia eucalyptifolia* Mold.		
*21	*Aegiceras corniculatum* Blanco		
22	*Aegiceras floridum* Roem. & Schult.		
*23	*Aegialitis rotundifolia* Roxb.		
*24	*Sonneratia apetala* Buch. Ham.		
*25	*Sonneratia grifithii* Kurz.		
*26	*Sonneratia caseolaris* Engler		
*27	*Sonneratia alba* Smith		
28	*Sonneratia ovata* Backer		
*29	*Xylocarpus granatum* Koen.		
*30	*Xylocarpus mekongensis* Pierre		
*31	*Aglaia cucullata* Pellegrin		
*32	*Heritiera fomes* Buch. Ham.		
*33	*Heritiera littoralis* Dryand		
*34	*Lumnitzera racemosa* Willd.		
35	*Lumnitzera littorea* Voight		
36	*Camptostemon philippinensis* Becc.		
37	*Camptostemon schultzii* Masters		
38	*Osbornia octodonta* Muell.		
*39	*Scyphiphora hydrophyllacea* Gaertn.f.		
*40	*Nypa fruticans* Wurmb.		

* Collected & Reported from the Indian Sundarbans.

Since the term mangroves has been defined differently by different experts, accordingly the total number of mangrove species is also different among the different published works. According to Tomlinson (1986) the major and minor mangrove species of the Old World is 40 and only 8 mangrove species are reported from the New World (Table 3).

1.3. Distribution of Mangroves in India

In the Indian sub-continent the mangrove ecosystem are found within the inter tidal, tidal and supra-tidal deltaic zones, estuarine mouths, sheltered coasts and edges of the islands of both East Coast facing the Bay of Bengal and West Coast facing the Arabian Sea; besides the coastal inter tidal zones the mangals are also spread over the Bay Islands of Andaman and Nicobar and in the Lakshadweep and Minicoy Coral Atolls (Naskar and Mandal, 1999). In the nine different maritime states of India the mangals are distributed in the deltaic regions, *viz.,* in West Bengal, they are restricted in the estuarine regions of Sundarbans between the river Hugli (Ganga) and the river Harinbhanga; in Orissa at Bhitarkanika, between the rivers Devi and Dharma at the mouth of the river Mahanadi; the Coringa mangrove area between the Krishna and the Godavari rivers in Andhra Pradesh; in the Pennar delta of Tamil Nadu at the mouth of the river Cauvery, all these accounting for the major mangrove areas of the East Coast facing Bay of Bengal. In the West Coast the mangals are distributed in the State of Gujarat in the Indus delta in five different regions, out of which, the two Gulfs, *viz.,* Gulf of Kachchh and Gulf of Cambay form the major mangrove zones and the other three, *viz.,* Rann of Kachchh, Saurashtra Coast and South Gujarat Coast constitute the minor mangrove zones; in Maharashtra, the mangals are distributed along the coastal belt of the districts Sindhudurg, Ratnagiri, Raigad, Mumbai and Thane; in Goa the mangrove areas are distributed between the Mandovi and Zuari estuaries and other five minor estuaries and Cambarjooa canal; in Karnataka, the major mangrove areas are in Coondapur and Malpe estuarine complex and in Kerala, a very negligible mangrove area is reported from the Cochin estuary (Naskar and Mandal, 1999) (Map 1). Out of the total mangrove areas in the Indian sub-continent, Sundarbans accounts for 65% of the total distribution followed by Andaman and Nicobar Islands accounting for 18% of the total distribution (Naskar, 1999). The important mangrove regions along with their total area coverage, overall species composition and important rivers, estuaries and deltas associated have been tabulated in Table 4.

1.4. The Mangroves

Mangroves are coastal tropical forest formation encircled or inundated by the tidal rivers and/ or sea-water, flooded frequently by the tidal water (Naskar and Guha Bakshi, 1987). The mangroves are extraordinary trees and shrubs capable of tolerating wide ranges of salinity and tidal influences. They grow and survive in the frequently inundated saline coastal/ estuarine soil in the tidal, inter tidal and supratidal regions in tropical and sub- tropical regions of the world.

Mangroves are a group of distinct plants with characteristic morphological, anatomical, physiological and physiognomical adaptations and succession mechanisms. These characteristic features include leaves, which are generally succulent or fleshy, flattened, shiny, large sized and dark green. To check the high rate of transpiration due to bright sunlight and to cover the stomata for reducing water loss the leaves have a waxy coating (cuticle) and or hair (tomentum), salt glands for excreting excess salt, sunken stomata associated with sub - stomatal cavity and hypodermal aqueous tissues for maintaining succulence by response to the presence of chlorine ion (Cl^-). The trees have an erect, solid and stout trunk base and an umbrella like spreading canopy. Number of vessels are more per square milimeter occupying large cross sectional areas, though the vessels are small. This is for uplifting plenty of water through these vessels and in case, any of the conductive columns get damaged there is a substitute for it. The vessels have small scalariform perforations, vascicentric parenchyma, fine celled multiseriate rays and libriform fibres to protect the vessels from any collapse under high tension. The major mangrove species have typical aerial root systems in the forms of pneumatophores, stilt roots, root buttresses, pneumatothodes all with lenticels and air cavity for respiration purposes and for providing mechanical support. Another interesting adaptive feature observed in mangroves is their mode of germination. The seeds germinate and hypocotyles develop while still hanging on the mother plant to avoid at first, the highly saline water in which they grow. When the hypocotyles are long and emerge out of the pericarp, the condition is referred to as vivipary as in the case of members of family mangrove Rhizophoraceae and when the hypocotyles are small and hidden within the pericarp, the condition is called pseudo - vivipary or crypto - vivipary present in the members of family Avicenniaceae (Naskar & Mandal, 1999).

Table 4 : Indian Mangrove Zones/Areas, Species Composition & Important Associated Rivers/Deltas/Estuaries

Mangrove Area/ Zone	*Total Area (sq. km.)*	*Overall Species Composition*		*Important Rivers/Deltas/Estuaries/Bays Associated*
		True Mangroves	*Mangrove Associates*	
EAST COAST				
West Bengal – Sundarbans	4267	32	69	Ganga-Brahmaputra delta, Rivers-Hooghly, Baratala, Saptamukhi, Thakuran, Matla, Goasaba and Herobhanga, Bidya, Kalindi, Raimangal, Kalagachi.
Andaman and Nicobar Islands	1152	32	36	
Orissa – Bhitarkanika	120	30	35	Baitarini-Brahmani-Mahanadi delta, Balasore Coast
Andhra Pradesh– Coringa, Kakinada Bay,	184	17	22	Godavari-Krishna delta, Rivers - Gautami Godavari, Vashishti Godavari, Nizampatam Bay
Tamil Nadu–Pichavaram, Muthupet, Chatram	26	17	17	Cauvery delta, Estuaries - Vellar, Vaigai, Adyar

WEST COAST				
Kerala	44 ha	18	15	Estuaries - Cochin (Kannamali), Veli, Quilon, Vembanad, Calicut, Rivers - Chetwipuzha, Karanjirapuzha, Akla, Anjerakndi, Valapatanam, Puramba
Karnataka – Coondapur, Malpe,	60	12	6	Estuaries - Sita - Swarna, Mulki, Kopal; Rivers - Chatra, Kolluru, Haladi, Kalinadi, Belumdarnadi, Ganga Velli, Angashini
Goa – Mandovi, Zuari	20	14	13	Terakhol, Chapora, Sal, Talpona, Galgibag, Cambarjua
Maharashtra – Ratnagiri, Raigad, Mumbai, Thane, Sindhudurg	330	19	10	Mahim creek, Elephanta island, Waghotan, Rajapur, Vikhroli; Estuaries - Savitri, Shastri, Kundalika; Rivers - Vashishti, Vasai, Vaitarana, Thane, Panuel-Karanja-Dharmatar creek complex
Gujarat – Gulf of Kachchh, Gulf of Cambay, Rann of Kachchh, Saurashtra	260	13	9	Rivers - Luni, Mahi, Narmada, Tapti, Sabarmati, Ambica, Auranga, Kolak, Daman Ganga, Kori creek, Patre creek
Lakshadweep and Minicoy Islands	25 ha	1	not reported	

(Source : Naskar, 1999)

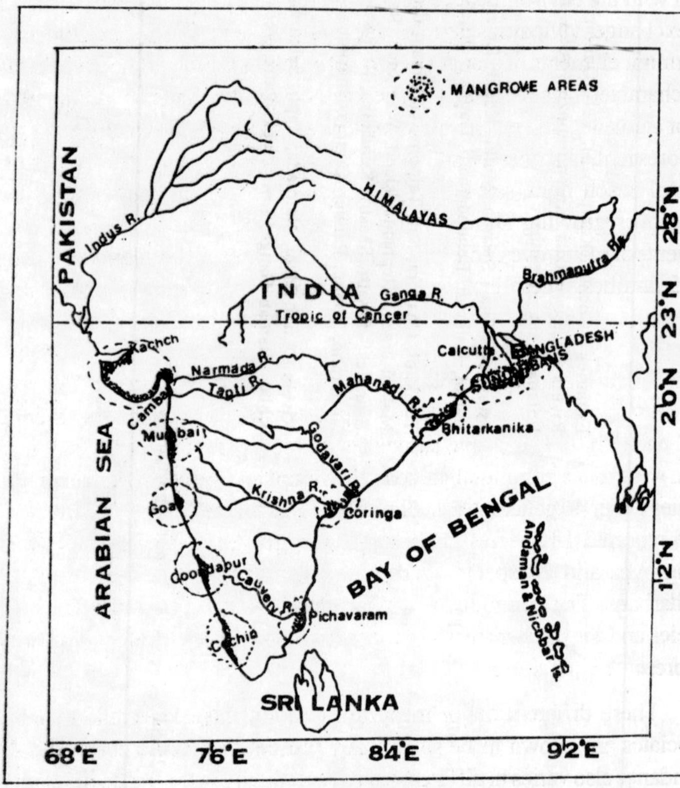

Map 1 : Indian Mangrove Zones

Based on the presence or absence of these characters the mangroves are divided into 'major elements of mangroves', 'minor elements of mangroves' and 'mangrove associates'. Tomlinson (1986) has characterized these major elements of mangroves or the true mangroves as "those having complete fidelity to the mangrove environment, *i.e.,* only growing in the mangrove areas and not existing in terrestrial communities; morphological specializations fitted with the environment or habitat, where they grow, *viz.,* aerial roots for gas exchange, viviparous germination and with salt tolerance mechanisms". The minor elements of mangroves do not exhibit the typical adaptive features and characteristics described above but occupy the mangals alongwith the major elements. The mangrove associates are those plants which grow as non - arborescent, herbaceous, sub - woody and climbers of the tidal zones or as tropical beach flora, salt marsh flora, wet coastal communities or as back mangroves growing above the normal high tide level. Among the major elements of mangroves Tomlinson (1986) has listed 34 species under 9 genera and 5 families. These include 8 species under *Rhizophora,* 6 species under *Bruguiera,* 2 species under *Ceriops,* 1 species under *Kandelia,* 8 species under *Avicennia,* 5 species under *Sonneratia,* 1 species under *Nypa,* 2 species under *Lumnitzera* and 1 species under *Laguncularia.* Among the minor elements of mangroves there are 20 species under 11 genera and under the third category of mangrove associates including back mangals, salt marsh flora, wet coastal communities, coastal swamps and swamp forest there are 60 species under 46 genera. Naskar and Guha Bakshi (1987) and Naskar (1983) have reported 18 species under major mangroves, 8 species under minor mangroves and 28 species under mangrove associates from the Indian Sundarbans. The common and major mangrove species, minor mangrove species and the back mangals or mangrove associates of Indian Sundarbans are presented in Table 5.

These different major mangroves, minor mangroves and mangrove associates are known to be specifically present in specific zones and their abundance also varies in different zones accordingly in the Indian Sundarbans. (Anonymous, 1999) (Fig. 1).

In these mangrove habitats of intertidal zones the soil and water phases receive organic nutrients after decomposition of the mangrove litter. The nutrient enriched and mineral rich tidal water and coastal soil of the intertidal zones in the coastal areas are noted as very suitable substrate for good number of very economic and potentially important algal species under the groups of Chlorophyta, Rhodophyta, Chrysophyta, Phaeophyta and Cyanophyta.

Table 5 : Major and Minor Elements of Mangroves and Mangrove Associates of Indian Sundarbans

	Families & Genera	Species Name	Vernacular Name
I	**MAJOR ELEMENTS OF MANGROVES IN INDIAN SUNDARBANS**		
A	**RHIZOPHORACEAE**		
1.	*Rhizophora* L.	*R. apiculata* Blume	Bhora, Garjan
		R. mucronata Lamk.	Khamu, Garjan
2.	*Bruguiera* Lamk.	*B. gymnorhiza* (L.) Lamk.	Kakra
		B. sexangula (Lour.) Poir	Kakra
		B. cylindrica (L.) Blume	Bakul Kakra
		B. parviflora W. & A. ex. Griff.	Bakul Kakra
3.	*Ceriops* Arnold	*C. decandra* (Griff.) Ding Hou	Jhamti Goran
		C. tagal (Perr.) Robin.	Mat Goran
4.	*Kandelia* W. & A.	*K. candel* (L.) Druce	Goria
B	**AVICENNIACEAE**		
5.	*Avicennia* L.	*A. alba* Blume	Kal Bani
		A. officinalis L.	Jat Bani
		A. marina (Forssk.) Vierh.	Peyara Bani
C	**SONNERATIACEAE**		
6.	*Sonneratia* Linn. f.	*S. caseolaris* (L.) Engle	Chak Keora
		S. griffithii Kurz.	Ora
		S. apetala Buch. Ham.	Keora
D	**COMBRETACEAE**		
7.	*Lumnitzera* Willd.	*L. racemosa* Willd.	Kripal
E	**ARECACEAE**		
8.	*Nypa* Steck.	*N. fruticans* (Thunb.)	Golpata
II	**MINOR ELEMENTS OF MANGROVES IN INDIAN SUNDARBANS**		
A	**MELIACEAE**		
1.	*Xylocarpus* Roem.	*X. granatum* Koenig.	Dhundul
		X. mekongensis Pierre	Pasur
2.	*Aglaia* Lour.	*A. cucullata* (Roxb.) Pelle	Amur
B	**MYRSINACEAE**		
3.	*Aegiceras* Gaertn.	*A. corniculatum* (L.) Blanco	Khalsi
C	**EUPHORBIACEAE**		
4.	*Excoecaria* L.	*E. agallocha* L.	Gneoa
D	**AEGIALITIDACEAE** *(nom. alt. PLUMBAGINACEAE)*		
5.	*Aegialitis* Brown	*A. rotundifolia* Roxb.	Tora

Contd...

	Families & Genera	Species Name	Vernacular Name
E	**STERCULIACEAE**		
6.	*Heritiera* Aiton	*H. fomes*	Sundari
F	**RUBIACEAE**		
7.	*Scyphiphora* Gaertn.	*S. hydrophyllacea* Gaertn. f.	Tagri Bani
G	**PTERIDACEAE**		
8.	*Acrostichum* L.	*A. aureum* L.	Hudo fern
H	**ACANTHACEAE**		
9.	*Acanthus* L.	*A. ilicifolius* L.	Harkoch Kata
		A. volubilis Wall.	Lata - Hargoja
I	**RUTACEAE**		
10.	*Atalantia* M. Roem.	*A. correa* M. Roem.	Ban Lebu
J	**TILIACEAE**		
11	*Brownlowia* Roxb.	*B. tersá* (L.) Kosterm.	Lata Sundari
III	**MANGROVE ASSOCIATES OF INDIAN SUNDARBANS**		
A	**MALVACEAE**		
1.	*Hibiscus* L.	*H. tiliaceus* L.	Ban Kapas
		H. tortuosus Roxb.	Ban Kapas
2.	*Thespesia* Soland ex Corr	*T. populnea* (L.) Sold.	Paras
		T. populneoides (Roxb.) Kostel	Paras
		T. lampus (Cav.) Dalz & Gibs.	Paras
B	**FABACEAE**		
a	**CAESALPINIACEAE**		
3.	*Cynometra* L.	*C. ramiflora* L.	Gila
4.	*Caesalpinia* L.	*C. bonduc* (L.) Roxb.	Nata Karanja
		C. crista L.	Shingri Lata
b	**PAPILIONACEAE**		
5.	*Dalbergia* L. f.	*D. spinosa* Roxb.	Chulia Kata
6.	*Derris* Lour.	*D. scandens* Roxb.	Noa Lata
		D. trifoliata Lour.	Pan Lata
		D. indica Bennet	Karanja
C	**ARECACEAE (*nom. alt.* PALMAE)**		
7.	*Phoenix* L.	*P. paludosa* Roxb.	Hental
D	**LECYTHIDACEAE**		
8.	*Barringtonia* Forst.	*B. asiatica* (L.) Kurz.	Hijal
		B. racemosa Roxb.	Sumudra

Contd...

Families & Genera		*Species Name*	*Vernacular Name*
E	**BIGNONIACEAE**		
9.	*Dolichandrone* Seem.	*D. spathaceae* (L. f.) Sch.	Gorshingiah
F	**EBENACEAE**		
10.	*Diospyros* L.	*D. ferrea* (Willd.) Roxb.	Ban Gab
G	**CLUCIACEAE (*nom. alt.* GUTTIFERAE)**		
11.	*Calophyllum* L.	*C. inophyllum* L.	Kath Champa
H	**CAPPARIDACEAE**		
12.	*Capparis* L.	*C. sepiaria* L.	Kata Gurkamai
13.	*Crataeva* L.	*C. roxburghii* R. Br.	Barun
I	**SAPINDACEAE**		
14.	*Allophyllus* L.	*A. cobbe* (L.) Bl.	
J	**PANDANACEAE**		
15.	*Pandanus* Parkin.	*P. tectorius* Sol. ex Park.	Keya
		P. foetidus Roxb.	Keya
K	**SAPOTACEAE**		
16.	*Manilkara* Adans.	*M. hexandra* (Roxb.) Dub.	Bilati Bakul
L	**ASCLEPIADACEAE**		
17.	*Sarcolobus* R. Br.	*S. carinatus* Wall.	Baoli Lata
		S. globosus Wall.	Baoli Lata
18.	*Pentatropis* Wight & Arn.	*P. capensis* (Linn. f.) Bullock	
19.	*Hoya* R. Br.	*H. parasitica* Wall.	Pargacha
M	**CHENOPODIACEAE**		
20.	*Suaeda* Forssk.	*S. maritima* Dumort.	Gire Sak
		S. nudiflora Moq.	Gire Sak
21.	*Salicornia* L.	*S. brachiata* Roxb.	
N	**AMARYLLIDACEAE**		
22.	*Crinum* L.	*C. defixum* Kar-Gawl.	Sukh - darshan
O	**ARACEAE**		
23.	*Cryptocoryne* Fisch.	*C. ciliata* (Roxb.) Fisch.	Kerali
P	**VERBENACEAE**		
24.	*Clerodendrum* L.	*C. inerme* (L.)	Ban Jui
Q	**RUPPIACEAE**		
25.	*Ruppia* L.	*R. maritima* L.	Nona Jhanji
R	**TAMARICACEAE**		
26.	*Tamarix* L.	*T. dioca* Roxb.	Nona Jhau
		T. gallica L.	Nona Jhau

Contd...

Families & Genera	Species Name	Vernacular Name
S BORAGINACEAE		
27. *Heliotrophium* L.	*H. curassavicum* L.	Nona Hati Sur
T POACEAE		
28. *Porteresia* Tak.	*P. coarctata* (Roxb.) Tak.	Dhani Ghas
29. *Myriostachya* Hook. f.	*M. wightiana* Hook. f.	Nalai
30. *Aeluropus* L.	*A. lagopoides* (L.) Trin.	Nona Durba
U OPUNTIACEAE		
31. *Opuntia* L.	*O. dillenii* Haw.	Nag Phana
V SOLANACEAE		
32. *Solanum* L.	*S. trilobatum* L.	Nona Begun
W LAURACEAE		
33. *Cassytha* L.	*C. filiformis* L.	Akash Bel
X LORANTHACEAE		
34. *Dendrophthoe* Mart.	*D. falcata* (Linn. f.) Etting.	Baramanda
35. *Viscum L.*	*V. monoicum* Roxb.	Manda
	V. orientale Willd.	Manda
Y RUBIACEAE		
36. *Hydrophyllax*	*H. maritima* L.	
Z CONVOLVULACEAE		
37. *Ipomoea* L.	*I. pes - caprae* (L.) Sweet	Chagal Kuri

Source : Anonymous (1999)

1.5. Algae in the Mangrove Habitat

The word 'algae' (sing. alga) has been derived from the Latin word '*alga*' meaning sea -weeds. The Greek word for algae or sea -weeds is *phykos*. Algae have been variously and quite facetiously referred to as 'frog spittle', 'pond scums', 'water mosses' and of course 'sea weeds'. Different Phycologists prefer to define 'Algae' in different ways, according to Fritsch (1935)– *"unless purely artificial limits are drawn, the designation alga must include all holophytic organisms (as well as their numerous colourless derivatives) that fail to reach the higher levels of differentiation characteristic of the archegoniate plants"*; Smith (1955) has defined algae as *"simple plants with an autotrophic mode of nutrition"*; Prescott (1969) has defined algae as *"those chlorophyll bearing organisms (and their colourless relatives) which are thalloid, i.e. having no true roots, stems and leaves or leaf - like organs"*.

Figure 1: Abudance of Mangroves in Different Parts of the Indian Sundarbans.

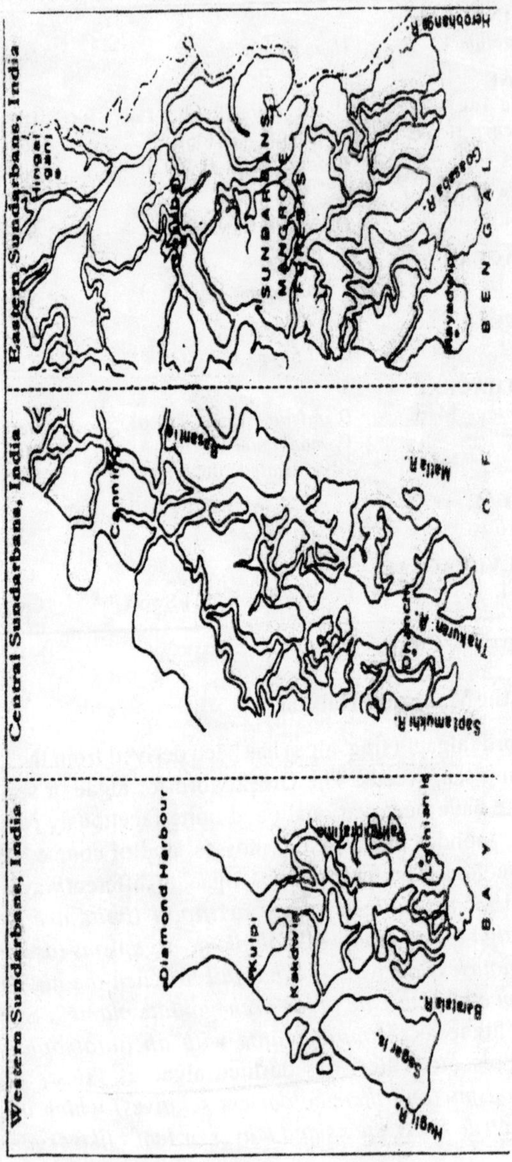

Map of Sundarbans Showing Western, Central and Eastern Parts & Dominant Plant Species

Western parts of Sundarbans Within Frequent Tidal Zone	Central parts of Sundarbans Within Frequent Tidal Zone	Eastern parts of Sundarbans Within Frequent Tidal Zone
Avicennia alba	*Avicennia alba*	*Sonneratia apetala*
Avicennia, marina	*Avicennia marina*	*Avicennia alba*
Avicennia officinalis	*Avicennia officinalis*	*Avicennia marina*
Acanthus ilicifolius	*Porteresia coarctata*	*Avicennia officinalis*
Porteresia coarctata	*Aegiceras corniculatum*	*Nypa fruticans*
In the Ridge Forest Zone	In the Ridge Forest Zone	In the Ridge Forest Zone
Excoecaria agallocha	*Ceriops decandra*	*Ceriops decandra*
Acanthus ilicifolius	*Excoecaria agallocha*	*Excoecaria agallocha*
Phoenix paludosa	*Phoenix paludosa*	*Phoenix paludosa*
Lumnitzera racemosa	*Ceriops tagal*	*Ceriops tagal*
Avicennia officinalis	*Avicennia alba*	*Avicennia alba*
Beyond the Tidal Reaches	Beyond the Tidal Reaches	Beyond the Tidal Reaches
Thespesia populnea	*Phoenix paludosa*	*Excoecaria agallocha*
Acanthus ilicifolius	*Nypa fruticans*	*Phoenix paludosa*
Brownlowia lanceolata	*Ceriops decandra*	*Ceriops decandra*
Excoecaria agallocha	*Excoecaria agallocha*	*Heritiera fomes*
Phoenix paludosa	*Heritiera fomes*	*Aglaia cucullata*
5 Most Dominant Species	5 Most Dominant Species	5 Most Dominant Species
Avicennia alba	*Avicennia alba*	*Sonneratia apetala*
Avicennia marina	*Avicennia marina*	*Heritiera fomes*
Phoenix paludosa	*Phoenix paludosa*	*Nypa fruticans*
Ceriops decandra	*Sonneratia apetala*	*Excoecaria agallocha*
Excoecaria agallocha	*Excoecaria agallocha*	*Avicennia marina*

The algae play a substantial role in the environment that they inhabit, specially the aquatic environment, firstly as the primary producer group of organisms in the food web and contrariwise as pollutants when uncontrolled growth occurs. As the algae require light, inorganic salts and a nitrogen source for their rapid growth, they are potential producers of proteins, lipids and carbohydrates for feeding humans and animals (Kumar & Singh, 1995). With phenomenal increase in population during the last few decades, a few of the major problems faced by human society are those of quantity and quality of food, fodder, drinking water, disposal of sewage and industrial wastes and proper conservation of soil and water. Properly regulated algal growth can provide substantial assistance in tackling these problems. The other uses that the algae are commercially exploited for are production of agar, carrageenin, alginic acid, antibiotics, minerals, funori, diatomite and also in biological research works.

The algae exhibit a wide range of thallus structure in nature, their diversity ranges from being unicellular to complex multicellular forms. The various thallus organizations encountered in this group of simple autotrophic plants are: motile unicellular, non-motile unicellular, multicellular flagellated or colonial, multicellular non-flagellated or non-flagellated colonies or coccoid, palmelloid, dendroid, filamentous–unbranched or branched, heterotrichous, siphonaceous, uniaxial, multiaxial and parenchymatous (Sharma, 1986). Due to the presence of such diversity of forms and structures and with nothing more than their characteristic oxygen evolving type of photosynthesis in common, the algae are a difficult group to classify or scientifically categorize into a hierarchical series of groups. Different criterion have been adopted by various Phycologists in classifying the group 'algae', the most important one being pigment type, their relative amount, chemical composition and their complements, alongwith the pigment the shape of the chromatophores are also taken into account during classification. The second criterion adopted is the external form of the algae, *i.e.,* their size, shape, appendages if any and other significant structures. The other factors taken into account are the reserve food material, the flagellar structure, number, position and anatomy, cell wall, nucleus (presence or absence) and number, arrangement and shape of chromosomes. Type of life – history, method of reproduction, shape of reproductive bodies, presence or absence of sexual reproduction is also considered during classification. Another important criterion considered during classification is the algal habitat ecology. Generally a combination of a few of the above mentioned characters are considered by algal taxonomists.

The International Code of Botanical Nomenclature (Lanjouw, 1956) has recommended the following suffixes for the different categories of algae:

Division:	-phyta
Sub-division:	-phytina
Class:	-phyceae
Sub-class:	-phycidae
Order:	-ales
Sub-order:	-inales
Family:	-aceae
Sub-family:	-oideae
Tribe:	-eae
Genus:	*-generally a Latin name*
Species:	*-generally a Latin name*
Variety:	*- A Latin word*
Form:	*- A Latin word*

Two different trends are observed in classification of algae. According to one thought, since there are different pigments and the product of photosynthesis is different in different groups of algae, they should be first divided into several "phyta" (*i.e.* divisions such as Chlorophyta and Rhodophyta) and each "phyta" should further be divided into different "phyceae" (classes such as Chlorophyceae, Rhodophyceae); this is supported by Pascher, 1914, 1931; Smith, 1933, 1955; Papenfuss, 1946; Prescott, 1969; all of them are among the noted phycologists. The other school of thought consider the facts that Chlorophyll-a is common to all algae and secondly product of photosynthesis is different but the process involved is same, so a common origin from a common source is indicated, therefore algae is itself equivalent to a division and can only be further divided into classes "phyceae", there is no "phyta" in the classification proposed by this trend. This school of thought is supported by Fritsch (1935, 1945) and his followers. Papenfuss (1951) proposed that to establish the link with algae the word "phyco" must be added before "phyta", thus the divisions in algae should be named as Chlorophycophyta, Rhodophycophyta, etc.

The classification proposed by Smith (1955) has been followed during taxonomic enumeration of the algal species described in this book. The basic outline of Smith (1955) classification is as follows :

Serial No.	Division	Class
1.	* **Chlorophyta**	* Chlorophyceae * Charophyceae
2.	**Euglenophyta**	Euglenophyceae
3.	**Pyrrophyta**	Cryptophyceae Desmokontae Dinophyceae
4.	* **Chrysophyta**	* Xanthophyceae Chrysophyceae *Bacillariophyceae
5.	* **Phaeophyta**	Isogeneratae Heterogeneratae * Cyclosporae
6.	* **Cyanophyta**	* Myxophyceae
7.	* **Rhodophyta**	* Rhodophyceae

* shown against the Division and Class indicates their presence, as the algal flora collected and reported during this present investigation from the Indian Sundarbans mangals.

In the mangrove habitats inundation by saline water is a regular feature. During high tide the mangrove areas get inundated with tidal river water, thus receiving sufficient moisture alongwith frequent exposure to sunlight during low tide, which suitably provides for ideal habitat for different groups of algal flora, *viz.,* benthic and planktonic. In the intertidal mangrove habitat, generally the aerial roots of major mangroves and low trunk regions of the mangrove trees and shrubs act as the substratum for different groups of filamentous or branched algal species, *viz.,* *Bostrychia* spp., *Caloglossa* spp., *Catenella* spp., *Polysiphonia* spp. and some others; as these grow in the periphery of aerial roots or trunk regions these groups of algae are sometimes referred to as *periphyton* (Naskar and Mandal, 1999). These periphytic algae form a compact matrix which receive moisture from the tidal water and exposure to sunlight during low tide phases. The estuarine, in shore and off shore fin-fish and shell-fish species very often graze on these periphytic algae; as such, these densely growing forest floor serve as the ideal grazing ground for these shell - fish and fin - fish fauna. Most of these algal species are noted as good food materials, because of the presence of minerals, vitamins, carbohydrates, fats and proteins either in the cell wall or in the cytoplasm (Anonymous, 1999). Moreover, in the mangrove ecosystem the photosynthetic activity of algae continuously oxygenates the surrounding environment during high tide, which is further beneficial to other small aquatic animals. Till date, in the Indian part of Sundarbans the algal flora is not consumed as human food though some of the genera have potential food value, particularly members of Rhodophyta, Chlorophyta and Cyanophyta; this may be due to lack of awareness and ignorance among the local people. Even now, though some algae in the Indian Sundarbans have important commercial and industrial use they are still not cultured or utilised as food or industrial raw material.

Besides the direct function in the mangrove habitats the Cyanophyta members also help to fix the atmospheric nitrogen in the initial silted up river flat mud in the accretion stages of alluvium soil, where Chlorophyceae group of algae gradually turn it into more consolidated soil for succession of higher groups of plants (Mac Nae, 1968; Naskar and Mandal, 1999). Due to the difficult terrain and inaccessibility of these hostile mangrove forests of Indian Sundarbans these algal species have received less attention by researchers.

1.6. Algal Habitats in the Indian Sundarbans

In Indian Sundarbans, algae are generally found to occupy a wide range of habitats, rather, they are ubiquitous because of their presence in nearly all types of habitats. They are found in fresh water, sea water and brackishwater either as phytoplankton, as free floating or attached masses on dry or wet soil as benthic algae, on bricks, dilapidated jetties, under water concrete construction and other hard substratum as lithophytes, on sandy beaches, on other plant parts as periphytic algae, on other algal species as epiphytic algae, on animals as epizoic algae and in association with fungi in the form of lichens. The Indian coastal Sundarbans inter tidal mangrove forest zone is devoid of any rocky substrate; as such, certain groups of rock attached benthic forms of algae are absent from this Indian Sundarbans, as in contrast to the similar ecological conditions prevailing in the Chakaria Sundarbans region of Bangladesh, where several benthic forms are present. Islam (1986) has stated that in depth ecological studies of the tropical benthic flora is little known in comparison to the temperate benthic ecosystem flora.

In the Sundarbans of West Bengal, India all the forms of algae mentioned above are encountered. The coastal environment of Sundarbans in India basically provides two different types of environment for the rich and abundant algal flora found here. First, the enormous open water systems in the form of vast river expanse of the major estuarine rivers, other minor rivers, creeks, canals and the Bay of Bengal and second, the shallower water by the edge. Only the free floating algal forms and the phytoplankton are found to survive in the open water. At the edge, the water becomes shallow enough for algal attachment to the bottom soil or on the exposed soil, as well as, to various other hard substrata in the forms of pneumatophores, pnematothodes, knee-roots, stem-buttress, plank roots, stilt roots, prop-roots, blind root suckers, bark and other aerial parts of mangroves, bricks, wooden and bamboo poles, concrete jetties, etc., frequently inundated by tidal water. The second type of estuarine/deltaic habitat is more complex in terms of survival because of the tides. Daily as the tides move in and out, the amount of water covering the algae changes from totally inundated to completely exposed, thus affecting the amount and quality of light reaching the algae, exposure to salinity regimes and current by varying degrees. Yet, we find a large number of algae that grow in these inter tidal regions and others that can survive in the subtidal regions, where regular tidal flow exist. This tidal flow helps to exchange gases and nutrients and also check eutrophication in this sheltered coast; as such, algal bloom is not formed in this above zone.

Another major niche for the algal flora of Sundarbans is the brackish water fisheries in the inland areas. The brackish water fisheries/ wetlands (natural or man - made) with wide range of salinity regimes (6 mmhos/cm^2 to 30 mmhos/cm^2) offer habitat for a number of specific periphytic, free floating and planktonic algal forms, which are found to restrict themselves to such conditions only (Naskar, 1985). In this estuarine wetland fisheries, the common algal species are *Enteromorpha intestinalis, E. clathrata, E. prolifera, Chaetomorpha gracilis, C. brachygona, C. aerea, Lola tortuosa, L. implexa, L. capillaris* and *Spirogyra* spp. These macrophytic forms of brackish water loving algae are very often noted to grow gregariously and often choke the estuarine wetland fisheries. As such, the exotic tilapia species, *viz., Eriochromis mossambicus* and *E. nilontica* are introduced for controlling these algal growth. But profuse growth of these algae at times turns out to be uncontrollable and manual clearing of these dense algae becomes inevitable. But on decomposition, these algal bio-mass release large quantity of organic nutrients and enhance the natural productivity of these estuarine wetland fisheries. Besides this, the half decomposed algal bio - mass is often preferred to the fresh algae by the brackish water fish and prawn species as natural food.

1.7. Fauna and Faunal Habitats in Indian Sundarbans

The Sundarbans mangals have a diverse range of habitats and niches in terrestrial, intertidal and aquatic environments, thus harbouring very rich and diverse animal communities, right from the unicellular protozoans to the mighty mammals. There have been a lot of publications on the fauna of the Sundarbans region (Nandi and Misra, 1987; Mandal and Nandi, 1989; Mandal and Ghosh, 1989; Nandi, *et. al.*, 1993; Chaudhuri and Choudhury, 1994, Nandi and Pramanik, 1994; Annonymous, 1995; Das and Nandi, 1999). Being the sea land interphase, mangals include habitats in the forms of mangrove dominated forests, mangrove swamps, tidal flats (mud flats/ sandy - mud flats) and estuarine water bodies, *viz.,* rivers, channels, creeks and canals. These niches include tree canopy, higher aerial parts of mangroves, root zone of mangroves, forest floor of supratidal

Table 6 : The Faunal Habitats in the Indian Sundarbans Mangals

Sl.No.	*Habitat*	*Inhabiting Faunal Species/ Groups*
1	Tree canopy	Monkey, bats, birds, insects and mites
2	Higher parts of mangroves	*Littorina scabra*
3	Lower parts and root zone of mangroves	*Cerithidea obtusa, Nerita articulata, N. semirugosa, Onchidium spp., Sesarma spp.*
4	Forest floor of supratidal zone	Soil litter inhabiting microarthropods, mud shrimps
5	Mud flats	*Assiminea* spp., *Telescopium telescopium, Cerithidea cingulata, Uca* spp., *Sesarma* spp., *Macrophthalmus* spp., *Metaplas* spp.
6	Sandy mud flats	*Ocypode macrocera, Dotilla blanfordi, Uca* spp.
7	Dead trees and stumps	Marine wood borers, sessile forms (barnacles, oyesters, sea-anemones)
8	Muds of rivers and channels near mangrove edge	Polychaetes, mud burrowing bivalves *(Tellina sp., Dosinia sp., Meretrix sp., Crassostrea gryphoides), Scylla serrata*
9	Edges of rivers and channels	*Periopthalamus* spp., *Boleopthalamus* spp., *Cerberus rhynchops*
10	Estuarine water	Shrimps, prawns and fishes

(Source : Das and Nandi, 1999).

Table 7 : Distribution of Mangrove Inhabiting Fauna in Relation to Their Habitats

Sl. No.	Phylum/Class	Total no. of Species	Number of Species in the Specific Zone		
			Supra tidal	Inter tidal	Estuarine
I	**Kingdom: Protista**				
	Subkingdom: Protozoa				
1.	Sarcomastigophora	45	26	6	14
2.	Aplicomplexa	24	19	1	4
3.	Myxozoa	4	—	3	1
4.	Ciliophora	31	2	19	10
II	**Kingdom: Animalia**				
1.	Porifera	1	—	1	—
2.	Cnidaria	33	—	13	20
3.	Ctenophora	2	—	—	2
4.	Platyhelminthes	41	—	3	38
5.	Nemathelminthes	68	—	68	—
6.	Acanthcephala	3	—	—	3
7.	Nemertinea	2	—	2	—
8.	Rotifera	4	—	—	4
9.	Mollusca	142	12	116	14
10.	Sipuncula	2	—	2	—
11.	Echiura	3	—	3	—
12.	Annelida	78	9	69	—
13.	Arthropoda	476	234	90	152
(a)	Crustacea	240	—	90	150
(b)	Insecta	201	201	—	—
(c)	Arachnida	33	33	—	—
(d)	Merostomata	2	—	—	2
14.	Entoprocta	1	—	—	1
15.	Bryozoa	3	—	—	3
16.	Brachiopoda	1	—	1	—
17.	Chaetognatha	4	—	—	4
18.	Echinodermata	20	—	14	6
19.	Hemichordata	1	—	1	—
20.	Chordata	445	184	88	173
(a)	Chondrichthyes	22	—	—	22
(b)	Osteichthyes	154	—	36	118
(c)	Amphibia	8	8	—	
(d)	Reptilia	58	40	—	18
(e)	Aves	163	101	52	10
(f)	Mammalia	40	35	—	5
TOTAL		1434	486	499	449

[Sources: Mukherjee (1975); Mandal and Nandi (1969); Nandi et. al., (1993); Chaudhuri and Choudhury (1994): Anon (1995).]

zone, mud flats, sandy beaches, dead trees and stumps, mud/silt/clay of rivers near edge of mangroves, estuarine waters, etc. (Table 6).

A total of 1,434 animal species have been reported so far from the Indian parts of Sundarbans (Das and Nandi, 1999). These animals comprise 989 species of invertebrates, one species of hemichordate and 445 species of vertebrates. According to their habitats 486 species have been reported from the supra littoral zone, 499 species from tidal flats and 449 species from estuarine waters (Table 7).

Among the 40 species of mammals, 5 species of dolphins and porpoises are aquatic and the rest are terrestrial. The Royal Bengal tiger *(Panthera tigris tigris)* is the key stone species of Sundarbans, Spotted deer *(Cervis axis),* Wild boar *(Sus scrofa)* and Rhesus monkey *(Macaca mullata)* are quite abundant. Among the 163 species of birds 110 species are resident and 53 species are migratory (Sanyal, 1999). The reptilian fauna include 1 species of marsh crocodile *(Crocodilus porosus),* 3 species of monitor lizards *(Varanus bengalensis, V. salvator* and *V. flavescens),* 12 species of turtles and terrapins and 40 species of snakes and lizards. Out of the 8 species of amphibia, 1 is a tree frog, *Rhacophorus maculatus.* Terrestrial invertebrates are mostly insects followed by spiders and mites. The aquatic invertebrate fauna include zooplankton, benthic and pelagic communities. Zooplankton include copepods, amphipods, ostracods, chaetognaths, mysids and hydromedusae and diverse planktonic larval forms. Benthic communities include infauna and epifauna. Among the epifauna gastropods are predominant. Sessile animals include barnacles, bivalves and sea anemones. The infauna comprise of mud burrowers and wood borers. Among the mud burrowers brachyuran crabs, polychaetes and bivalves are common and 6 species of molluscan wood borer have been reported till date (Das & Nandi, 1999). The water bodies of Sundarbans harbour about 200 shell-fish and fin-fish species of which at least 100 are commercially important (Table 8).

Habitat destruction is posing a major threat to the rich biodiversity of Sundarbans. Over exploitation and unplanned developmental activities are the other major threats. Several species of vertebrate are enlisted as extinct or endangered to a point of extinction (Table 9).

1.8. Primary Productivity in the Sundarbans Mangals

All chlorophyll bearing organisms, *viz.,* phytoplankton, periphyton, benthic algae and other macrophytic forms, viz., the major mangroves,

Table 8 : Shell - fish and Fin - fish Species of Economic Importance from the Estuaries and Off-shores of Indian Sundarbans
Commercially Important Fin - Fish Species

Sl. No.	Fin - Fish Species	Sl. No.	Fin - Fish Species
1	*Hilsa ilisha* (Hamilton)	2	*Hilsa toli* (Cuvier)
3	*Ilisha elongata* (Bennet)	4	*Gadusia chapra* (Hamilton)
5	*Corica soborna* (Hamilton)	6	*Anodontostoma chacunda* (Hamilton)
7	*Nematolosa nasus* (Bloch)	8	*Pellona ditchela* Valenciennes
9	*Raconda russeliana* (Grey)	10	*Coilia dussmieri* Valenciennes
11	*Coilia ramcartii* (Hamilton)	12	*Setipinnia phasa* (Hamilton)
13	*Setipinnia taty* (Valenciennes)	14	*Thryssa hamiltoni* (Gray)
15	*Thryssa purava* (Hamilton)	16	*Megalops cyprinoides* (Broussouuet)
17	*Anguilla bengalensis* (Gray)	18	*Murenesox cinereus* (Forsskal)
19	*Chanos chanos* (Forsskal)	20	*Aorichthys aor* (Hamilton)
21	*Mystus cavasius* (Hamilton)	22	*Mystus gulio* (Hamilton)
23	*Mystus vittatus* (Bloch)	24	*Silonia silondia* (Hamilton)
25	*Pangasius pangasius* (Hamilton)	26	*Arius sona* (Hamilton)
27	*Arius gagora* (Hamilton)	28	*Arius caelatus* Valenciennes
29	*Arius jella* (Day)	30	*Osteogeniosus militaris* (Linn.)
31	*Plotossus canius* (Hamilton)	32	*Harpadon nehereus* (Hamilton)
33	*Lates calcarifer* (Bloch)	34	*Chanda nama* (Hamilton
35	*Parambasis baculis* (Ham.)	36	*Parambassis ranga* (Ham.)
37	*Terapon jarbua* (Forsskal)	38	*Sillaginopsis panijus* (Ham.)
39	*Sillago sihama* (Forsskal)	40	*Caranx carangus* (Bloch)
41	*Megalaspis cordyla* (Linn.)	42	*Leiognathus blochii* (Val.)
43	*Leiognathus equalus* (Forsskal)	44	*Leignathus fasciatus* (Lacepede)
45	*Gazza minuta* (Bloch)	46	*Lutjanus argentimaculatus* (Forsskal)
47	*Lutjanus johni* (Bloch)	48	*Datniodes quadrifasciatus* (Sovas.)
49	*Gerres oyena* (Forsskal)	50	*Gerreomorpha setifer* (Ham.)
51	*Pomadasys argenteus* (Forsskal)	52	*Pomadsys hasta* (Bloch)
53	*Dasysciaena albida* (Cuvier)	54	*Johnius sina* (Cuvier)
55	*Johnius vogleri* (Bleeker)	56	*Nibea soldado* (Lacepede)
57	*Otolithoides biauritus* (Cantor)	58	*Pama pama* (Ham.)
59	*Drepane punctata* (Linn.)	60	*Scatophagus argus* (Linn.)
61	*Etroplus suratensis* (Bloch)	62	*Liza microlepis* (Smith)
63	*Liza parsia* (Ham.)	64	*Liza tade* (Forsskal)
65	*Mugil cephalus* (Linn.)	66	*Rhinomugil corsula* (Ham.)
67	*Valamugil cunnesius* (Val.)	68	*Valamugil speigleri* (Bleeker)
69	*Eleutheronema tetradactylum*	70	*Polydactylus indicus* (Shaw)
71	*Polynemus paradiseus* (Linn.)	72	*Xenentodon cancila* (Ham.)
73	*Glossogobius guiris* (Ham.)	74	*Boleophthalamus boddaerti* (Val.)
75	*Pseudapocryptes lanceolatus*	76	*Odontamblyopus rubicundus* (Ham.)
77	*Lepturacanthus savala* (Cuvier)	78	*Lepturacanthus gangeticus* (Gupta)

Contd...

Sl. No.	Fin - Fish Species	Sl. No.	Fin - Fish Species
79	*Enpleurogrammus muticus* (Gray)	80	*Trichiurus pantului* (Gupta)
81	*Pampus argenteus* (Euphrasen)	82	*Pampus chinensis* (Euphrasen)
83	*Stromateus sinensis* (Day)	84	*Scomberomorus commerson* (Lacepede)
85	*Euryglossa pan* (Ham.)	86	*Cynoglossus cynoglossus* (Ham.)
87	*Chirocentrus dorab* (Forsskal)	88	*Bagarius bagarius* (Ham.)
89	*Carcharhinus melanopterus* Quoy	90	*Rhizoprionodon acutus* (Rupell)
91	*Scoliodon laticaudus* (Muller & Henle)	92	*Aetobatus narinari* (Blainville)
93	*Aetomylaeus nichofii* (Schneider)	94	*Himantura marginata* (Blyth)

Commercially Important Shell - Fish Species

	Prawns		*Crabs*
1	*Palaemon styliferus* Milne - Edw.	1	*Scylla serrata*
2	*Macrobrachium lamarrei* (Milne-Edw.)	2	*Varuna litterata* (Fab.)
3	*Penaeus monodon* Fabricius	3	*Matuta planipes*
4	*Penaeus indicus* Edwards	4	*Matuta lunaris*
5	*Metapenaes monoceros* Fabricius	5	*Portunus pelagicus*
6	*Metapenaeus brevicornis* (Milne-Edw.)	6	*Portunus sanguinolentus*
7	*Parapenaeopsis sculptilis* (Hellar)	7	*Charybdis feriata*
8	*Parapenaeopsis stylifera* (Milne-Edw.)	8	*Charybdis rostata* (Milne - Edw.)
9	*Palaemon rude* (Hellar)		**Mollusca**
10	*Macrobrachium rosenbergii* (Fab.)	1	*Anadara granosa* (Linn.)
11	*Acetes indicus* Milne - Edw.	2	*Crassotrea gryphoides* (Schlotheium)
12	*Acetes erythraeus* Nobili	3	*Saceostrea cuculata* (Born)

(Source: Das and Nandi, 1999).

Table 9 : List of Extinct and Threatened/Endangered Fauna in the Indian Sundarbans.

Sl. No.	*Groups/ Common Name*	*Scientific Name*
A	**EXTINCT FAUNA**	
1.	Swamp Deer	*Cervus duvaucelli*
2.	Wild Buffalo	*Bubalus bubalis*
3.	Javan Rhinoceros	*Rhinoceros sondaicus*
4.	Great One-horned Rhinoceros	*Rhinoceros unicornis*
B	**THREATENED / ENDANGERED FAUNA**	
I	**REPTILES**	
1.	Estuarine Crocodile	*Crocodylus porosus*
2.	Olive Ridley Turtle	*Lepidochelys olivacea*
3.	Batagur Turtle	*Batagur baska*
4.	Indian Flapshelled Turtle	*Lissemys punctata*
5.	Indian Softshelled Turtle	*Trionyx gangeticus*
6.	Peacock - marked Softshelled Turtle	*Trionyx hurum*
7.	Indian Tent Turtle	*Kachuga tecta*
8.	Common Indian Monitor	*Varanus bengalensis*
9.	Yellow Monitor	*Varanus flavescens*
10.	Water Monitor	*Varanus salvator*
11.	Indian Rock Python	*Python morulus*
II	**BIRDS**	
1.	Giant Heron	*Ardea goliath*
2.	Dalmatian Pelican	*Pelecanus philippensis*
3.	Lesser Adjutant Stork	*Leptotilos dubius*
4.	Whitebellied Sea Eagle	*Haliaeetus leucogaster*
5.	Osprey	*Pandion haliaetus*
6.	Shaheen Falcon	*Falco peregrinus*
III	**MAMMALS**	
1.	Leopard Cat	*Felis bengalensis*
2.	Fishing Cat	*Felis viverrina*
3.	Royal Bengal Tiger	*Panthera tigris tigris*
4.	Gangetic Dolphin	*Platenista gangetica*
5.	Irrawady Dolphin	*Orcaella brevirostris*
6.	Little Indian Porpoise	*Neophocaena phocaenoides*
7.	Chinese Pangolin	*Manis pentadactyla*

(Source: Red Data Book of Indian Animals).

Note: Barking Deer *Muntiacus muntjac* (Zimmerman) was reported to be extinct but have been recently recorded from Holiday Island, Sundarbans, West Bengal.

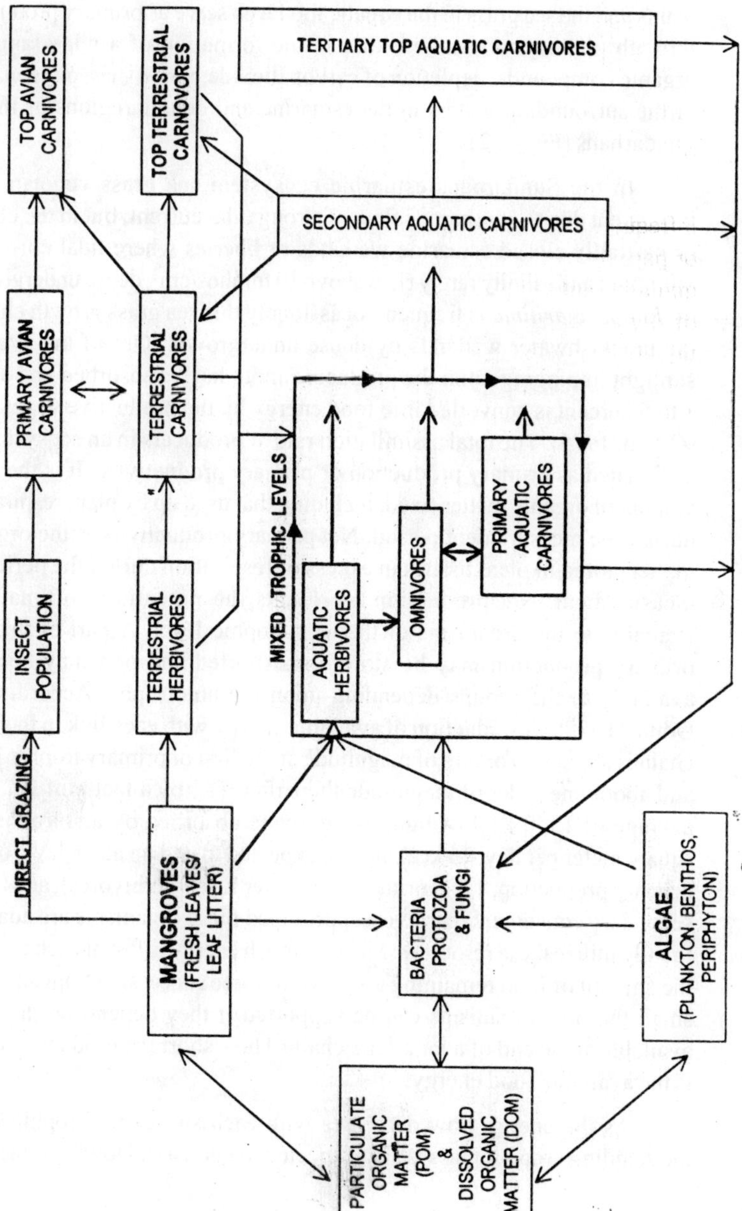

Figure 2 : Compartmental Model of the Food Chains Operating in the Indian Sundarbans Mangals

minor mangroves, mangrove associates, other terrestrial forms of green plants and the sea grass in the aquatic food web serve as primary producers, through photosysnthesis resulting in the formation of a wide range of organic compounds, depletion of carbon dioxide, *vis - a - vis,* oxygenation of the surrounding waters in the estuarine and deltaic regions of Indian Sundarbans (Figure 2).

In the Sundarbans estuarine ecosystem sea grass vegetation is infrequent due to continuous flow of strong tidal current; but in the closed or partiallly closed estuarine wetlands or bheries where tidal current is minimum and salinity rarely rises above 10 mmhos/cm^2, dense undergrowth of *Ruppia maritima* is frequent; ocassionaly this sea grass growth chokes the brackishwater wetlands by dense undergrowth. Out of the average sunlight impinging upon the producers about half is absorbed and about 1 to 5 percent is converted into food energy by the productive vegetation (Odum, 1975). The total assimilation rate of producers in an ecosystem is designated as primary production or primary productivity. It is the total amount of organic matter fixed, including that used up by plant respiration during the measurement period. Net primary productivity is the organic matter stored in plant tissues in excess of respiration during the period of measurement. Net production represents the potential food material available to the organisms of the next trophic level. A part of this net primary production may be stored or exported and part may become available to the groups depending upon the autotrophs. According to Odum (1975), the reduction of available energy with each link in the food chain is about two orders of magnitude at the first or primary trophic level and about one order of magnitude thereafter, *i.e.,* by a factor of 10. If an average of 1500 kcal of light energy were absorbed by autotrophs per square meter per day, 15 kcal may be expected to end up at the level of net primary production, 1.5 at primary consumer level (herbivores), and 0.3 at secondary consumers (carnivores) provided these organisms are adapted to fully utilize these resources. Since so much energy is lost at each transfer the amount of food remaining after two or three successive transfers is so small that few organisms can be supported if they depend on the food available at the end of a long food chain. Thus, shorter a food chain more is the available food energy.

As the energy flow decreases with each successive trophic level the standing crop biomass is also expected to decrease. However, the size of the individual organisms making up a trophic level has influence on the standing crop biomass. Hence, in an aquatic system the producers are composed largely of very small organisms and the consumers are large and by harvesting at frequent intervals an equivalent amount of food (net production) from mass cultures of small algae may be obtained as from a grain crop harvested after a long interval of time (Odum, 1975). Rates of photosynthesis in water bodies are related to phytoplankton / algal abundance and light intensity, therefore, the photosynthetic production will increase with increased abundance of phytoplankton and algae.

The optimum factors for healthy primary production from any area depends on adequate availibility of Sun light as radiant energy and its optimum transmission in the water column and also adequate and continuous availability of nutrients to maintain a steady flow of productivity. In the Sundarbans tidal estuaries the availability of the solar radiation and adequate supply of nutrients from the dense mangrove forest zones is beyond any question. As such, it is definitely referred as the highly productive ecosystem among the tropical and subtropical zones of the world (Naskar & Mandal, 1999).

1.9. Interaction in the Mangrove Ecosystem

A major characteristic of all ecosystem in general is that the net production is either directly consumed as living material or is consumed later as dead material, likewise, it can be generalized that two different types of food chains are found to operate in nature–the Grazing Food Chain and the Detritus Food Chain. The food chains operating in the mangals of Sundarbans are of a complex type, with influx from both Grazing Food Chain and Detritus Food Chain. In the estuarine water it is practically a Detritus Food Chain operating, which begins with dead mangrove leaves (leaf litter) and other detritus that falls in, is blown in or washed down from the forest floor. These are transported by tides and currents over large areas of the bay. The leaf fragments, a very small amount of which is directly grazed upon by insects are acted upon by bacteria, protozoa and fungus and along with the algal forms of phytoplankton, benthos and periphyton are eaten and re-eaten by a key group of small animals, *viz.*, zooplankton (including rotifers, copepods, cladocerons and larval forms of molluscs and insects), nematodes, crabs, shrimps, small herbivorous fin - fish and molluscs. These animals though small in size make upto a large number of individuals. These are eaten by small carnivorous fish, fish eating birds, fishing cats, snakes, water monitors etc. The top consumer in the estuarine water of Sundarbans is the estuarine crocodile *(Crocodylus porosus).* The faecal matter and the organic matter of all these organisms go back into the ecosystem as detritus. On land the Grazing Food Chain operates, wherein the grazing herbivores including deer, wild boar, rhesus monkeys etc. feed on mangrove leaves and terrestrial top consumer tiger *(Panthera tigris tigris)* in turn feeds on these herbivores (Fig. 3).

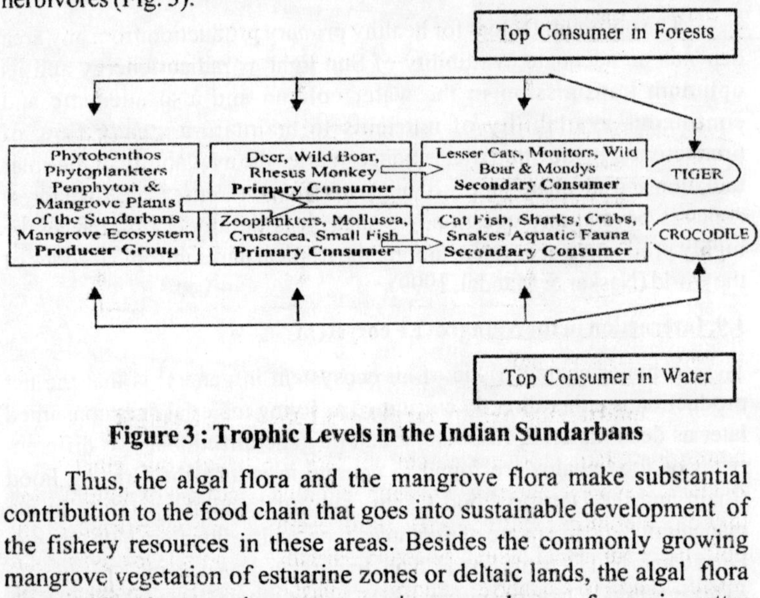

Figure 3 : Trophic Levels in the Indian Sundarbans

Thus, the algal flora and the mangrove flora make substantial contribution to the food chain that goes into sustainable development of the fishery resources in these areas. Besides the commonly growing mangrove vegetation of estuarine zones or deltaic lands, the algal flora are considered as most important as primary producers of organic matter in the estuarine environment because of their active photosynthetic ability. This view is also supported by Bold and Wynne (1968). Most animal life in the estuarine environment mainly depend on algae because of their primary source of energy and natural food for them. In the mangrove ecosystem major algal flora constitute the main part of the food chain (Naskar and Mandal, 1999).

Chapter 2

Geology and Geography of the Indian Sundarbans

2.1. Geology of the Indian Sundarbans

The present day status of the coastal plains of the lower Ganga delta is the outcome of the combined effects of the deltaic deposits of the Ganga- Brahmaputra rivers, the tidal deposits from the Bay of Bengal in a basement of the Bengal basin and an overall southerly tilt during the recent times (Bhattacharya, 1999). The coastal area has a gentle seaward slope and major portion of it lies 7 - 8 m above mean sea - level (M.S.L.). The meso-macrotidal regime inundates most areas twice daily. Erosion in the coastal area is led by gradual denudation, as well as, by catastrophic events. Gradual erosion results from: (i) flood and ebb tidal currents, (ii) high tidal range and (iii) strong littoral drift; catastrophies include : (a) cyclones and hurricanes, (b) storm - induced waves, (c) tidal bores and (d) tectonic (neo-tectonic) movements (Bhattacharya & Das, 1994; Bhattacharya, 1999). Inspite of all these, the continuous carrying of silt load on the overall deltaic regions of lower Ganga delta are in prograding condition (Sanyal, 1999).

Out of 9,630 sq. km. of the Indian Sundarbans deltaic region, towards the south-eastern part of the *Dampier & Hodges line* (1831), more than 5,000 sq. km. area has already been reclaimed for human habitation and conversion to agricultural fields, fisheries, roadways, railways, ports and harbours (Table 1). Out of about 110 deltaic islands of the Indian Sundarbans more than 50% area in 54 islands is now under human settlement with a total estimated population as on the beginning of the 21st Century of about 40 lakhs, who are solely dependent on these deltaic

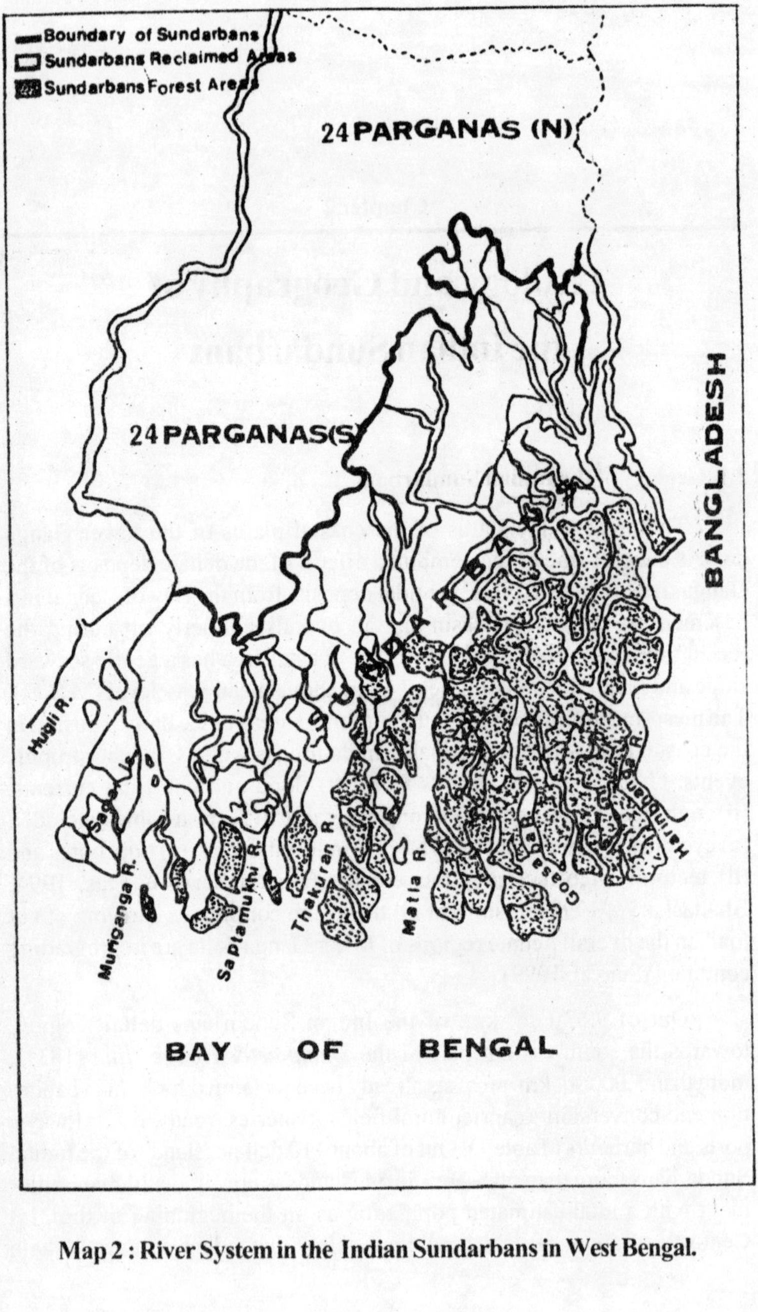

Map 2 : River System in the Indian Sundarbans in West Bengal.

regions, primarily on rain-fed agriculture, secondarily on inland brackish water fisheries and fishing and thirdly on forest resources. This reclamation of Sundarbans forest zones was initiated since the later half of the 18th Century and till date about 56 islands within the inter tidal limits exist in the Sundarbans mangrove ecosystem covering about 55% of the total area; remaining 45% area in the inland of Indian Sundarbans are 6 major estuarine rivers, *viz.*, Hugli mouth (main Hugli river and Baratala river), Saptamukhi, Thakuran or Jamira, Matla, Goasaba, Harinbari or Harinbhanga and several other wide to narrow rivers, creeks, canals and marshy areas, which form the criss-cross network system (Map 2). During the solar phases, *i.e.*, first moon and full moon most of the islands get inundated with the high tide water diurnally and get exposed during low tide. Due to choking of the distributaries of the Ganga (Bhagirathi - Hugli) in the Sundarbans with upstream, water presently freshwater discharge in these rivers of Sundarbans are negligible or mostly absent except during monsoon months when downpour in the southern parts of Sundarbans occur. Monsoon precipitate water wash and flow through the rivers of the southern parts of Bengal in Sundarbans during the middle of June to the middle of October (Table 10).

Table 10 : Climate of the Sundarbans in West Bengal

Sl. No.		***Winter*** *15th Nov.- 14th Feb.*	***Summer*** *15th Feb.- 14th June*	***Monsoon*** *15th June- 14th Nov.*
A.	**Temperature (°C)**			
(i)	Highest Temperature	23.0°C	31.2°C	31.1°C
(ii)	Lowest Temperature	13.0°C	28.8°C	26.0°C
B.	**Average Rainfall (mm)**			
(i)	Average rainfall/month	20.7 mm	111.2 mm	281.1 mm
(ii)	Average rainy days/month	1.7 days	7.5 days	19.3 days
(iii)	Total rainfall during the season	82.8 mm	44.8 mm	1124.4 mm
C.	**Relative Humidity (%)**			
(i)	Average humidity at 8.30 hrs.	0.77	0.83	0.84
(ii)	Average humidity at 17.30 hrs.	0.72	0.83	0.82
(iii)	Highest humidity	0.94	0.96	0.96
(iv)	Lowest humidity	0.50	0.69	0.70

(Source : Naskar, 2000)

2.2. Geographical Limits of the Indian Sundarbans

The Indian Sundarbans falls within the geographical limits of the *Dampier & Hodges* line (an imaginary line drawn in 1831) in the north-west, the river Hugli in the west, the river Icchamati-Kalindi -Raimangal-Harinbari in the east, which is also the International boundary between India and Bangladesh and the Bay of Bengal in the south. This total 9,630 sq. km. is restricted within the latitudes 21° 31′ N and 22° 30′ N and longitude between 88° 10′ E and 89° 51′ E, spread over the districts 24 Parganas South and southern parts of 24 Parganas North. Since the latter part of the 18th Century about 5,364 sq. km. area of the north - western parts of these deltaic and intertidal lands of Sundarbans have been cleared for human habitation and settlement, development of agricultural fields and brackish water fisheries (Naskar and Guha Bakshi, 1987). These Sundarbans mangrove reclaimed zones are densely populated and include 19 rural block, 13 blocks under the jurisdiction of South 24 Parganas and 6 blocks under North 24 Parganas (Table 11; Map 3).

The remaining 4,266.6 sq. km. area is the present day Sundarbans Reserve forests; out of this 4,266.6 sq. km. area, about 2,300 sq. km. is land or forest area, 1,750 sq. km. is under water and about 216.6 sq. km area includes denuded sea shore and sand char areas (Naskar & Guha Bakshi, 1987). Total length of the coast-line of the Indian Sundarbans is about 150 km and is spread over about 50 -70 km. inside the main land. The river systems include several distributaries of the river Hugli, the 6 major estuarine rivers, which meet the Bay are Muriganga or Baratala river, Saptamukhi river, Thakuran river, Matla river, Goasaba river and Raimangal/ Harinbari/ Herobhanga river. Besides these major rivers and distributaries several other minor but important rivers also exist, these are rivers Bidya, Raimangal, Ichhamati, Jhilla, Gomor and Panchamukhani. Also, there are a large number of heavily choked up or silted up rivers or canals namely Ghugudanga gang or Hatania Duania river, Deogra khal, Piyali river, Kharatya khal, Durgaduani river and Terobanki river. Unfortunately, since the 16th Century, all these rivers/ distributaries have lost their earlier connections with the Hugli river. Besides the rivers, canals and creeks mentioned here, there are innumerable small and narrow creeks or canals, which derive their names from the local names of the nearby villages or forest areas they pass through.

Table 11: Population in the Rural Blocks of Mangrove Reclaimed Areas of Indian Sundarbans.

Sl.No.	*Name of the Blocks*	*Area (hectare)*	*Estimated Population (1991)*
	Thirteen Blocks Under the District 24 Parganas South		
1	Gosaba	33,725	2,00,514
2	Basanti	29,000	2,26,974
3	Canning - 1	20,568	2,06,100
4	Canning - 2	22,528	1,51,635
5	Joynagar - 1	12,711	1,85,271
6	Joynagar - 2	17,518	1,77,335
7	Kultali	23,948	1,56,450
8	Mathurapur - 1	14,838	1,41,888
9	Mathurapur - 2	23,050	1,72,982
10	Patharpratima	46,950	2,45,601
11	Namkhana	22,723	1,34,354
12	Sagar	47,080	1,54,202
13	Kakdwip	26,110	1,90,088
	Six Blocks Under the District 24 Parganas North		
1	Minakhan	20,095	1,37,361
2	Haroa	15,675	1,51,100
3	Sandeshkhali - 1	17,679	1,20,539
4	Sandeshkhali - 2	19,500	1,18,895
5	Hasnabad	14,688	1,51,115
6	Hangalgunj	23,040	1,42,291
	TOTAL	4,13,568	31,64,695

Source : Sundarbans Development Board, West Bengal).

* Population during 2001 is estimated about 4.2 million.

In 1989, the entire 9,630 sq. km. area under Sundarbans of West Bengal, India was declared under the Man and Biosphere Reserve Project. This area has further been categorised under the following management practices :

(1) **Core area in the Sundarbans Tiger Reserve,** covering an area of 1,700 sq. km., which also includes the 1,300 sq. km. area of the National Park of the Sundarbans; this area is densely covered with the mangrove species, which are protected under strict conservation practices.

(2) **Restoration Zone** covering an area of 350 sq. km. These degraded forest areas include river banks and the degraded non - forested mud -

flats of 200 sq. km., these areas are being afforested with suitable mangrove species.

(3) **Manipulation Zone - I** (Forestry) covering an area of 2,564 sq. km. of the mangrove forests; in this zone developmental activities are allowed and the rural population is allowed to harvest the forest products and fishing is also permitted in these tidal rivers, canals, creeks and lagoons.

(4) **Manipulation Zone - II** (Non - forested area or agricultural lands) includes an area of about 4,916 sq. km. situated outside the mangrove forest zones. This area needs to be developed by implementation of scientific and modern techniques of agriculture, aquaculture and other resource harvesting practices for the benefit and upliftment of the rural population, residing in these areas.

The important physical and meteorological factors of this dynamic coastal zones are

Wind : Two dominant prevailing winds during South-West and North-East monsoons lead to the Southwesterly and the Northeasterly wind (Banerjee, 1972). The normal maximum wind velocity is 16.7-50 km/h (April - June) and minimum wind velocity is 10.7 - 11.8 km/h (December - February).

Cyclones : Bay of Bengal is cyclone prone. On an average 3 - 4 cyclones may occur in a year. Maximum wind velocity during cyclones range from 80 - 140 km/h. Cyclones initiate large - scale littoral drift, severe coastal erosion and underwater accumulation of eroded material apart from causing heavy loss of life and property. Since 1582, there have been about 90 severe cyclone, strong wind or tidal surges partly or fully devastating the Sundarbans area again and again and the human settlement suffered most; among these the cyclone at the rate of 250 km/hour, tidal thrust at the rate of 41 feet (about 13 m) along with earth quake completely devastated the Indian and Bangladesh part of Sundarbans during 1737. During 1582, there is report of a severe cyclone which took the lives of more than 2 lakh people inhabiting the Sundarbans area (Table 12).

Waves : Wave height ranges from 0 - 0.6 m with a wave period of 5 - 7 seconds during the clear winter season, whereas, these become 1.8-2.4m and 12 - 14 seconds during the rough summer seasons. During periods of cyclonic storms wave height goes above 2.4 m and wave period above 14 seconds (Banerjee, 1972).

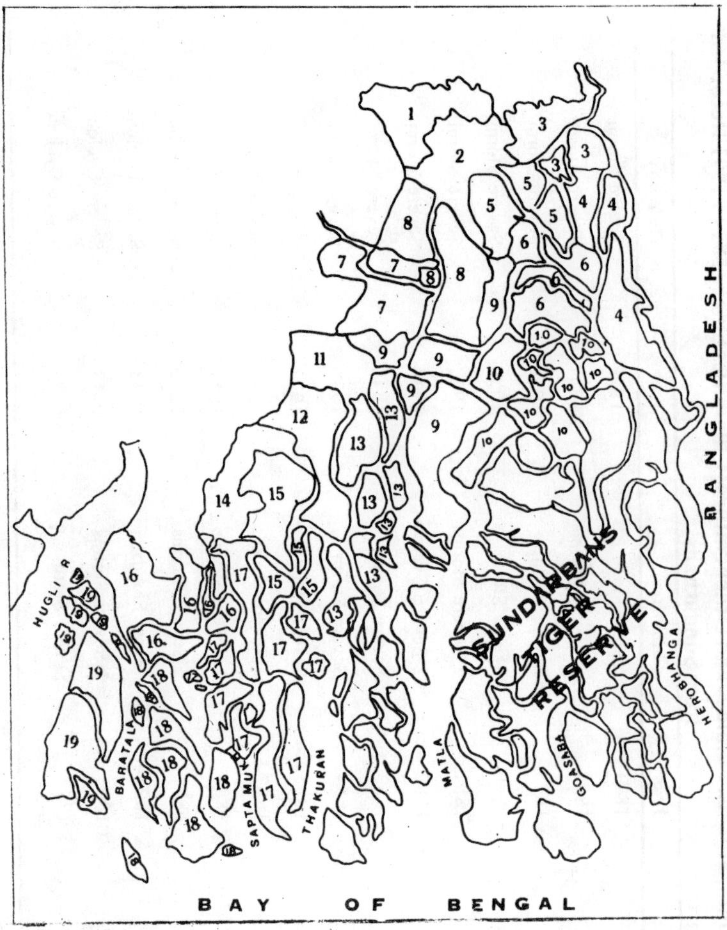

Map 3: Different Rural Blocks in Indian Sundarbans, West Bengal

Six Rural Blocks in the District 24 Parganas (North) of West Bengal :

1. Haroa; 2. Minakhan; 3. Hasnabad; 4. Hingalganj; 5. Sandeshkhali I; 6. Sandeshkhali II.

Sixteen Rural Blocks in the District 24 Parganas (South) of West Bengal: 7. Canning I; 8. Canning II; 9. Basanti; 10. Gosaba; 11. Joyngar I; 12. Joynagar II; 13. Kultali; 14. Mathurapur I; 15. Mathurapur II; 16. Kakdwip; 17. Patharpratima; 18. Namkhana; 19. Sagar.

Table 12: Devastating Cyclones, Storms, Famines and Other Natural Calamities Since 1582 to 1991 Recorded from Sundarbans

Year	*Calamity*	*Year*	*Calamity*	*Year*	*Calamity*
1582	Severe Cyclone	1871	Severe Flood	1922	Severe Storm
1688	Severe Cyclone	1877	3 Severe Storms	1927	Severe Storm
1707	Severe Cyclone	1878	Severe Storm	1928	Severe Storm
1737	Cyclone +Earthquake	1880	Severe Storm	1929	Severe Storm
1742	Severe Cyclone	1881	2 Severe Storms	1932	Severe Storm
1762	Earthquake	1882	2 Severe Cyclones	1934	Severe Storm
1770	Famine	1883	Severe Storm	1935	Severe Storm
1791	Famine	1884	Severe Storm	1936	Severe Storm
1823	Severe Flood	1885	Severe Flood	1937	Severe Storm
1830	Severe Storm	1887	2 Severe Storms	1940	Severe Storm
1832	3 Severe Storms	1888	3 Severe Storms	1941	Severe Storm
1833	Severe Cyclone	1889	2 Severe Storms	1942	Severe Famine
1838	Severe Flood	1890	Severe Flood	1943	Severe Cyclone
1839	Severe Storm	1893	2 Severe Storms	1948	Severe Cyclone
1840	Severe Storm	1894	Severe Storm	1956	Severe Cyclone

1844	Severe Storm	1897	Earthquake	1961	Severe Cyclone
1848	Severe Storm	1898	4 Severe Storms	1962	Severe Cyclone
1850	Severe Storm	1899	Severe Storm	1965	Severe Storm
1852	Severe Storm	1900	Severe Flood	1966	Severe Flood
1856	Severe Flood	1901	4 Severe Storms	1968	Severe Flood
1858	Severe Storm	1904	2 Severe Storms	1970	Severe Storm
1859	2 Severe Storms	1907	2 Severe Floods	1976	Severe Flood
1862	Severe Cyclone	1909	Severe Cyclone	1978	Severe Flood
1864	Flood + Cyclone	1913	2 Severe Storms	1981	Severe Storm
1865	Famine	1916	2 Severe Storms	1982	Severe Storm
1867	Severe Cyclone	1917	Severe Storm	1985	Severe Storm
1868	Severe Flood	1919	Severe Storms	1988	Cyclone + Storm
1869	Severe Flood	1921	2 Famines	1991	Cyclone + Storm

(Source : Naskar, 1999)

Tide : Tidal length in different inlets ranges from 60 - 80 km, spring tidal range is 4.87 and the neap is 2.13 m.

Rainfall : Average annual rainfall is 1,750 - 1,800 mm of which 80% of precipitation takes place between May and September (Naskar & Mandal, 1999) Table 1- 10).

Temperature : Monthly average air temperature ranges from 34°C- 26°C during April - June, the minimum 25°C - 21 °C during December - June (Table 10).

Relative humidity : Relative humidity is over 80% from June - September and over 75% from October - May (Table 10).

2.3. Land Forms in the Indian Sundarbans Mangal

The various land forms encountered in the Sundarbans areas in the three different geo-morphological settings, *viz.,* estuarine swamps, deltaic swamps and lagoon swamps are mudflats, salt-pans, salt-flats, lower swamps, higher swamps and natural levees (Paul, 1999).

Mudflats: Mudflats are produced in estuarine and sheltered coastal areas by flocculation process of an expanse of fine clay or silt deposition. The seaward and channel-ward sides of mangrove swamps or salt marshes are dominated by mudflats.

Salt-pans: High tides usually submerge the mature mangrove swamps. The seepage channels carry silts mainly from interior of the tidal flood plains by scouring caused by ebb- tides. This continuous scouring without compensatory silting during the flood tends to lower the interior of the flood plains and gradually form pan areas. In the tidal swamps of the abandoned delta where the influence of freshwater is almost nil in the dry period the rapid evaporation of water between two high tides leads to formation of salt encrusted flat in the landward zone. Such areas with hypersaline environment remain as salt blanks.

Lower Swamp Surface: These areas are situated at the edges of the islands and riverside bars within the height under frequent tidal flooding. These inter tidal flats comprise of smooth seaward and channel-ward dipping flats dissected by large and small - scale tidal channels. Fresh silt accumulation takes place during overflow of channels in each high tide.

Higher Swamps: These are certain plains dissected by a network of tidal channels or creeks. During high state of tides (specially high springs) salt water enter the channels, overtop the channel banks and inundate the adjacent swamp surface. After a period of slack water stage, tide water drain through the creeks or channels and re-expose the swamp surface.

Natural Levees: These are formed on either bank of tidal creeks. If a channel overflows its banks during the flood tide, the water does not remain confined to the channel but flows over the land surface in a broad sheet. This flow has reduced water velocity and some of the suspended sediment settles out and is deposited close to the channel building up an embankment known as natural levee.

Since the later part of the 18th century, the newly settled human population had converted these natural landscape into agricultural fields, brackish-water fisheries, habitation areas, road ways, railway tracks, market place, ports, harbours and very recently as the most profitable brick fields.

2.4. Water Characteristics in the Indian Sundarbans

The most distinctive feature of the Sundarbans mangrove forests areas is its criss-cross tidal river and creek systems in the estuarine environment. The physico-chemical conditions of these water bodies in Sundarbans is conducive for survival and growth of brackish-water fishes, shrimps, prawns and crabs. The important physico-chemical features as observed by Nath and De (1999) are :

Temperature : Water temperature ranges between 20° C and 34° C. Maximum temperature is observed during May - June and minimum during January - February.

Transparency : Transparency values fluctuate both seasonally and diurnally. Maximum transparency was recorded during winter, when the weather was calm and sunny and minimum transparency during monsoon due to turbulence and flooding.

Dissolved Oxygen : The dissolved oxygen in the water bodies ranges between 4.1 and 7.5 ppm.

pH : The water reaction is slightly alkaline (pH 7.5 - 8.5). Higher pH values were observed during winter, while lower values during monsoon. In general the water has good buffering action so that only minor fluctuation of pH was recorded.

Total Alkalinity : The total alkalinity ranged between 90 and 150 ppm. Alkalinity was maximum during summer and minimum during post - monsoon.

Free Carbon di oxide : Free carbon di oxide in water ranges between 0 and 9 ppm.

Salinity : According to Nath & De (1999), the salinity of the Hugli estuary has dropped considerably after the construction of Farakka Barrage during 1975, due to discharge of more freshwater from Ganga riverine system.

Phosphate : Phosphate concentration in the water bodies ranges between 0.03 and 0.15 ppm. Phosphate is found to be maximum during winter and minimum during April.

Nitrate : Nitrate concentration ranges between 0.06 and 0.504 ppm. The organic substances decompose and release nitrate into the system.

Silicate : Silicate concentration ranges between 1.4 and 10.0 ppm. At higher salinity, the silicate content was usually low and *vice versa.*

Calcium and Magnesium – Moderately high content of calcium (40-393 ppm) and magnesium (12 - 216ppm) have been recorded. Maximum concentration of calcium and magnesium was found during early summer and minimum after monsoon.

Free Ammonia : Free ammonia is an index of water pollution and this was usually absent from the water samples from the estuaries.

Potassium : Potassium content ranges between 16.4 and 718.8 ppm.

Sulphate : Sulphate content ranges between 13 and 390 ppm.

But due to choking and blocking of earlier connecting rivers of the Sundarbans from the main flow of the upstream Hugli River (branch of Ganga) the salinity in the mid part of the Sundarbans, *i.e.,* in between the river Saptamukhi to the river Matla have raised considerably Naskar (1999). Naskar (1999) has also reported the chemical properties of the saline water of the Sundarbans, based on the collection and analysis of water samples from the different parts of the Sundarbans estuaries, covering all the seasons of the year (Tables 13, 14 & 15).

Table 13 : Chemical Properties of Estuarine Water in Sundarbans (Zone - I. Eastern Part of the Sundarbans Mangal)

Parameters	*Summer (March - May)*	*Post monsoon (Sept. - Nov.)*	*Winter (Dec. - Feb.)*
pH	7.39 - 7.97	7.02 - 7.12	7.27 - 7.50
Ec (mmhos/ cm)	20 - 36	12 - 14	12.0 - 14.5
Na^+ (ppm)	8000 - 10000	3000 - 4000	5000 - 8000
K^+ (ppm)	200 - 500	190 - 300	280 - 530
Ca^{++} (me/l)	8.8 - 15.0	13.2 - 20.4	15.2 - 22.4
Mg^{++} (me/l)	40.0 - 91.4	35.0 - 37.6	78.0 - 82.9
NH_4^+ (ppm)	12.57 - 18.32	20.38 - 25.20	1.28 - 2.60
NO_3^- (ppm)	12.36 - 16.46	12.57 - 18.00	0.64 - 1.28
CO_3^{2-} (me/l)	1.04 - 1.50	nil - 1.00	nil - 0.94
HCO_3^- (me/l)	1.56 - 2.00	1.248 - 1.872	4.8 - 7.36
Cl^- (ppt)	10.13 - 11.61	5.26 - 7.28	8.17 - 8.52
Salinity (ppt)	18.83 - 21.00	9.71 - 13.17	14.79 - 15.40
SO_4^{2-} (ppm)	460 - 490	410 - 430	420 - 440
PO_4^{3-} (ppm)	Trace - 0.17	Trace - 0.09	Trace - 0.90

(Source: Naskar, 1999)

Table 14: Chemical Properties of Estuarine Water in Sundarbans (Zone - II. Central Part of the Sundarbans Mangal)

Parameters	*Summer (March - May)*	*Early monsoon (June - Aug.)*	*Late monsoon (Sept. - Nov.)*
pH	7.3 - 7.6	7.1 - 7.8	7.2 - 7.7
Ec (mmhos/ cm)	12 - 18	1.4 - 9.0	5.5 - 9.0
Na^+ (ppm)	9000 - 10500	3500 - 9000	6000 - 7500
K^+ (ppm)	405 - 425	55 - 295	245 - 270
Ca^{++} (me/l)	14 - 16.8	4.8 - 5.2	6.8 - 9.2
Mg^{++} (me/l)	64 - 72.8	2.8 - 51.8	12.7 - 59.8
NH_4^+ (ppm)	1.0 - 1.4	2.9 - 4.7	3,0 - 9.2
NO_3^- (ppm)	0 - 0.8	0.7 - 1.2	1.2 - 2.5
Total N (mg/l)	0.2 - 0.6	3.6 - 5.9	4.2 - 7.7
CO_3^{2-} (me/l)	0 - 0.32	0 - 0.36	0 0.36
HCO_3^- (me/l)	1.1 - 1.8	0.03 - 1.9	1.4 - 1.9
Cl^- (ppt)	13.0 - 14.5	0.6 - 7.3	3.6 - 6.1
Salinity (ppt)	24.00 - 26.24	1.18 - 13.35	6.68 - 11.10
SO_4^{2-} (ppm)	148 - 153	85 - 144	135 - 138
DO (ppm)	4.0 - 7.6	3.0 - 8.2	5.8 - 9.2
PO_4^{3-} (ppm)	Trace - 0.1	Trace - 0.1	Trace - 0.2
Temperature- air (°C)	24.0 - 26.0	20.0 - 22.0	19 - 21
Temperature - water (°C)	14.0 - 16.0	11.0 - 12.0	10 - 13

(Source: Naskar, 1999)

Table 15 : Chemical Properties of Estuarine Water in Sundarbans (Zone - III. Western Part of the Sundarbans Mangal)

Parameters	*Summer (March - May)*	*Post monsoon (Sept. - Nov.)*	*Winter (Dec. - Feb.)*
pH	7.46 - 7.82	7.36 - 7.87	7.24 - 7.91
Ec (mmhos/ cm)	14.5 16.0	1.7 - 10.0	4.8 - 19.5
Na^+ (ppm)	5500 - 12000	4000 - 6000	8500 - 14000
K^+ (ppm)	490 - 505	270 - 290	500 - 540
Ca^{++} (me/l)	16.8 - 33.2	5.0 - 12.8	10.2 - 22.0
Mg^{++} (me/l)	51.6 - 100.5	5.4 - 38.0	84.2 - 91.2
NH_4^+ (ppm)	3.5 - 24.5	2.2 - 4.8	1.1 - 4.7
NO_3^- (ppm)	2.63 - 23.8	0.82 - 1.40	0.8 - 2.5
Total N (mg/l)	6.13 - 58.4	3.0 - 6.2	1.9 - 7.2
CO_3^{2-} (me/l)	0 - 0.2	0 - 0.72	0.30 - 0.91
HCO_3^- (me/l)	1.0 - 1.2	4.00 - 4.05	2.0 - 3.3
Cl^- (ppt)	9.07 - 13.86	8.00 - 8.49	8.62 - 9.63
Salinity (ppt)	16.40 - 28.05	14.47 - 15.27	15.58 - 17.42
SO_4^{2-} (ppm)	147 - 153	127 - 130	100 - 108
DO (ppm)	4.0 - 8.3	2.2 - 4.0	3.0 - 7.8
PO_4^{3-} (ppm)	Trace - 0.17	Trace - 0.16	Trace - 0.27
Temperature-air (°C)	23.0 - 25.0	19.5 - 21.0	17.0 - 18.5
Temperature-water (°C)	20.0 - 20.5	15.0 - 17.0	10 - 12

(Source: Naskar, 1999)

Chapter 3

Factors Governing Habitats of Algae

Besides the climatic factors, like optimum temperature ranging between 15°C and 30°C, rainfall between 1,000mm and 3,000mm, humid climate and aridity, the biotic interference and geo-morphological, geological and the interactive biological factors determine the regeneration, growth, development and survivability of the mangrove ecosystem (Naskar & Mandal, 1999). Along with these factors, the nutrient requirements, tidal amplitude of the said area and salinity also have an effect on the health, survivability and regeneration of mangroves and mangrove ecosytem in its true perspective: these are narrated in brief below :

3.1. Nutrients

As in case of all other plant communities, algal flora also require nutritious or high NPK content in the water bodies in which they grow for vigorous growth, regeneration and distribution. These algae are the smallest or simplest forms of chlorophyllous. photosynthetic plants, but for the long chain of protein and fat synthesis and for other mineral requirements they depend on nutrient rich water phase and soil substratum. In the Sundarbans mangals the long flowing river mouths of the mighty river Ganga and Brahmaputra have immense nutrient carrying capacities and transport these nutrients to the deltaic regions after gradual siltation and accumulation of organic matter from the long upstream stretches. Generally in such deltaic lands potash is very high and nitrogen is derived from decomposed organic matter, whereas phosphate is very high in sea water. The silicate, iron and other macro - and micro - nutrients are also rich in saline sea water and deltaic areas developed on gradual silt deposition. Phosphate reducing and nitrogen fixing bacteria also play very important role in nutrient enrichment in these inter tidal brackish water regions of Sundarbans.

3.2. Tidal Influx - Current

Tidal water periodically inundates the entire deltaic areas of the mangal habitats along with the changing lunar phases and during the ebb tide period the substrate and the trunk or aerial roots of the mangrove species get the chance for light exposure during low tide. All these help in increasing the moisture levels in the atmosphere and conducive light intensity helps in photosynthetic activation of the autotrophs. So, these alternate high and ebb tides play very dominant roles in establishment, growth and regeneration of algae in the mangrove habitats.

3.3. Salinity

The different algal species or the different groups of algae prefer and grow in different degrees of salinity. Some algae grow in marine water, some algae also prefer either fresh water or brackish water. As such, the salt tolerant species prefer these mangrove habitats for their ideal habitat, but some freshwater forms of algae, *viz., Oedogonium* sp., *Uronema* sp., *Triplastrum* sp., *Chlorella* sp. *Volvox* sp. and *Closterium* sp. may be obtained in these mangrove habitats during monsoon or post - monsoon months, when the salinity comes down by some extent. On the contrary some of the marine forms of algae, *viz., Gelidiella* sp., *Dictyota* sp. and *Boodleopsis* sp. may also grow in these mangrove habitats during the summer months when the degree of salinity of these mangrove habitats goes slightly high.

Chapter 4

Phycological Studies : India & Bangladesh

4.1. Earlier Studies in the Indian Subcontinent

The first paper on Indian algae was published by William Jones in 1768 based on the collections made by Koening from the eastern parts of Indian sub continent. In the 19th century, Sir William Roxburgh initiated exploratory works in both cryptogamic and phanerogamic plants. Griffith (1844) published the work on cryptogamic plants from the manuscripts of Roxburgh. Griffith reported four species of *Chara* from Serampore, West Bengal, which was probably one of the earliest works on the Eastern Indian algae. Wallich (1860) gave the first systematic account of desmids of this sub - continent based on his collections from Ramgunge in West Bengal. Turner (1892) published his monumental work on the freshwater algae in East India based on Wallich's works. Lagerheim (1888) recorded about 52 species of desmids from Bengal but did not mention the sites of collections. Borge (1899) described the tropical and subtropical green algae from Bengal. Prain (1905) reported 84 species of algae from Hugly, Howrah and 24 Pergunnahs districts in India. From the first quarter of the 20th century a number of pioneer Indian Algologists initiated systematic studies of algae and described a number of new taxa from this sub continent. Leading among them was Prof. M. O. P. Iyengar of Madras University regarded as the Father of Indian Phycology. The other noted phycologists include K. Biswas, B. P. Pal, B. C. Kundu, M. S. Randhawa, Y. Bharadwaja, S. P. Agharkar, J. C. Banerjee, R. N. Singh and P. R. Parukutty.

Prof. Iyengar has described a large number of new algal genera and species of which mention must be made of his publications on *Fritschiella, Characiosiphon, Cylindrocapsopsis, Triplastrum etc.*, but his major contribution remains his monumental work on the order Volvocales, in which he described many new genera and species (Iyengar and Desikachary, 1981). Bharadwaja (1934) did extensive studies on the *Scytonema* and *Tolypothrix* species from India and Sri Lanka. Randhawa (1959) worked and published works mainly on order Zygnemales. Kundu (1929, 1934, 1935, 1937, 1941, 1951) studied extensively the Charales from Bengal. Biswas (1925, 1926, 1930, 1932, 1934, 1936, 1937, 1942, 1949) made immense contribution to the studies on blue- green, green and diatom algae from Bengal, Assam, Chilka Lake and Burma. Boergessen, a Danish phycologist studied the marine algae of the western and southern coasts of India, Burma and Sri Lanka, between 1930 and 1938. Fritsch also made some significant contribution to the field of phycology in India by studying some algae from Bengal and Sri Lanka and more importantly by training a large number of Indian phycologists. Banerjee (1935, 1936, 1938) reported algae from rice fields of Faridpur (presently Bangladesh) and the blue green algae from lower Bengal.

No further phycological research activity was undertakn for nearly a decade or so with regards to algal taxonomic and floristic studies after Biswas, Kundu, Banerji and their contemporaries in West Bengal and other eastern states. Later, Das (1963) studied the Compsopogonales in India. Gupta (1965, 1975) recorded some blue green algae as new records from West Bengal. Sinha & Mukherjee (1975a, 1975b) recorded a number of blue - green algae from the paddy fields of Bankura, West Bengal. Santra (1984) made extensive studies on the blue-green algae from Sikkim Himalayas. Santra & Adhya (1973, 1976a, 1976b, 1977) did a lot of work on members of Zygnemataceae and Vaucheriaceae from Darjeeling in eastern Himalayas. Das & Santra (1982) recorded several diatoms from Senchal lake, Darjeeling. Adhya & Santra (1977) recorded *Draparnaldiopsis indica* from Jalpaiguri in West Bengal for the first time. Pal & Santra (1982, 1984, 1985, 1986) and Maity & Santra (1985) did extensive research on the blue - green algae from different districts of West Bengal. Naskar & Santra (1986) recorded *Enteromorpha tubulosa* in brackishwater mixed with sewage - fed fisheries from Sundarbans, West Bengal. Sarma (1971) recorded *Cladophorella* for the first time in West Bengal. Sarma and Mukherjee (1985, 1986) have reported two new species of *Bulbochaete* from West Bengal. Later Kargupta & Sarma (1987) recorded six species and Sarma *et al.* (1988) recorded seven new species of *Bulbochaete* for the first time from West Bengal. Sarma & Kargupta (1986) added three new species to *Zygnemopsis* and for the first time (1987) reported 14 species of

Stigeoclonium from West Bengal. Sarma & Mustafa (1981) reported the occurrence of *Fritshiella tuberosa* in Burdwan, West Bengal for the first time. Kargupta & Sarma (1991) reported eleven taxa of *Coleochaete* and Sarma & Keshri (1991) reported the presence of a rare alga *Charciosiphon rivularis* Iyeng., for the first time from West Bengal. Pal (1990) has done extensive work on the floristics of the algal flora of Midnapore district of West Bengal. Chatterjee (1975, 1976, 1979) added to the list of Characeae from different districts of West Bengal. Chatterjee (1980) also made a new record of *Aphanochaete repens* A. Br. from West Bengal. Maity & Santra (1985) have given an account of the blue green algae of the 24 Parganas, West Bengal. Santra, *et. al.* (1988) have extensively studied the blue-green algae from the saline habitats of West Bengal. Santra & Pal (1988) gave an account of the benthic marine algae of the mangrove delta regions of West Bengal. Pal, *et. al.* (1988) gave a check list of the algal flora of Sundarbans delta. Mandal & Naskar (1994) made studies on the periphytic algae from the Sundarbans mangals, India and most recently in 1999, Naskar and Mandal have given a systematic account of 50 algae belonging Cyanophyceae, Xanthophyceae, Chlorophyceae and Rhodophyceae from various habitats of Indian Sundarbans.

The discussion reveals the fact that inspite of the seemingly substantial work done with regards to the algal floristics of West Bengal, there is very little published work on the algal flora of the Sundarbans mangals in particular and whatever exists is quite sporadic.

4.2. Earlier Studies in Sundarbans of Bangladesh

In Bangladesh, after the works of Kundu (1929, 1934, 1935, 1937, 1941, 1951), Banerjee (1935), De (1939) and Fritsch (1949) research on the algal flora had to wait for more than a decade during which it was completely neglected. It was only in 1952, that Prof. A.K.M.N. Islam initiated phycological studies in the University of Dhaka and started publishing his works since 1960. Prof. Islam, his co-workers and students have made wide range and also extensive studies on the algal flora of Bangladesh and particularly the Sundarbans. Besides reporting innumerable known taxa both benthic and planktonic for the first time from varied habitats, nearly 150 new species, varieties and one new genus have been described by Islam and his associates from Bangladesh, vide his publications Islam (1960, 1961, 1962, 1963, 1964, 1965, 1966, 1969a, 1969b, 1969c, 1970a, 1970b, 1972a, 1972b, 1972c, 1973a, 1973b, 1973c, 1974a, 1974b, 1974d, 1975a, 1975b,

1976, 1976a, 1976b, 1977, 1978, 1979a, 1980a, 1980b, 1984, 1985, 1986), Islam & Ahia (1964), Islam & Alam (1980), Islam & Aziz (1975, 1977, 1979, 1980, 1982, 1987a, 1987b), Islam & Begum (1970, 1981a, 1981b, 1987), Islam & Haroon (1978, 1980), Islam & Hossain (1978, 1979), Islam & Khair (1978b), Islam & Khatun (1966), Islam & Mannan (1986), Islam & Morshed (1985), Islam & Nahar (1967), Islam & D. Sarma (1968, 1976), Islam & P. Sarma (1963, 1964a, 1964b, 1965, 1976), Islam & Uddin (1969, 1973, 1978a, 1978b), Islam & Moniruzzaman (1981), Islam & Zaman (1974), Islam *et al.* (1976, 1991), Khan & Islam (1988).

Other than the algal floristics, Islam has published reports on the algal ecology, limnology, hydrobiology, physiology and biochemistry, as well, vide his publications Islam (1970c, 1974c), Islam & Chowdhury (1979), Islam & Haroon (1975), Islam & Mendes (1976a, 1976b, 1977a, 1977b), Islam & Saha (1975), Islam & Sobhan (1976), Islam & Zaman (1975), Islam *et al.* (1974, 1980), Khondker *et al.* (1985, 1988, 1990).

Thus in Bangladesh there exist a good number of very important, informative and valuable scientific publications and large number of papers, bulletins, books and reports, etc. on the algal flora of Bangladesh in general and Sundarbans in particular. This type of effort and publications on the algal flora is lacking or absent in the Indian Sundarbans.

Chapter 5

Field and Laboratory Methodologies for the Investigation of Algae

5.1. Morphometric Investigation on Algae

5.1.1. Collection and Preservation Methods

During this study (since April, 1996), seasonal collection of algal samples were made from the mangrove forest areas, river flat areas, tidal rivers and also from brackish water fisheries in the vicinity of the rivers in Sundarbans, West Bengal in the eastern, western and central parts (Map 4), covering all the four dominant seasons of the year in Sundarbans, *viz.,* Summer (March-June), Monsoon (June-September), Post monsoon (October-November) and Winter (November-February). The samples of the periphytic algal forms, *i.e.,* algae growing on mangrove bark, pneumatophores, other aerial root systems, bricks, wooden and bamboo poles, concrete jetties, on the sides of boats and launches were scraped with a scalpel and were collected during all the tidal phases. Free-floating filamentous algae were collected from the natural, as well as, man-made wetlands in the brackish water areas employed for brackish water fish production. The phytoplankton were collected from the brackish water areas, rivers, canals and creeks during all the tidal phases with the help of a truncated plankton net of 25 μ mesh size, bolten silk. Reed stems, twigs of submerged weeds, shells of animals, small stones and soil samples with algal growth were collected in polythene bags with just enough water to ensure a saturated atmosphere when tied, were brought to the laboratory for further examination. The samples thus collected were thoroughly washed with distilled water a number of times arid then

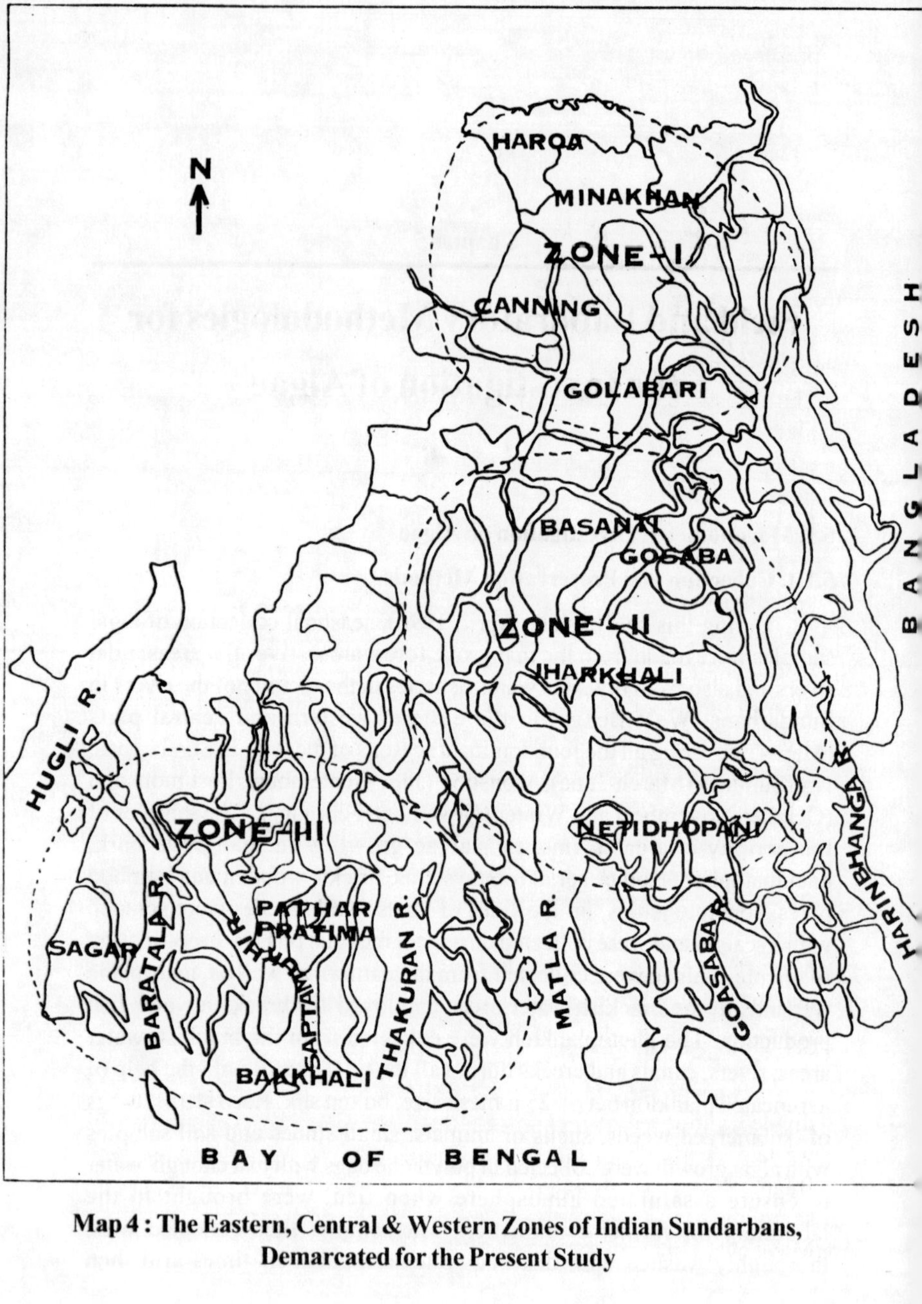

Map 4 : The Eastern, Central & Western Zones of Indian Sundarbans, Demarcated for the Present Study

preserved in formalin acetic alcohol (FAA) solution prepared in the following way :

50 % ethyl alcohol	90 ml
40 % formalin	4 ml
Glycerol	3 ml
Glacial acetic acid	3 ml
Cupric chloride ($CuCl_2$)	9.5 gm

The permanent slides of the algae were prepared for morphological study and for identification. The alga was placed on a slide in a drop of distilled water. A small drop of 40% formalin was added to fix the material. The excess water was drained away. A small lump of glycerine jelly was placed on the alga and the slide was kept in the oven for a few minutes at 60° C for the jelly to melt and to let the alga spread out. A cover glass was placed on this, the excess jelly wiped off and the slide was sealed and labelled.

The samples containing diatoms were suspended in distilled water and washed by centrifugation, the supernatant was removed and the pellet was re-suspended in 10 ml of distilled water, to this 10 drops of hydrogen peroxide was added and kept for few hours. A drop of this hydrogen peroxide–cleaned material was dried on a cover-glass and this was mounted in Canada Balsam for further observation.

A total number of 150 algal species were identified from the Sundarbans, which are described along with the photos and sketches during this present investigation (Tables 16, 17, 18 & 19). The camera lucida sketches were prepared and photo-micrographs were taken of the slides thus prepared. The measurements were noted using ocular and stage micrometer. These data were compared with Algae Identification Manuals, published Monographs, reports and papers in different National and International Journals for identifying the collected specimens.

5.1.2. Sites of Collection

The Sundarbans, West Bengal for the present study, had been divided into three different zones, *viz.,* (a) eastern, (b) central and (c) western parts (Map 4). The sampling sites in the eastern zone included Bagna, Marichjhapi, Kalitala, Hasnabad, Jhilla and Burirdhabri; the sampling sites in the upper central zone included Haroa, Minakhan, Malancha, Canning and Golabari; the sampling sites in the lower central zone included Basanti, Purandar, Ramgopalpur, Gosaba, Pakhiralay, Jamespur and Jharkhali in the human habitation areas and Sajnekhali, Pirkhali, Panchmukhani and Sudhanyakhali in the Tiger Reserve Area. The sampling sites in the western zone included Sagar, Namkhana, Bakkhali, Frazerganj, Patharpratima, Harwood Point, Dhanchi Island and Lothian Island (Map 5).

5.2. Investigation on Algal Habitat

5.2.1. Dissolved Oxygen

Other than the direct effect that Dissolved Oxygen has on the respiration of different biotic organisms, it controls many other limnological properties of a water body, which ultimately govern the productivity of the ecosystem. The Winkler or iodometric method (Winkler, 1888) was adopted to determine the dissolved oxygen of the water bodies. This method is a titrimetric procedure based on the oxidizing property of dissolved oxygen and involves the addition of divalent manganese solution, followed by a strong alkali to the samples. The dissolved oxygen in the sample rapidly oxidizes an equivalent amount of the dispersed manganous hydroxide. In the presence of iodide ions in an acidic solution the oxidized manganese reverts back to its original divalent state liberating iodine of an amount equal to the dissolved oxygen. This is titrated with a starch indicator to get the dissolved oxygen content (Tables 20, 21, 22 & 23).

5.2.2. Salinity

Salinity was originally conceived as a measure of the mass of dissolved salts in a given mass of solution. The experimental determination of the salt content usually involves indirect methods like measuring conductivity, density, sound speed, or refractive index. From an empirical relationship of salinity and the physical properties, salinity can be calculated. However, in the present study argentrometric method has been adopted for determining salinity of the water samples. Potassium chromate indicates the end point of titration by Silver nitrate solution, which is used for calculating the chloride content and further estimation of salinity of the water sample (Tables 20, 21, 22, 23 & 29).

5.2.3. Alkalinity

Alkalinity of a water sample is its acid-neutralizing capacity. It is the sum of all the titratable bases. The alkalinity in water can be

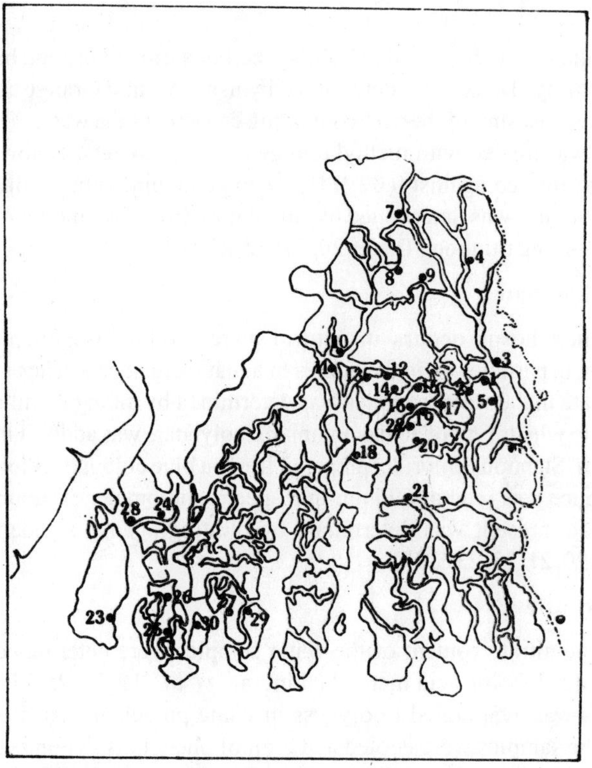

Map 5 : Sampling Sites in the Indian Sundarbans

Sites of Collection from the Indian Sundarbans :

Eastern Part– **1.** Bagna; **2.** Marichjhapi; **3.** Kalitala; **4.** Hasnabad; **5.** Jhilla; **6.** Burirdabri

Central Part – **Human Habitation Area–7.** Haroa; **8.** Minakhan; **9.** Malancha; **10.** Canning; **11.** Golabari; **12.** Basanti; **13.** Purandar; **14.** Ramgopalpur; **15.** Gosaba; **16.** Pakhiralay; **17.** Jamespur; **18.** Jharkhali.

Sundarbans Tiger Reserve Area – 19. Sajnekhali; **20.** Pirkhali; **21.** Panchmukhani; **22.** Sudhanyakhali.

Western Part – 23. Sagar; **24.** Namkhana; **25.** Bakkhali; **26.** Frazerganj; **27.** Patharpratima; **28.** Harwood Point; **29.** Dhanchi Island; **30.** Lothian Island.

differentiated into hydroxide alkalinity, carbonate alkalinity and bicarbonate alkalinity. These were determined by using Methyl Orange as indicator to give measure of the acid combining capacity of the water. The water sample was treated with methyl orange indicator to get a yellow colour. This was titrated against 0.02 N H_2SO_4 to get a faint orange colour. The total alkalinity was determined by calculating from the amount of H_2SO_4 used up during titration (Tables 20, 21, 22, 23 & 29).

5.2.4. Phosphate

Phosphorus occurs in natural water bodies as phosphates in solution, in particles or in the bodies of aquatic organisms. The amount of phosphate in the water samples was determined by taking 25 ml of water sample to which 1 ml of acid ammonium molybdate was added. On adding 1 drop of Stannous chloride, prepared fresh, a blue colour developed. The absorbance was read at 690 nm in a spectrophotometer. The amount of phosphate present was determined by plotting against standard curve (Tables 20, 21, 22, 23 & 29).

5.2.5. Nitrate

The nitrate content of the water samples were determined by the Phenol disulphonic acid method (Jhingran, *et. al.,* 1988). 25 ml of water samples was evaporated to dryness in white porcelain basins on water bath. The samples were cooled and 2 ml of phenol disulphonic acid was added drop by drop and rubbed with a glass rod. Then 2 ml of $Al_2(SO_4)_3$ solution was added, to this 12N NaOH was slowly added with constant stirring until the solution turned alkaline. After filtering this solution to remove $Al(OH)_3$ the yellow colour developed was read at 410 nm in a Spectrophotometer. The amount of nitrate present was determined by plotting against standard curve (Tables 20, 21 22, 23 & 29).

5.2.6. Light Intensity

Light is essential for photosynthesis, but algae differ markedly in respect to their tolerance to light intensity. Some algae are indifferent to the intensity of light while others are very particular. Aquatic algae of unshaded ponds are usually considered as growing in full sunlight, but the more deeply submerged ones are not so strongly illuminated as those, at the surface, because there is gradual geometrical decrease in intensity of illumination with an arithmetic increase in depth. Despite this, many at a certain depth than they do at the surface (Manning, *et. al.,* 1930).

As is well known, there is qualitative penetration of light in water and a greater absorption at the red end of the spectrum. The vertical distribution of attached algae is directly correlated with the intensity of illumination at different levels (Oberdorfer, 1927).

In the present study, the light intensity was measured with help of a Lux Meter, Model - LX - 101 (Tables 24, 25, 26 & 27).

5.2.7. Primary Productivity

Primary productivity in a water body can be determined by various ways as described by A.P.H.A. (1980). The Oxygen method by measuring the changes in oxygen and CO_2 concentration (light and dark bottle method) was adopted in the present investigation. The advantage of this method is that it provides estimates of gross and net productivity and respiration. The DO concentration is estimated at the beginning and end of incubation period. Productivity is calculated on the assumption that one atom of carbon is assimilated for each molecule of oxygen released. Samples collected from each pre-selected depth were taken in triplicate-light bottle, dark bottle and initial analysis bottle. DO of the initial bottle was determined by Winkler's method and the light and dark bottles were suspended at the depth from which the collections were made for 3 hours. At the end of incubation period the DO of both the bottles were determined by Winkler's method (Tables-24, 25, 26 & 27).

$$\textbf{Gross primary production} = \frac{LB\text{-}DB}{T} \times \frac{12}{32} \times \frac{1000}{PQ} \text{ mg C/m}^3\text{/hr}$$

$$\textbf{Net primary production} = \frac{LB\text{-}IB}{T} \times \frac{12}{32} \times \frac{1000}{PQ} \text{ mg C/m}^3\text{/hr}$$

LB = light bottle; IB = initial bottle; DB = dark bottle; T = time (hrs.);

PQ = Photosynthetic coefficient (1.2)

5.3. Bio-Chemical Investigation of Algae

Eleven major macrophytic algae were selected and collected for the bio–chemical estimation of carbohydrates, proteins, mineral and lipid content, due to their noted abundant growth and abundance in the Sundarbans and for their potential food value for the estuarine and brackish water fish prawns, shrimps, crabs and molluscs, *viz.*, *Enteromorpha* spp., *Chaetomorpha* spp., *Lola* spp., *Spirogyra* spp.,

Rhizoclonium spp., *Ulva* spp., *Catenella* spp., *Polysiphonia* spp. *Pterosiphonia* sp., *Dictyota* sp., *Gelidiella* sp.

5.3.1. Total Carbohydrate Content of the Major Algae

Carbohydrates are important components of storage and structural materials in all autotrophic organisms. The basic units of carbohydrates are the monosaccharides, which can not be further split by hydrolysis into simpler sugars. The total carbohydrate content of the collected macrophytic algal species were thus determined by hydrolysing the polysaccharides into simple sugars by acid hydrolysis and estimating the resultant monosaccharides. The algae collected from the estuarine and brackishwater environments of Indian Sundarbans were cleaned, segregated and air-dried for further investigations. The moisture content of this air-dried biomass was first determined. In a boiling tube 100 mg air dried biomass of the algal sample was taken. this was hydrolysed with 5 ml of 2.5 N Hydrochloric acid in a boiling water bath for 3 hours. After cooling to room temperature the solution was neutralized with solid sodium carbonate. The volume was made to 100 ml and then centrifuged. The supernatant was used for analysis. To the aliquots of samples taken anthrone reagent was added and again heated in a boiling water bath for 8 minutes. The sample was cooled rapidly and the green to dark green colour was read at 630 nm in the spectrophotometer. Amount of carbohydrate present was determined by plotting on a standard curve. The amount of carbohydrate present was expressed as percentage of dry weight. The results are highlighted in Table 30.

5.3.2. Total Protein Content of the Major Algae

Proteins form the structural and functional basis of the cells of all living organisms. The algae collected from the estuarine and brackishwater environments of Indian Sundarbans were cleaned, segregated and air-dried for further investigations. The moisure content of this air-dried biomass was first determined. The total protein content of the samples were determined by Lowry's method. The air-dried biomass (500 mg) was extracted with 0.2M Phosphate buffer of pH 7. The extract thus prepared was centrifuged and the supernatant collected was treated with Trichloro Acetic Acid, the precipitated protein was dissolved in 1N NaOH. To the aliquots 5 ml of Alkaline Copper Solution was added, and allowed to stand for 10 minutes. Then 0.5 ml of Folins-Ciocalteau reagent was added, mixed well and incubated in the dark for 30 minutes. The blue colour developed was read at 660 nm in a spectrophotometer. The amount of protein present in the sample was determined by plotting in the standard curve prepared with Bovine Serum Albumin (Fraction V). The protein content was expressed as percentage of dry weight (Table 30).

5.3.3. Total Lipids of the Major Algae

The above mentioned algae collected from the estuarine and brackishwater environments of Indian Sundarbans were cleaned, segregated and air-dried for further investigations. The moisture content of this air-dried biomass was first determined. The sample was then extracted with chloroform : methanol : water (2: 1: 0.8, vol./ vol./ vol.) according to Bligh and Dyer (1959). The homogenized sample was transferred to a separatory funnel and the chloroform layer containing the lipid was collected. This chloroform was completely evaporated to obtain the total lipid of the sample. Total lipid content was determined gravimetrically and expressed as a percentage on dry basis (Table 30).

5.3.4. Total Mineral Content of the Major Algae

The mineral content of the above mentioned algae was determined by charring known amount of dry bio - mass in crucibles in a Blast Furance at 350-400°C. The total mineral content (ash content) was then determined gravimetrically and expressed as percentage on dry basis (Table 30).

5.3.5. Fatty-Acid Composition of the Major Algae

To understand the quantity and composition of the extracted lipids from the macrophytic algal species from Indian Sundarbans, identification and quantification of fatty acids present was carried out. The polyunsaturated fatty acids may be grouped as (i) Omega 3 fatty acids and (ii) Omega 6 fatty acids. Linolenic acid, eicosapentaenoic acid and docosahexanoic acid belong to the Omega -3 family. No higher animals including man are able to introduce a double bond between the terminal methyl group and Omega-9 carbon atom. Hence, there is need for supplying these fatty acids to the body system through our diet. The ω -3 fatty acid, α-linolenic acid and the ω -6 fatty acid, linoleic acid are therefore essential fatty acids. These are important dietary precursors for a number of essential long-chain fatty acids including arachidonic acid ($C_{20:4,}$ ω-6) eicosapentaenoic acid ($C_{20:5,}$ ω -3) and docosahexanoic acid ($C_{22:6,}$ ω-3). Fatty acids were made volatile by converting the lipids extracted earlier into methyl esters. Fatty acid composition of the lipids was determined by gas-liquid-chroatography of the methyl esters. A Hewlett-Packard Model 5890 Gas Chromatograph was equipped with a stainless steel column (i.d. 3/8 inch, length 6 feet) packed with 10% diethylene glycol succinate (DEGS) adsorbed on Chromosorb WHP and a Flame-Ionization Detector. The oven, injector and detector temperatures were maintained at 190, 230 and 240°C, respectively. Peaks were identified by comparison with relative retention times of the standard fatty acids and methyl esters (Tables 31a, 31b & 31c; GLC 1-9).

5.3.6. Chlorophyll Content of Plankton in Water Bodies

Several types of chlorophyll occur in the algal communities of which the most common is chlorophyll - a, which is the main centre of photosynthetic activities in autotrophic organisms. It is also reported that primary productivity has a close correlation with fish yield in a water body (Wolny and Grygierel, 1972; Sreenivasan, 1968). Chlorophyll - a content may be used as an index of productivity in a fish pond. 50 litres of water was filtered through a plankton net to collect 50 ml of sample. The sample was collected in dark bottles of 50 ml volume and immediately transferred to an ice box and brought to the laboratory. The sample was centrifuged to let the sample settle at the bottom of the tube. The supernatant was decanted and the plankton was rinsed repeatedly with small volumes of ice cold acetone and transferred to a 50 ml volumetric flask. This flask was stoppered, wrapped with dark paper and kept in the refrigerator for 24 hours. The absorbance of the solution was measured at 665 and 750 nm, respectively in a spectrophotometer. The concentration of chlorophyll -a in the original water sample was calculated according to Vollenweider, 1969 (Tables 24, 25, 26 & 27).

5.4. Impact Assessment of Algae on Fisheries

5.4.1. Collection and Preservation of Fin-fish & Shell-fish

The growth and development of fish species is dependent on the availability and optimum utilization of fish food resources. The gut content analysis throws light on the food preferences of the fish species examined and also the feasibility of sustainable development in aquaculture of the particular species studied based on the environmental conditions and food availability. In the present study six commonly available fin-fish species, *viz., Scatophagus argus, Liza parsia, Liza tade, Mugil cephalus, Rhinomugil corsula* and *Valamugil cunnesius;* two most high priced, as well as, common prawn species, *viz., Penaeus monodon* and *Macrobrachium rosenbergii;* two common crab species, *viz., Scylla serrata* and *Uca* sp. and one molluscan species, *viz., Telescopia* sp. were collected from the estuarine and brackishwater habitats of Sundarbans, West Bengal. The fish caught during different seasons, different catches and different sites were immediately weights and segregated according to their respective weighed and preserved in 4% formalin and kept in separate groups and brought to the laboratory for further microscopical examination of the gut contents.

5.4.2. Fish Gut Content Analysis

The gut content analysis of the fish species can be carried out by three different methods as described by Jhingran, *et. al.* (1988), *viz.*, volumetric, gravimetric and numerical methods. The gravimetric method was followed in the present study. The guts dissected out from the fish species were weighed and the contents were sorted into different groups to determine the food and feeding habits of the fish species studied. The gut content was examined under miscroscope to determine the nature of preferred food of the individual species.

5.4.3. Plankton Collection

Plankton collection was carried out with the help of a truncated cone shaped plankton net. The plankton net is made of bolting silk No. 25 standard grade. This has an aperture size of 0.064 mm (79 meshes per linear cm). Approximately 50 litres of water was passed through the plankton net to get a sample or 50 ml volume. This sample with fixed with 4% formalin and brought to the laboratory for further quantitative and qualitative estimation. In the laboratory the plankton samples were transferred to plankton tubes of 10 ml volume after letting the samples to settle down and removing the supernatants. In the plankton tubes the samples were allowed to settle overnight to get the settling volume.

5.4.4. Species Diversity Index

Natural communities contain a bewildering number of species, so much so, that it is difficult to identify and enlist all the species of plants, animals and microbes found in any large area, be it a forest or an ocean. Much of this impressive diversity of species observed in the environment result in mixtures or gradients; however, even if small samples from an apparently homogenous, uniform habitat are selected to restrict oneself to a limited taxon, the pattern which emerges is, a few dominant species associated with many rare ones (Odum, 1975). Although, the pattern of *many species most of which are rare* almost seems to be an ecological "law", the actual number of rare species and hence the total diversity is quite variable within and between ecosystems, even if we stick to the same taxonomic and ecological grouping (Odum, 1975). It has been observed that the older and more stable a community happens to be, the more will its species diversity tend to be (Misra, 1974). In fact, nature prefers and simultaneuosly favours high species diversity, since natural communities with high species diversity are less vulnerable to environmental vagaries, while man made communities are open to greater damages by environmental hazards or epidemics and when such an incident takes place the total community is destroyed.

Species diversity is a very useful parameter for comparison of two communities specially to study the influence of biotic disturbances or to know the state of succession and stability in the community. A convenient way to express and compare diversity is to calculate the diversity indices based on the ratio of parts to the whole, or n_i/N where n_i is the number or other *importance value, viz.,* bio-mass, productivity, surface coverage, etc. of each component, *viz.,* species and N is the total of importance values.

If "N" represents some specified importance value (number, biomass, energy flow, etc.), number in case of the present study and designate n_i as the importance value for each component (species) and capital N as the total of importance values-then we can say that the ratio n_i/N is the probability function, p_i, for each of the parts of the whole, using these symbols the calculations for Shannon Weaver Diversity index has been formulated (Margalef, 1963).

$$H = -\Sigma (n_i/N) \log_2 (n_i/N)$$

or

$$H = -\Sigma (p_i) \log_2 (p_i)$$

The sigma (Σ) stands for "the sum of," *i.e.,* one calculates the ratio for each component and multiplies it by the log of the ratio for computing H which represents Shannon index of general diversity. Since log of fractions are negative values a minus sign is inserted in the formula to convert the sum to a positive number.

The function is widely used these days for a variety of population data to obtain species diversity index and explain the habitat characteristics particularly in relation to pollution following the explanations of Wilhum and Dorris (1966) and Staub *et. al.*, (1970).

Wilhum & Dorris (1966) explanation :

Diversity Index	**Pollution Level**
>3	Clean water
1 - 3	Moderately polluted
<1	Heavily polluted

Staub, *et. al.* (1970) explanation :

Diversity Index	**Pollution Level**
3.4 - 4.5	Slightly polluted
2.1 - 3.0	Light pollution
1.0 - 2.0	Moderate pollution
0.0 - 1.0	Heavy pollution

Though this index is widely used for estimating species diversity, but this index is not considered as a real assessment of species diversity in community but relative importance value of each species is taken into account (Whittaker, 1977). Therefore the index obtained by this function is a good indicator of water quality of the habitat and forms the basis of bio-monitoring (Sinha, *et. al.* 1993).

5.4.5. Phytoplankton

Phytoplankton form the pastures of all water bodies on the surface of the earth wherever sunlight reaches and provided environmental conditions are favourable. These organisms are the basis of all the food- webs in the aquatic systems. The phytoplanktonic organisms vary in size, shape and diversity of groups. The groups include (1) diatoms, (2) green flagellates, (3) green algae (single cells/ colonies/filaments), (4) dinoflagellates and (5) blue - green algae. The phytoplankton floating in open water are referred to as Euplankton, in rivers are called Potamoplankton and in ponds are called Heleoplankton. Prescott (1969) classified the plankton on the basis of their size as organisms more than 2000 μ in diameter- megaplankton; organisms 200-2000μ in diameter-mesoplankton; organism 20-200μ in diameter-microplankton;

organisms 2-20μ in diameter-nanoplankton and organisms less than 2 μ in diameter-ultranannoplankton. The phytoplankton community has further been classified according to their habitat by Kumar (1999) as the permanently floating community of the open water, referred to as euplankters, some accidental plankters which may be washed into the plankters are called pseudoplankton or tychoplankton, the floating community at the water - air interface is called the neuston, those floating organisms which develop on suspended particulate organic matter or detritus are called trypton or obioseston and seston refers to both the living plankton and the trypton.

The growth and multiplication of phytoplankton is primarily dependent on temperature, solar illumination and availability of certain essential nutrients. The phytoplankton quickly absorb the nitrates and phosphates available in the water. On the death of these organisms the nutrients locked up in their body are released through bacterial decomposition. The dead remains generally sink beyond the euphotic zones and reach the bottom, where the nutrients get released and accumulate. These nutrients are utilised only if they get transported to the euphotic zone through vertical movement of water. This vertical movement is usually dominant in two places in the marine environment, *viz.,* the coastal areas and regions of upwelling. Upwelling usually leads to bloom formation, phytoplankton blooms are often predictable features of marine and freshwater habitats. Blooms are observed after certain hydrobiological changes, *viz.,* enhanced terrigenous runoff, water column turnover, upwelling, increased thermal stratification accompanied by temperature and light conditions conducive to growth and multiplication of biomass (Fogg, 1969: Reynolds & Walsby, 1975). The environmental conditions which elicit the blooms have more or less become analogous to the bloom forming communities like coastal marine (dinoflagellate dominated), estuarine (dinoflagellate and cyanobacteria dominated) and freshwater (cyanobacteria dominated) ecosystems (Paerl, 1988). The primary nuisance created by such blooms include (1) water quality deterioration, (2) chronic or intermittent health hazards like toxicity and (3) loss of aesthetic and recreational value of affected waters.

Grazing of phytoplankton by zooplankton is often a major factor determining phytoplankton abundance and it at times determines the timing and magnitude of phytoplankton peaks. Cells of different types or sizes may be grazed at different rates and the species composition of a phytoplankton community can change as consequence of grazing. The grazing rate depends on abundance of phytoplankton because many grazers remove algal cells from water by filteration. This indicates that changes in filteration rate reflect the feeding strategy of the grazers (Vareschi, 1977).

The mean concentration of chlorophyll -a gives a good indication of the biomass of phytoplankton in the euphotic zones of water bodies and are commonly estimated by limnologists to get an idea of the phytoplankton in aquatic systems. Studies on seasonal cycles and succession of phytoplankton usually reveals three important phases based on algal abundance and species succession, *viz.,* mid February-April, when a low biomass and stable species composition is found, intermediate biomass and species replacement among chrysophytes and diatoms during May to mid September and highest biomass and species diversity during mid September to November (Heckey & Kling, 1981).

5.4.6. Zooplankton

The zooplankton are small animals with independent powers of movement but virtually at the mercy of currents as because of their size and drifting about passively. They consist of mostly invertebrates and larvae and immature stages of both invertebrates and vertebrates. The planktonic community shows considerable variety as it is composed of members of every group from protozoa to chordata, though coelentrates and crustaceans dominate in any perceivable environment. The interesting feature of zooplankton is that though many are planktonic throughout their life, others are so only during part of their life. As suitable food in the form of phytoplankton is available only in the surface water most of the benthic animals including some deep sea first liberate their larvae in the surface waters. The same rule should apply to other zooplankton forms also, since the whole zooplankton community is dependent on the phytoplanktonic forms for their food.

Depending on the nature of the animals that form the community, plankter is divided into holoplankton and meroplankton. The holoplankton are those which are planktonic throughout their lives like the siphonophores, ctenophores and chaetognaths. These are generally found in oceanic water and are referred to as off - shore plankton. The plankton of the neretic waters exhibit a diversity in young stages of the animals inhabiting these areas. These are referred to as in-shore plankton or meroplankton and vary in composition, depending on the breeding season of the animals of the shelf region. The zooplankton collected during the present investigation are presented in Table 28.

Chapter 6

Taxonomy and Morphology

The Sundarbans, West Bengal has been divided into three different zones, *viz.*, (a) eastern, (b) central and (c) western parts (Map-4). The sampling sites in the eastern zone inlude Bagna, Marichjhapi, Kalitala, Hasnabad, Jhilla and Burirdhabri; the sampling sites in the upper central zone included Haroa, Minakhan, Malancha, Canning and Golabari; the sampling sites in the lower central zone included Basanti, Purandar, Ramgopalpur, Gosaba, Pakhiralay, Jamespur and Jharkhali in the human habitation areas and Sajnekhali, Pirkhali, Panchmukhani and Sudhanyakhali in the Tiger Reserve Area. The sampling sites in the western zone included Sagar, Namkhana, Bakkhali, Frazerganj, Patharpratima,Harwood Point, Dhanchi Island and Lothian Island (Map-5).

Seasonal collection of algal samples were made from the mangrove forest areas, river flat areas, tidal rivers and also from brackish water fisheries in the vicinity of the rivers in Sundarbans, West Bengal in the eastern, western and central parts, covering all the four dominant seasons of the year in Sundarbans, *viz.,* Summer (March-June), Monsoon (June-September), Post Monsoon (October-November) and Winter (November-February). The sample of the periphytic algal forms, *i.e.,* algae growing on mangrove bark, pneumatophores, other aerial root systems, bricks, wooden and bamboo poles, concrete jetties, on the sides of boats and launches were scraped with a scalpel and were collected during all the tidal phases. Free-floating filamentous algae were collected from the natural, as well as, man-made wetlands in the brackish water areas employed for brackish water fish production. A total number of 150 algal species were identified from the Sundarbans, which are described along with the photos and sketches in this treatise (Tables–16, 17, 18 & 19). These algae belong to different groups, *viz.*, Cyanophyta (50 species), Chlorophyta (39 species), Chrysophyta (44 species), Phaeophyta (2 species) and Rhodophyta (15 species).

In Pie Charts-IIa, IIb & IIc, the distribution of different algal groups in different zones (northern, central and southern parts of Indian Sundarbans) have been highlighted. In Zone-I, *i.e.,* the northern parts of Indian Sundarbans, the Chlorophyta contribute to 40%, the Chrysophyta 21%, Rhodophyta 12% and Cyanophyta 34% of the total algal flora. In Zone-II, *i.e.,* the central parts of Indian Sundarbans, the Chlorophyta contribute to 40%, the Chrysophyta 21%, Rhodophyta 12% and Cyanophyta 35% of the total algal population. In Zone-III, *i.e.,* southern parts of the Indian Sundarbans, the Chlorophyta contributes to 13%, the Chrysophyta 45%, the Phaeophyta 4%, the Rhodophyta 13% and the Cyanophyta 25% of the toal algal population encountered here.

6.1 Division-Cyanophyta

The Division Cyanophyta is unique as it includes algae with procaryotic cell organization and hence they are usually called bacteria as their cell structure is very similar to eubacteria. The cells lack an organized double-membraned nuclei, chromatophores and mitochondria, in addition they posses characteristic photosynthetic pigments, *viz.,* biliproteins and myxoxanthophyll along with the usual chlorophyll-a and β-carotene. Also the Cyanophyta group of organism are morphologically, *i.e.,* cell structure, heterocysts and reproduction-wise quite conservative but metabolically versatile, *i.e.,* in the perspective of nitrogen fixation and photosynthetic metabolism. This is just the opposite in case of other eucaryotic algae which are structurally versatile but metabolically conservative (Kumar, 1999).

The identified Cyanophyta members from the Indian Sundarbans include 50 species which belong to 5 orders, *viz.,* Chroococcales, Chamaesiphonales, Pleurocapsales, Nostocales and Stigonematales, 10 families, *viz.,* Chroococcaceae (5 genus, 11 species), Entophysalidaceae (1 species), Chamaesiphonaceae (1 species), Dermocarpaceae (2 genus, 3 species), Hyellaceae (1 genus, 2 species), Oscillatoriaceae (8 genus, 21 species), Nostocaceae (1 species) (Table-17, Pie charts-IIa, IIb, IIc). Out of these 50 species, 5 are epiphytic on green or other blue green algae, 13 occur as planktonic forms and the others occur as benthic flora in various habitats river beds, forest floor, agricultural fields and dried up or damp brackishwater fisheries in the Indian Sundarbans.

6.2 Morphometric Features and Taxonomic Enumeration

CYANOPHYTA

Class - CYANOPHYCEAE

Unicellular or multicellular algae without true nucleus or chromatophores. Protoplast differentiated into peripheral zone with photosynthetic pigment and a central colourless portion.

Key to the Orders

1. Plants unicellular or colonial, sometimes forming a pseudofilamentous colony, never with a trichome organisation, no differentiation into base and apex, endospores not formed in sporangia, no exospores, nannocytes present. **-Chroococcales**

2. Plants unicellular, attached, typically differentiated into base and apex, reproduction by endospores or exospores. **-Chamaesiphonales**

3. Plant more or less filamentous, attached, arrangement uniform, chroococcaceous structure, often forming parenchymatous thalli with prostrate and erect filaments, without differentiation into trichome and filaments, no hormogones, no heterocysts, endospores in sporangia. **-Pleurocapsales**

4. Plants filamentous, with trichome and filament organised, hormogones present, often with heterocysts, akinetes, exospores or endospores, pseudo-hormogonia present.

   (a) Without true branching, unbranched or with false branching **- Nostocales**

   (b) With true branching or dichotomous branching, heterotrichous condition **- Stigonematales**

Order-CHROOCOCCALES

Unicelluar or colonial at times pseudofilamentous colony, never with trichome organisation, plants not differentiated into base and apex, sporangia do not have endospores, no exospores present.

Key to the Families

1a. Cells unicellular or form colonies, no filament like growth -**Chroococcaceae**
1b. Cells show filament like growth in a linear series forming pseudofilamentous growth -**Entophysalidaceae**

Family - CHROOCOCCACEAE

Key to the Genera

1a. Cells many in a single colony - 2
1b. Cells single or few in shapeless colonies with distinct vesicular individual sheaths - *Gloeocapsa*
2a. Cells without any regular or definite arrangement, a few distinct sheaths round individual cells may or may not be present - 3
2b. Cells in regular transverse and longitudinal rows in flat colonies - *Merismopedia*
3a. Cells typically packed into microscopic colonies of definite shapes, mostly planktonic - *Microcystis*
3b. Cells loosely arranged, mostly not planktonic, forming macroscopic colonies - 4
4a. Cells spherical - *Aphanocapsa*
4b. Cells ellipsoidal to cylindrical - *Aphanothece*

Microcystis Kützing

Linnaea : 8, 372 pp., 1833 em Kützing, Tab. Phycologicae, 1 : 7, 1846; Species Algarum, 208, 1849.

Cells spherical or elongated, many in spherical, ellipsoidal or net-like colony, homogenous mucilage, individual envelopes absent.

Key to the species

1a. In salt water, round colonies - *M. litoralis*
1b. In fresh water, irregular colonies - *M. bengalensis*

1. *Microcystis litoralis* (Hansg.) Forti

De Toni, Sylloge Algarum, 5 : 89, 1907, Geitler, Kryptogamentflora, 134, 1932; Frémy, *Cyano.* cotes d' Eur., 11, 1933; Desikachary, Cyanophyta 85, 1959; Santra *et. al.*, Biol. Mem. 14 (1) : 86 fig. 12, 1988.

Colony round with thick gelatinized mucilage, 55 - 70 μ broad, cells spherical or ellipsoidal, 3 - 5 μ broad, closely arranged, pale bluish - green in colour, without gas vacuoles.

Habitat: Planktonic in a ditch at Ramgopalpur, Sundarbans, 24 Parganas (S) West Bengal.

Specimen No. : N.S. 10. (Plate 1, Fig. 1).

2. *Microcystis bengalensis* Banerjee

Myxo. of Lower Bengal, 1. J. Indian bot. Soc., 15 : 295, figs. 2A and B, 1936: Desikachary, Cyanophyta, 89, pl. 19, figs. 5, 6, 1959; Santra *et. al.*, Biol. Mem. 14 (1) : 87, fig. 13, 1988.

Colonies irregularly arranged, branched or lobed irregularly, distinct sheath present, thickly stratified, cells spherical with gas vacuoles, 2.7 - 5.5μ diameter.

Habitat : Planktonic in brackishwater fishery at Haroa, Sundarbans, 24 Parganas (N), West Bengal.

Specimen No. : N. S. (P) 8. (Plate 1, Fig. 2).

***Gloeocapsa* Kützing**

Phyco. generalis, 174, 1843.

Colonies single or held together forming expanded mass, cells spherical, 2 - 8 in each colony, at times many with concentric rows of envelope around the cells, lamellated or unlamellated sheaths around individual cells.

Key to the species

1a.	Sheath colourless	- 2
1b.	Sheath coloured	- 4
2a.	Sheath unlamellated	- 3
2b.	Sheath lamellated	- *G. decorticans*
3a.	Cells without sheath, thalamus blackish	- *G. punctata*
3b.	Cells with sheath, thalamus bluish - green	- *G. aeruginosa*
4a.	Sheath lamellated	- *G. rupestris*
4b.	Sheath unlamellated	- *G. kuetzingiana*

3. *Gloeocapsa decorticans* (A. Br.) Richter

ex Wille, Nyt. Mag. Naturvid., 62: 186,1925; Geitler, Kryptogamenflora, 184, fig. 83b, 1932; Desikachary, Cyanophyta, 114, pl. 24, fig., 9, 1959.

Cells spherical or oval, bluish - green coloured, single or in groups of upto 4 cells together, sheath colourless, thick and distinctly lamellated, cells upto 20μ diameter.

Habitat : On wet soil near a fresh water pond at Jharkhali, Sundarbans, 24 Parganas (S), West Bengal.

Specimen No. : N. S. 52. (Plate I, Fig. 3).

4. *Gloeocapsa punctata* Näg. ex Kützing

Species Algarum, 22, 1849; Nageli, Gatt. eizell. Algen, 51, pl. 1F, fig. 6, 1849; Forti in De Toni, Sylloge Algarum, 5 : 55, 1907; Frémy, Myxo d' Afr. équat. franc., 47, 1929; Geitler, Kryptogamenflora, 189, 1932; Desikachary, Cyanophyta, 115, pl. 23, fig. 2, 1959.

Thallus gelatinous, dark blue to blackish, cells without sheath, 1.0-2.3μ diameter, in groups or colonies upto 20μ diameter.

Habitat : On the bricks and cemented stairs of a freshwater pond at Jharkhali, Sundarbans, 24 Parganas (S), West Bengal.

Specimen No. : N.S. 30. (Plate I, Fig. 4).

5. *Gloeocapsa aeruginosa* (Carm.) Kütz

Phyc. gene., 174, 1843; Phyc. germ., 151, 1845; Tab. Phycologicae, I : pl. 21, fig. 2, 1846; Species Algarum, 218, 1849; Forti in De Toni,, Sylloge Algarum, 5: 55, 1907; Biswas, 361, 1924; Desikachary, Cyanophyta., 115, 1959.

Thallus crustaceous, mucilaginous, cells 2 - 4μ diameter, bluish - green coloured, unlamellated, colourless sheath, forming spherical colonies 30 - 45μ diameter.

Habitat : Blue-green crusts on bricks and walls alongside Basanti canal, Sundarbans, 24 Parganas (S), West Bengal.

Specimen No.: N.S. 31. (Plate I, Fig. 5).

6. *Gloeocapsa rupestris* Kütz.

Tab. Phycologicae, I : 17, pl. 22, fig. 2, 1846; Species Algarum, 223, 1849;

Forti in De Toni, Sylloge Algarum, 5 : 46, 1907; Frémy, Myxo d' Afr. équat. Franc., 50, fig. 1929; Geitler, Kryptogamenflora, 194, figs. 88e, 89, 1932; Desikachary, Cyanophyta, 117, 1959; Santra *et. al.*, Biol. Mem. 14 (1), 83, fig., 4, 1988.

Thallus brownish, sheath yellow to brown, lamellated, colonies 30-5μ in diameter, cells 7.8 - 10μ in diameter, cells 7.8 - 10μ in diameter.

Habitat : From stagnant water at Ghusighata, Sundarbans, 24 Parganas (N), West Bengal.

Specimen No.: N. S. 19. (Plate I, Fig. 6).

7. *Gloeocapsa kuetzingiana* Näg.

Gatt. einzell. Algen, 50, 1849; Rabenhorst, Fl. Eur. Alg., 2 : 46, 1865; Forti in De Toni, Sylloge Algarum, 5 : 47, 1907; Geitler, Kryptogamenflora, 194, fig. 91a, 1932; Desikachary, Cyanophyta, 118, pl. 23, fig. 4, pl. 24, fig. 12, 1959; Santra *et. al.,* Biol Mem. 14 (1), 83, fig. 3, 1988.

Thallus brownish to blackish, cells densely aggregated in colonies upto 150μ diameter, cells with unlamellated sheath, 5 - 7.5μ diameter.

Habitat: On moist soil at Ghusighata, Sundarbans, 24 Parganas (N), West Bengal.

Specimen No.: N. S. 24. (Plate I, Fig. 7). Photo Plate I.

***Aphanocapsa* Näg.**

Gatt. einzell. Algen, 52, 1849.

Cells nearly spherical, loosely and irregularly arranged forming a gelatinous mass, sheath homogenous and colourless.

Key to the species

1a. In saline water - *A. littoralis*
1b. In freshwater - *A. pulchra*

8. *Aphanocapsa littoralis* Hansgirg

Beitr. zur Kenntnis der Meeresalg. und Bact. Fl., Sitzb. K. Böhm Ges. Wiss., Math. Nat. CI., 229, 1892; Forti in De Toni, Sylloge Algarum, 5 : 70,1907; Geitler, Kryptogarmenflora, 153, fig. 66b, 1932; Frémy, Cyano. cotes d' Eur., 15, pl. 3, fig. 1, 1933; Desikachary, Cyanophyta, 131, pl. 21, fig. 1, 1959.

Thallus (colony) without any definite shape, mucilaginous, green or yellow cells, spherical to sub spherical, single or in twos, densly, as well as, sparesely aggregated. 3 - 4μ diameter.

Habitat : Planktonic in a brackishwater unmanaged pond at Kumirmari, Sundarbans, 24 Parganas (S), West Bengal.

Specimen No. : N. S. 2 (S). (Plate I, Fig. 8).

9. *Aphanocapsa pulchra* (Kütz.) Rabenh

Fl. Eur. Alg., 2 : 49, 1865; Forti in De Toni, Sylloge Algarum, 5 : 73, 1907; Frémy, Myxo. d'Afr. équat franc., 22, fig. 22, 1929; Geitler, Kryptogamenflora, 155, fig. 69, 1932; Frémy, Cyano. cotes d' Eur., 14, pl. 2, fig. 3, 1933; Desikarchary, Cyanophyta, 132, pl. 21. fig. 2, 1959; Santra *et. al.* Biol. Mem., 86, pl. 1, fig. 11, 1988.

Thallus bluish-green, homogenous, cells spherical, 2.5 - 4.5μ diameter, loosely arranged, single or in twos, individual sheaths indistinct.

Habitat : On wet soil during low tide at Gosaba Bazar, Sundarbans, 24 Parganas (S), West Bengal.

Specimen No. : N. S. 9. (Plate I, Fig. 9).

***Aphanothece* Näg.**

Gatt. einzell. Algen., 59, 1849

Cells ellipsoidal to cylindrical, straight or slightly bent, many in a more or less shapeless expanded thallus, mucilage homogenous, occasionally with individual envelopes.

10. *Aphanothece stagnina* (Spreng.) A. Br.

Rabenhorst, Fl. Eur. Alg., 2 : 66, 1865; Forti in De Toni, Sylloge Algarum, 5: 76, 1907; Boye - Peterson, Freshwater Cyanophyceae of Iceland, Bot. of Iceland, 2, 1923; Frémy, Myxo. d' Afr. équat. franc., 26, fig. 266, 1929; Geitler, Kryptogamenflora, 164, fig. 72, 75a, b, 1932; Frémy, cyano. cotes d' Eur 18, pl. 3, fig. 5, 1933; Desikachary, Cyanophyta, 137, 21, fig. 10.

Thallus spherical, ellipsoidal, dull brown, cells oblong, to cylindrical, 3-6μ broad, 5-9.5μ long, densely arranged in the colony without individual envelopes, homogenous mucilage.

Habitat : Free floating in freshwater pond at Jharkhali, Sundarbans, 24 Paraganas (S), West Bengal.

Specimen No.: N.S. S. 17. (Plate I, fig. 10).

***Merismopedia* Meyer**

N. Syst. d. pflanzen - Physiologie, 3: 440, 1839; Wiegmann, Arch. f. Naturg., 2 : 67, 1839.

Cells many in tabular colonies, arranged in a homogenous mucilage, generally in fours, arranged in a single plane, free floating.

11. *Merismopedia tenuissima* Lemm.

Beitr. Kennt. Planktonalg., 11, Bot. Central bl. 76, 154, 1898; Kryptogamenflora der Mark Brandenburg, 3 : 82, fig. 8, 1910; Forti in De Toni, Sylloge Algarum, 5 : 108, 1907; Frémy, Myxo. d' Afr. équat. frac., 11, fig. 7, 1929; Geitler, Kryptogamenflora, 263, fig. 129 a, b, 1932; Frémy, cyano. cotes d' Eur. 6, pl. 1, fig. 3, 1933; Desikachary, Cyanophyata, 154, pl. 29, fig. 7, pl. 30, figs. 8,9, 1959; Santra *et. al.* Biol. Mem. 14 (1), 85, fig. 8, 1988.

Cells closely packed in groups of 16 - 100 cells, pale bluish - green, sub-spherical, 1.5 - 2μ broad.

Habitat : On wet soil during low tide at Gosaba, Sundarbans, 24 Paraganas (S), West Bengal.

Specimen No. : N. S. 9. (Plate I, Fig. 11).

Family - ENTOPHYSALIDACEAE Geitler

Synopt. Darst. Cyano., Beih. bot. Cbl., 4 (Abt. 2) : 235, 1925; Kryptogamenflora, 292, 1932.

***Johannesbaptistia* J. de Toni**

Notes. Nomencl. Algol., 1 : 6, 1936

Thallus microscopic, linear, elongated, cylindrical, straight or curved, cells discoid arranged in a single series, in narrow cylindrical hyaline mucilage.

12. *Johannesbaptistia pellucida* (Dickie) Taylor *et* Drouet

Bull. Torrey bot. cl., 65 : 285, 1938; Drouet and Daily, Rev. Coccoid Myxophyceae, Butler Univ. Stud. Bot., 12 : 85, figs. 182-184, 1956; Desikachary, Cyanophyta, 165, pl. 32, figs. 14-19, 1959; Islam, pl. 1, figs. 1-5, 1985; Santra, et al. Biol. Mem. pl. 2, fig. 16, 1988.

Elongated, pseudofilamentous thallus, straight or curved, cells discoid or semi circular, round at the tip of the filament, arranged in a single series in a cylindrical, hyaline, homogenous mucilage, occasionally dead cells found in between cells, cells 4 - 4.5μ broad and 2.2 - 3.2μ long.

Habitat : Planktonic in brackishwater fishery at Haroa, Sunderbans, 24 Parganas (N), West Bengal.

Specimen No. : N. S. (P) 8. (Plate I, Fig. 12).

Order-CHAMAESIPHONALES

Unicellular, often differentiated into base and apex, generally occur gregariously through endospores and exospores.

Key to the Families

la. Plants attached, differentiated into base and apex, with exospores - *Chamaesiphous*

lb. Plants attached, differentiated into base and apex, with endospores - *Dermocarpaceae*

Family-CHAMAESIPHONACEAE

***Chamaesiphon* A. Br., *et* Gurnow**

Rabenhorst, Fl. Eur. Alg., 2:148,1865; em. Geitler, Synopt. Darst. Cyano., Beih. bot. Cbl., 41 (2) : 256, 1925.

Sporangia with a short mucilage stalk or without such a stalk, at first spherical, later ellipsoidal, club-shaped or cylindrical, abstricting exospores at the apex, membrane lamellated, the wall opening at the rip, exospores sometimes remaining attached at the apex of the sporangium and germinate forming a filamentous or dendroid colony.

13. *Chamaesiphon curvatus* Nordst.

De, Algis et Chacaceis Sandwicensibus, 4, pl. 1, figs. 1,2, 1878; Frémy, Myxo. Afr. équat. franc., 69, fig. 74, 1929; Geitler, Kryptogarmenflora, 426, figs. 244, 245, 1932; Desikachary, Cyanophyta, 168, pl. 33, fig. 1, 1959; Santra *et. al.,* Biol.-Mem. pl. 2, fig. 17, 1988.

Club shaped sporangia, single or in groups, tips slightly bent, 3.2 - 4.8μ broad and 30-58μ long, spores 2.8-4μ diameter.

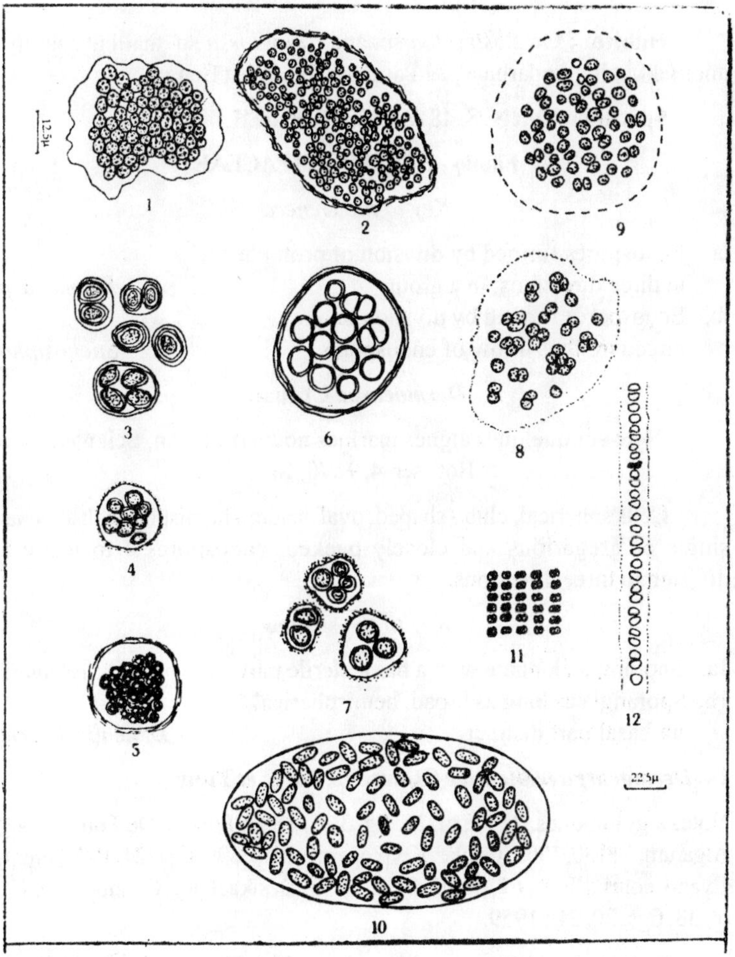

Plate I

1. *Microcystis litoralis* (Hansg.) Forti; 2, *Microcystis bengalensis* Banerjee; 3. *Gloeocapsa decorticans* (A. Br.) Richter; 4. *Gloeocapsa punctata* Näg. Ex Kützing; 5. *Gloeocapsa aeruginosa* (Carm.) Kützing; 6. *Gloeocapsa rupestris* Kützing; 7. *Gloeocapsa kuetzingiana* Näg.; 8. *Aphanocapsa littoralis* Hansgirg; 9. *Aphanocapsa pulchra* (Kütz.) Rabenh; 10. *Aphanothece stagnina* (Spreng.) A. Br.; 11. *Mersimopedia tenuissima* Lemm.; 12. *Johannesbaptistia pellucida* (Dickie) Taylor et Drouet.

Habitat : On *Bostrychia* sp. and *Caloglossa* sp. thalli at Jharkhali and Bakkhali, Sundarbans, 24 Parganas (S), West Bengal.

Specimen No.: N. S. 48 & N. S. 51 (Plate II, Fig. 13).

Family - DERMOCARPACEAE

Key to the Genera

la. Endospores formed by division of protoplast in three directions, in a group - *Dermacarpa*

lb. Endospores formed by division in one direction, thus a row of endospores - *Stichosiphon*

***Dermocarpa* Crouan**

Note sur quelques algnes marines nouvelles, Ann. Sci. nat., Bot., ser. 4, 9 : 70, 1858.

Cells spherical, club - shaped, oval, seldom hemispherical, attached, single or gregarious and closely packed, endospores 4 to many by division in three directions.

Key to the species

la. Sporangia elongate with a basal sterile part - *D. leibleiniae*

1b. Sporangia as long as broad, hemispherical, no basal part distinct - *D. hemisphaerica*

14. *Dermocarpa leibleiniae* (Reinsch.) Born. *et* Thur.

Notes algologiques, 2 : 73, pl. 26. fig. 3 - 5, 1880; Forti in De Toni, Sylloge Algarum, 5 : 130, 1907; Geitler, Kryptogamenflora, 399. fig. 224, 1932; Frémy, Cyano. cotes d'Eur., 61, pl. 17, fig, 3, 1933; Desikachary, Cyanophyta, 173, pl. 33, figs. 20, 21 ; 1959.

Sporangia elongate, oval, pyriform, 19 - 23μ broad, 47 - 63μ long, single bluish - olive green, thick lamellated membrane, endospores upto 20, 3 -4.3μ diameter.

Habitat : On *Lyngby heironymusii* from brackishwater pond at Ramgopalpur, Sundarbans, 24 Parganas (S), West Bengal.

Specimen No.: N. S. 4, N. S. 4 (S). (Plate II, Fig. 14).

15. *Dermocarpa hemisphaerica* Setchell *et* Gardner

In Gardner, New Pacific Coast Mar. Alg., 11, Univ. Calif. Publ. Bot., 6 : 438, pl. 37, fig. 21, 1918; Setchell and Gardner, the Marine Algae of the Pacific Coast of North America, *ibid.*, 8 (1); 22, pl. 3, fig. 21, 1919; Geitler, Kryptogamenflora, 391, fig. 215, 1932: Desikachary, Cyanophyta, 173, pl. 33, figs. 11, 12, 1959; Santra *et. al.*, Biol. Mem., 88, pl. 2, fig. 18, 1988.

Sporangia single, hemispherical in shape, attached by the flat side, base 18 - 20μ broad, bluish-green, wall colorless, not lamellated, endospores upto 20, 0.8-1.1 μ diameter.

Habitat : On *Cladophora* from regularly inundated mangrove forest floor at Dhanchi Island, Sundarbans, 24 Parganas, West Bengal.

Specimen No. : N. S. 80. (Plate II, Fig. 15).

Stichosiphon Geitler

Kryptogamenflora, 14 : 411, 1932.

Sporangia attached by means of a mucilaginous foot, elongate with a single row of endospores, 4 - 12 endospores.

16. *Stichosiphon sansibaricus* (Hieron.) Drouet *et* Daily

Butler. Univ..Studies, 10 : 223, 1952; Rev. Coccoid Myxophyceae, 125, fig. 253, 1956; Desikachary, Cyanophyta, 177, pl. 32, figs, 9-13, pl. 33, figs. 8 - 10, 1959; Santra *et. al.*, pl. 2, fig. 19, 1988.

Sporangia ellipsoidal, attached with the, help of a mucilaginous foot, epiphyte, 3-4 times long as broad, 2.6μ brad, endospores uniseriate, many, sheath hyaline, spores olive-green, spherical or ovoid or cylindrical.

Habitat : On *Lyngbya* sp. in a brackishwater fishery at Ramgopalpur, Sundarbans, 24 Parganas, West Bengal.

Specimen No. : N. S. 4. (Plate II, Fig. 16). Photo Plate I.

ORDER - PLEUROCAPSALES

Family - HYELLACEAE

Xenococcus Thuret

Essai clas. nostoch., Am. Sci. nat., 6 ser., Bot., 1 : 373, 1875; Bornet and Thuret, Notes Algologiques, 2 : 73, pl. 26, figs. 1, 2,1880; em. Geitler, Synopt. Darst. Cyano., Beih. bot. cbl., 41(2) : 244, 1925

Thallus forming hemispherical or crustose flakes by the formation of short erect branches or merely loosely arranged cells; erect filments when present short and few celled, dichotomously or tetrachotomously branched, cells in groups of four or eight; sporangia when present in the margins of disc at the tips of erect filaments.

Key to the species

la. Cells upto 45μ broad - *X. chaetomorphae*
1b. Cells upto 25μ broad - X. cladophorae

17. *Xenococcus chaetomorphae* Setchell *et* Gardner

New Pacific Coast Marine Algae, III, Univ. Calif, Publ., Bot., 6 : 436, pl. 36, figs 2 - 4, 1918; Geitler, Kryptoganmen flora, 336, fig. 173, 1932; Desikachary, Cyanophyta, 182, pl. 31, fig. 25, 1959; Santra *et. al.*, Biol. Mem., pl. 2, fig. 20, 1988.

Cells differently arranged in groups or colonies, usually 4 cells together of various shapes and sizes, ellipsoidal or angular, bright blue - green, 25 - 48μ diameter.

Habitat : On *Chaetomorpha* sp. filaments at Jharkhali, Sundarbans, 24 Parganas (S), West Bengal.

Specimen No. : N. S. 2, N. S. 4 (S). (Plate II, Fig. 17). Photo Plate-I.

18. *Xenococcus cladophorae* (Tilden) Setchell *et* Gardner

New Pacific Coast Marine Algae, III, Univ. Calif. Publ. Bot., 6 : 461, pl. 38, fig. 8, 1918; Geitler, Kryptogamenflora, 336, fig. 171, 1932; Desikachary, Cyanophyta, 183, pl. 31, fig. 26, 1959; Santra, *et. al.* Biol. Mem., pl. 2, fig, 21, 1988.

Cells irregulary arranged but more compact, in single layer, spherical or angular colonies, 8.8 - 13.6μ diameter, blue - green, wall colourless.

Habitat : On *Cladophora* sp. and *Chaetomorpha* sp. filaments, free floating in a brackishwater fish pond at Kumirmani, Sundarbans, 24 Parganas, West Bengal.

Specimen No. : N. S. 6 (S). (Plate II, Fig. 18).

Order- NOSTOCALES

Key to the Families

1a. Trichomes without false branching -2
1b. Trichomes with false branching -4

2a. Without heterocysts, spores commonly absent - *Oscillatoriaeae*
2b. With heterocysts and spores -3
3a. Trichomesi differentiated into base and apex - *Microchaetaceae*
3b. Trichomes not so differentiated - *Nostocaceae*
4a. With out a intercalary meristematic zone and generally terminal hair absent - *Scytonemataceae*
4b. With an intercalary meristarnatic zone and a terminal hair - *Rivulariaceae*

Family - OSCILLATORIACEAE

Key to the Genera

1a. Cylindrical trichomes, many in a sheath, sheath at the apex generally closing after hormogone formation - 2
1b. Cylindrical trichomes, without sheath or Single within a sheath, end /small of sheaths open - 4
2a. Trichomes many in a sheath, densely arranged, sheath slimy filaments twisted into rope-like bundle -*Microcoleus*
2b. Trichomes fewer, loosely arranged - 3
3a. Filaments branched, sheath slimy, end cells with calyptra - *Hydrocoleum*
3b. Filaments branched, sheath firm, end cell without a calyptra, many trichomes in a sheath, filaments richly branched - *Schizothrix*
4a. Trichomes with a prominent sheath - 5
4b. Trichomes without a sheath - 6
5a. Sheath firm, filaments not in bundles - *Lyngbya*
5b. Sheath mucilaginous, filaments forming a thallus with more or less confluent sheaths - *Phormidium*
6a. Trichomes more or less straight, not regularly spirally coiled, not in bundles - *Oscillatoria*
6b. Trichomes regularly spirally coiled - 7
7a. Cells of trichomes not visible or unicellular - *Spirulina*
7b. Cells of trichomes clearly visible - *Arthrospira*

***Microcoleus* Desmazieres**

Catal. pl. Botanagr. belg., 7, 1823; Gomont, Monogr. Oscillariées, 350, 1892.

Filaments rarely branched, sheath colourless, unlamellated, trichomes many in a sheath, densely aggregated, often coiled like a rope, ends straight, attenuated, end cell usually conical.

19. *Microcoleus chthonoplastes* Thuret ex Gomont

Thuret, Am. Sci. nat. Bot. 6th ser., 1 : 378, 1875; Gomont, Monogr. Oscillariées, 353, pl. 14, figs. 5 - 8, 1892; Frémy, Myxo. d' Afr. équat. frac., 78, 1929; Geitler, Kryptogamenflora, 1133, fig. 739, 1932; Desikachary, Cyanophyta, 343, pl. 60, fig. 7 - 9, 1959; Islam, Biblio. Phyco. pl. 73, fig. 431, 1976; Santra, *et. al.* Biol. Mem., pl. 2, fig. 23, 1988.

Filaments straight, unbranched, 3 - 6µ broad, many trichomes closely grouped, coiled, constricted at cross - walls, not granulated, 6 - 8µ long, about half broad as long, end cells pointed conical.

Habitat : On muddy substratum adjacent to Bakkhali beach Sundarbans, 24 Paraganas (S) West Bengal.

Specimen No.: N. S. 43. (Plate II, Fig. 19). Photo Plate II.

***Hydrocoleum* Kütz**

Phyc.gene., 196, 1843; Gomont, Monogr. Oscilariées, 332. 1892.

Few trichomes in gelatinous, mostly colourless sheath, filaments branched, forming a tuft, ends of trichomes straight, often attenuated, end cells with calyptra.

20. *Hydrocoleum lyngbyaceum* Kuetz. ex Gomont

Kuetzing, Species Algarum, 259, 1849; Gomont, Monogr. Oscillariées, 337, pl. 12, figs. 8 - 10, 1892; Geitler, Kryptogarnenflora, 1150, fig. 757, 1932; Frémy, pl. 19, fig. 1, 1933; Desikachary, Cyanophyta, 347, pl. 46, fig. 1,2,6, 1959; Santra *et. al.* Biol Mem., 92, figs. 27, 28, 1988.

Few trichomes in gelatinous and expanding, colourless sheath dark coloured trichomes, filaments unbranched at the base and richly branched above, spirally coiled, trichomes 7-8µ broad, cells often with calyptra.

Habitat : On muddy substratum at Bakkhali beach, Sundarbans, 24 Parganas ((S), West Bengal.

Specimen No. : N. S. 43. (Plate II, Fig. 20).

Schizothrix Kütz.

Phyc. gene., 230, 1843; Gomont, Monogr. Oscillaries, 292, 1892.

Trichomes generally many in large soft to firm thallus, enclosed in a sheath, sheath colourless to yellow - brown, filaments densely packed, forming a leathery stratum, thallus attached or free swimming.

21. *Schizothrix lamyi* Gomont

In Bornet, Algues du department de la Haute-Vienne etc., Bull. Soc. bot. Fr., 38 : 250, 1891; Monogr. Oscillariés, 323, pl. 11, figs. 1-3, 1892; Frémy, Myxo. d' Afr. équat. franc., 110, fig. 101, 1929; Geitler, Kryptogamen-flora, 1114, fig, 720, 1932: Desikachary, Cyanophyta, 330, pl. 56, fig. 1,4, 1959; Islam, Bangladesh J. Bot., pl. 9, figs. 2 - 5, 1973; Santra, Biol. Mem., pl. 3 fig. 31, 1988.

Many trichomes embedded in a common mucilaginous sheath forming brown thallus, sheath golden yellow to brown, firm lamellated, pointed at the ends, trichomes bent and much branched, with spread out branches, filaments also much branched, trichomes constricted at cross - walls, cells longer than broad, 3.2 - 4µ broad, 6.8 - 8µ long, end cell conical, slightly rounded.

Habitat: In a stagnant water pond on decomposed matter at Ramgopalpur, Sundarbans, 24 Paraganas(S), West Bengal.

Specimen No.: N. S. 9. (Plate II, Fig. 21). Photo Plate II.

Lyngbya Ag.

Syst. Alg., 25,1824 : Gomont, Monogr. Oscillariées, 118, 1892.

Trichomes single or free in a thin to massive thick, firm sheath mostly colourless, seldom yellow to brown or red coloured, filaments blue to purple red, sometimes spirally coiled or attached at the base or in the middle or the entire filament attached, mostly without such attachment or free-swimming or forming free thallus.

Key to the species

1a.	Filaments free floating	- 2
1b.	Filaments among other algae	- 3
2a.	Filaments 20 - 24µ broad	- *L. birgei*
2b.	Filaments 12 - 14µ broad	- *L. hieronymusii*

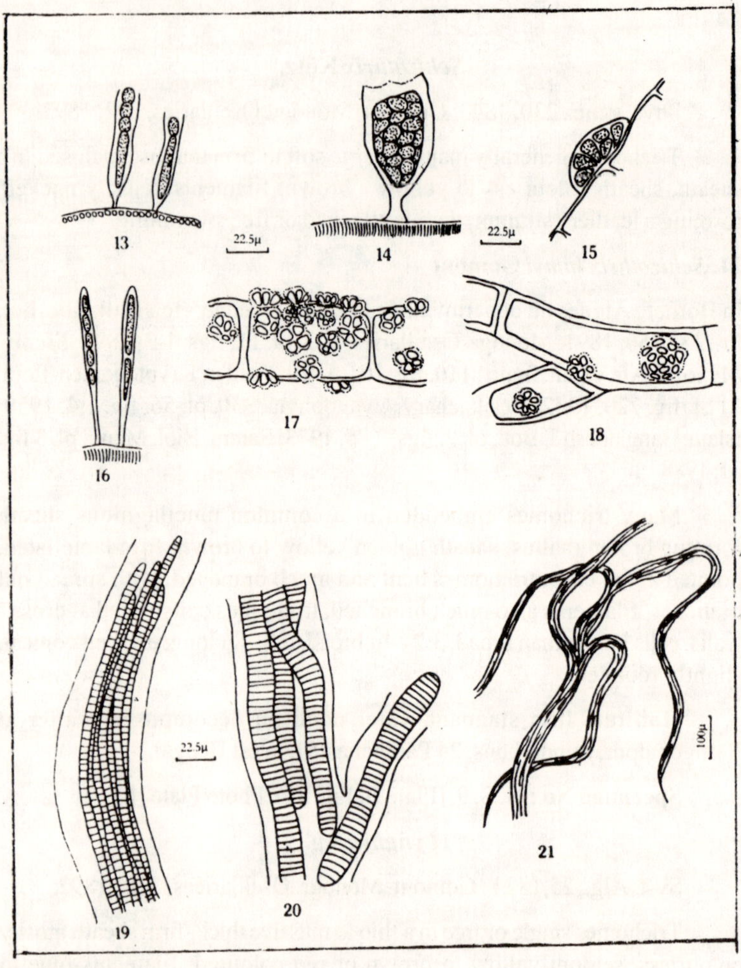

Plate II

13. *Chamaesiphon curvatus* Nordst.; 14. *Dermocarpa leibleiniae* (Reinsch.) Born *et.* Thur.; 15. *Dermocarpa hemispherica* Setchell *et* Gardner; 16. *Stichosiphon sansibaricus* (Hieron.) Drouet *et* Daily; 17. *Xenococcus chaetomorphae* Setchell *et* Gardner; 18. *Xenococcus cladophorae* (Tilden) Setchell *et* Gardner; 19. *Microcoleus chthonoplastes* Thuret ex Gomont; 20. *Hydrocoleum lyngbyaceum* Kütz. Ex Gomont; 21. *Schizothrix lamyii* Gomont.

3a. Filaments upto 5μ broad, trichomes 2 - 3μ broad end cells capitate - *L lutea*
3b. Filaments broader - 4
4a. Trichomes 26 - 32μ broad, without calyptra - *L. majuscula*
4b. Trichomes 8 - 15μ broad, without calyptra - *L. confervoides*
4c. Trichomes upto 20μ broad with calyptra - *L. semiplena*

22. *Lyngbya birgei* Smith G. M.

New or Interesting Algae from the lakes of Wisconsin, Bull, Torrey bot. 61; 43 : 482, pl. 26, fig. 28, 1916; Geitler, Kryptogamenflora, 1048, fig. 663, 1932; Desikachary, Cyanophyta, 296, pl. 50, figs. 7, 8, 1959.

Straight, long, brown filaments, 20 - 24μ broad, firm and unlamellated sheath, 2 - 3.5μ thick, trichomes not constricted, end cells rounded, not attenuated.

Habitat : Free floating in brackishwater fish pond at Chandipur, Sundarbans, 24 Paraganas (S), West Bengal.

Specimen No.: N. S. 3. (Plate III, Fig, 22). Photo Plate III.

23. *Lyngbya hieronymusii* Lemm.

Forschbers., Plön, 12 : 146, pl. 4, figs. 12, 13, 1905; Kryptogamenflora der Mark Brandenburg, 3 : 139, fig. 6 (on p. 102), 1910; Frémy, Myxo. d' Afr. équat. franc., 192, fig. 160, 1929; Geitler, Kryptogamenflora, 1047, fig. 656a, 1932; Desikachary, 1959, pl. 43, fig. 4; Islam, Biblio. Phyco., pl. 73, fig. 429, 1976; Santra *et. al.* Biol. Mem., pl. 3, fig. 32, 1988.

Filaments single, free floating, dark olive green, straight, 12 - 14μ broad, firm sheath, colourless, homogenous, not constricted at cross - walls, granulated, each cell 1.9 - 2.5μ long, end cells broadly rounded.

Habitat : From brackishwater fishery at Ramgopalpur, Sundarbans, 24 Parganas (S), West Bengal.

Specimen No. : N. S. 5. (Plate III, Fig. 23). Photo Plate III.

24. *Lyngbya lutea* (Ag.) Gom.

Essai class Nostoch. homocystees, Morots J. Bot., 4 : 354, 1890; Monogr. Oscillariées, 161, pl. 3, figs. 12, 13, 1892; Forti in De Toni, Sylloge Algarum, 5 : 275, 1907; Frémy, Myxo. d' Afr. équat. franc., 189, 1929; Geitler, Kryptogamenflora, 10577, fig. 670a, b, 1932; Frémy Cyano. cotes d' Eur.,

109, pl. 28, fig. 4, 1933; Desikachary, Cyanophyta, pl.52, fig. 9, 1959; Islam, Bangladesh J. Bot., pl. 8, fig. 18, 1973; Islam, Biblio. Phyco. pl. 72, fig. 416, 1976; Santra *et. al.* Biol. Mem. pl. 3, fig. 35, 1988.

Filaments yellow to brown to olive green, long, oiled and densely entangled 5-10µ broad, sheath colourless, 2-3µ broad, lamellated, trichomes not constricted at cross - walls, not attenuated, end cells with calyptra.

Habitat: On wet soil adjacent to brackishwater fishery at Ramgopalpur, Sundarbans 24 Parganas (S), West Bengal.

Specimen No.: N. S. 5. (Plate III, Fig, 24). Photo Plate IV.

25. *Lyngbya majuscula* Harvey ex Gomont

Harvey in Hooker, English Flora, 5 (Part 1) : 370, 1833; Gomont, Monogr. Oscillariées, 151, pl. 3, figs. 3,4, 1892, Forti in De Toni, Sylloge Algarum, .5: 268, 1907; Geitler, Kryptogamen flora, 1060, figs. 672c, d, 1932; Frémy, cyano. cotes d' Eur., 106, pl. 28, fig. 1, 1933; Desikachary, Cyanophyta, pl. 48, fig, 7, pl. 49, fig. 12 & pl. 52, fig. 10, 1959; Santra *et. al.* Biol. Mem., pl. 3, fig. 39, 1988.

Filaments very long, dull green to yellowish brown, slightly curved, 26 - 32µ broad, sheath colourless, lamellated, 4 - 6µ broad, trichomes not constricted at croos-walls, not attenuated at the ends, cross-walls not granulated, calyptra absent.

Habitat : Bottom biota of a brackishwater fishery at Chandipur I, Sundarbans, 24 Parganas (S), West Bengal.

Specimen No.: N. S. 3. (Plate III, Fig. 25). Photo Plate IV.

26. *Lyngbya confervoides* C. Ag. ex Gomont

C. Agardh, Syst. Alg., 73, 1824; Gomont, Monogr. Oscillariées, 156, pl. 3, figs. 5-6, 1892; Forti in De Toni, Sylloge Algarum, 5 : 271, 1907; Geitler, Kryptogamenflora, 1061, fig. 672 b, 1932; Frémy, Cyano. cotes d' Eur., 106, pl. 28, fig. 2, 1933; Desikachary, Cyanophyta, 314, pl. 49, fig. 9 & pl. 52, fig. 13, 1959; Islam, Bangladesh J. Bot., pl. 8, fig. 15 - 17, 1993; Santra *et. al.* Biol. Mem. pl. 3, fig. 36, 1988.

Straight filaments, bluish green, 13-16µ broad, sheath colourless, upto 6µ thick, tricomes 8 - 10µ broad, not constricted at cross-walls, cross-walls sparesely granulated not attenuated, calyptra absent.

Habitat : Bottom biota of a brackishwater water fishery at Chandipur I, Sundarbans, 24 Parganas (S), West Bengal.

Specimen No. : N. S. - 3. (Plate III, Fig. 26), Photo Plate IV.

27. *Lyngbya semiplena* (C. Ag.) J. Ag. ex Gomont

Agardh, J., Algae maris Meditterranei et Adriatici, 11, 1842; Gornont, Monogr. Oscillariées, 158, pl. 3, ags. 7 - 11, 1892; Forti in De Toni, Sylloge Algarum, 5 : 273, 1907; Geitler, Kryptogamenflora, 1061, fig. 672a, 1932; Frémy Cyano. cotes d' Eur., 108, pl. 28, fig. 3, 1933; Desikachary, Cyanophyta, 315, pl. 49, fig. 8, pl. 52, fig. 7, 1959; Santra *et. al.* Biol. Mem. pl. 3, fig. 37, 1988.

Filaments dark yellowish green, curved or flexuous, upto 20μ broad, sheath colourless, slimy, mucilaginous, slightly lamellated, 3 - 4μ thick, trichomes not cross - walls granulated, end slightly attenuated, end cells with a calyptra.

Habitat : Among other algae in a brackishwater fish pond at Ramgopalpur, Sundarbans, 24 Parganas (S), West Bengal.

Specimen No. : N. S. 5, N. S. 1 (S). (Plate III, Fig. 27).

***Phormidium* Kütz.**

Kützing, Phyc. gene. 190, 1843, Monogr. Oscillariées, 156, 1892.

Filaments many together forming a leathery stratum, thallus attached by the lower side, or floating in water with torn margins, more or less firm, sheath present, colourless, trichomes cylindrical, apex attenuated, straight or bent.

Key to the species

1a. Trichomes constricted at crosswalls, upto 2μ broad, end cells conical - *P. fragile*
1b. Trichomes not constricted at cross-walls broader than 3μ, end cells with calyptra - *P. stagnina*

28. *Phormidium fragile* (Meneghini) Gomont

Monogr. Oscillariées, 163, pl. 4, figs. 13 -15, 1892; Forti, in De Toni, Sylloge Algarum, 5 : 220, 1907; Frémy, Myxo. d' Afr, équat. franc., 140, fig. 123, 1929; Geitler, Kryptogamenflora, 99, fig. 636b, 1932, Frémy, Cyano. cotes d' Eur., 86, pl. 22, fig, 6, 1933; Desikachary, Cyanophyta, 253, pl. 44, figs. 1 - 3, 1959; Santra *et. al.* Biol. Mem. 96, fig. 41, 1988.

Thallus mucilaginous, lamellated, yellowish green, closely packed in a regular parallel manner, constricted at cross -walls, attenuated apices, 1.2 - 2.3μ broad, calyptra absent.

Habitat : On a twig in a water body at Jharkhali, Sundarbans, 24 Parganas (S), West Bengal.

Specimen No. : N. S. 52. (Plate III, Fig. 28). Photo Plate III.

29. *Phormidium stagnina* Rao, C. B.

The Myxophyceae of the Madras Presidency, I. J. Indian bot. Soc.,17: 93, figs. 4-7, 1938a; Desikachary, Cyanophyta, pl. 45, figs. 16-18, 1959; Santra *et. al.* Biol. Mem. 96, fig. 42, 1988.

Trichomes blue-green, single in each sheath, sheath very thin, firm and hyaline, straight, parallel filaments in typical bundles, densely aggregated or inter woven, thallus torn at the edges, filaments 12-15μ broad, end cells swollen with calyptra.

Habitat : Free floating along with other algae in a brackishwater fishery at Ramgopalpur, 24 Parganas (S), West Bengal.

Specimen No. : N. S. 7. (Plate III, Fig. 29).

***Oscillatoria* Voucher**

Hist. Conf., 165, 1803; Gomont. Monogr. Oscillariées, 198, 1892.

Trichome single or forming a flat free swimming thallus, sheath absent.

Key to the Species

1a.	Cells upto 1/3 as long as broad	- 2
1b.	Cells longer, trichomes constricted, cells 4 - 8μ long	- *O. chlorina*
2a.	Trichomes unconstricted	- 3
2b.	Trichomes constricted	- 6
3a.	Trichomes straight	- 4
3b.	Trichomes bent or spirally coiled	- 5
4a.	End cell with a thickened outer wall	- *O. limosa*
4b.	End cells without a thickened outer wall, 5 - 6μ broad	- *O. subbrevis*

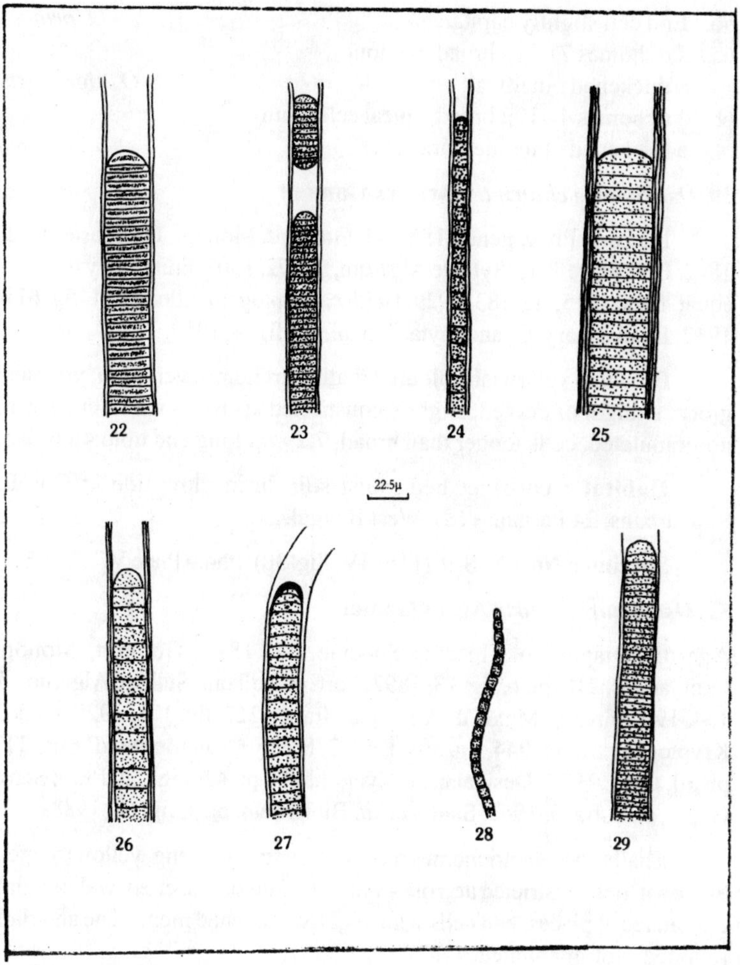

Plate III

22. *Lyngbya birgei* Smith; 23. *Lyngbya hieronymusii* Lemm; 24. *Lyngbya lutea* (Ag.) Gomont; 25. *Lyngbya majuscula* Hearvey ex Gomont; 26. *Lyngbya confervoides* C. Ag. Ex Gomont; 27. *Lyngbya semiplena* (C. Ag.) J. Ag. Ex Gomont; 28. *Phormidium fragile* (Meneghini) Gomont; 29. *Phormidium stagnina* Rao, C.B.

5a. End cell rounded - *O. curviceps*
5b. End cell slightly capitate - *O. princeps*
6a. Trichomes 7 - 11μ broad, without a thickened outer wall - *O. nigroviridis*
6b. Trichomes 4 - 10μ broad, apical cells with a thickened outer membrane - *O. tenuis*

30. *Oscillatoria chlorina* Kürtz. ex Gomont

Kützing, Phyc. gene., 185,143; Gomont, Monogr. Oscillariées, 223, 1892; Forti in De Toni, Sylloge Algarum, 5 : 172, 1907; Frémy, Myxo. d' Afr. équat. franc., 215, fig. 183, 1929; Geitler, Kryptogamenflora, 951, fig. 611c, 1932; Desikachary, Cyanophyta, 215, pl. 40, fig. 4, 1959.

Brown to yellowish coloured thallus, trchomes very thin, yellowish green, straight or curved, slightly constricted at cross - walls, cross walls notgranulated, cells longer than broad, 7.2 - 8μ long and upto 4μ broad.

Habitat : On river bed moist soil during low tide at Gosaba, Sundarbans, 24 Parganas (S), West Bengal.

Specimen No. : N. S. 9. (Plate IV, Fig. 30). Photo Plate V.

31. *Oscillatoria limosa* Ag. ex Gomont

Agardh, Dispositio Algarum Sueciae, 35, 1812; Gomont, Monogr. Oscillariées, 210, pl. 6, fig. 13, 1892; Forti in De Toni, Sylloge Algarum, 5 : 154, 1907: Frémy, Myxo. d' Afr. équat. franc., 212, fig. 178, 1929;.Geitler, Kryptogamenflora, 944, fig. 598 d, 1932; Frémy Cyano., cotes d' Eur., 117, pl. 30, fig. 4, 1933; Desikachary, Cyanophyta, pl. 42, Fig. 11, 1959; Sen & Gupta, pl. 1, fig. 9, 1987; Santra *et. al.* Biol. Mem. pl. 4, fig. 48, 1988.

Thallus brown, trichomes more or less straight, long, yellowish green to brown, not constricted at cross - walls, 12-20μ broad, cross-walls slightly granulated at places, end cells with slightly thickened membrane and flatly rounded, not attenuated.

Habitat : Attenuated to twigs in a water body at Jharkhali forming a yellow coloured coating on the twigs.

Specimen No. : N. S. 52. (Plate IV, Fig. 31). Photo Plate V.

32. *Oscillatoria subbrevis* Schmidle

Engler's Bot. Jaheb., 30, 243, pl. 4, fig. 7, 1901; Forti in De Toni, Sylloge Algarum, 5 : 208, fig. 174, 1907; Frémy, Myxo. d' Afr. équat. franc., 208, fig.

174, 1929; Geitler, Kryptogamenflora, 949, fig. 601b, 1932; Desikachary, Cyanophyta, pl. 37, fig. 2, pl. 40, fig. 1, 1959; Islam, Bangladesh J. Bot. pl. 8, fig. 12 - 13, 1973; Islam, Biblio. Phyco. pl. 72, fig. 414, 1976; Santra *et. al.* Biol. Mem., pl. 4, fig, 52, 1988.

Straight trichomes, not attenuated at apex, 5-6μ broad, not constricted, no granules at cross - walls, end cells rounded.

Habitat : On dried up river bed with *Porteresia coarctata* at Golabari, Sundarbans, 24 Parganas (S), West Bengal.

Specimen No. : N. S. 59. (Plate IV, Fig. 32). Photo Plate V.

33. *Oscillatoria curviceps* Ag. ex Gomont

Agardh, Syst. Alg., 68, 1824; Gomont, Monogr. Oscillariées, 213, pl. 6, fig. 14, 1892; Forti in De Toni, Sylloge Algarum, 5: 157, 1907; Geiser, Kryptogamenflora, 947, fig. 598 e, 1932; Frémy, Cyano. Cotes d' Eur., 117, pl. 30, fig, 5, 1933; Desikachary, Cyanophyta 209, pl. 38, fig. 2; Santra *et. al.* Biol. Mem., pl. 4, fig. 54, 1988.

Thallus dark blue green, trichomes straight, bent at the tip, tips not attenuated, not constricted at cross - walls, sometimes cross - walls granulated, 13.5 - 15.8μ broad end cells flat rounded.

Habitat : Planktonic in a brackishwater fishery at Haroa, Sundarbans, 24 Parganas (N), West Bengal.

Specimen No. : N. S. 14. (Plate IV, Fig. 33).

34. *Oscillatoria princeps* Voucher ex Gomont

Voucher, Histoire des conferves d'eau douce, 190, pl. 15, fig. 2, 1803; Gomont. Monogr. Oscillariées, 206, pl. 6, fig. 9,1892; Forti in De Toni, Sylloge Algarum, 5 : 150, 1907; Frémy, Myxo. d'Afr. équat. franc., 208, fig. 175, 1929; Geitler, Kryptogamenflora, 947, figs. 598a, 601-cg, 1932; Desiliachary, Cyinophyta, pl. 37, figs. 1,10, 11, 13, 14, 1959; Santra, *et. al.* Biol. Mem., pl. 4, fig. 56, 1988.

Trichomes blue green, straight, not constricted at the cross-walls, 28 - 36μ broad, slightly bent at the apex, end cells flatly rounded.

Habitat : Free floating in a brackish water fishery at Haroa, Sundarbans, 24 Parganas (N), West Bengal.

Specimen No. : N. S. 15. (Plate IV, Fig. 34).

35. *Oscillatoria nigroviridis* Thwaites ex Gomont

Thwaites in Harvey Phyc. Brit. Syn., 39, nr. 375, pl. 251A, 1846 - 1851; Gomont, Monogr., Oscillariées, 217, pl. 6, fig. 20, 1892; Geitler, Kryptogamenflora, 942, 597C, 1932; Frémy, Cyano. cotes d' Eur., 120, pl. 30, fig. 8, 1933; Desikachary, Cyanophyta, pl. 42, fig. 6 & pl. 49, fig. 2, 202.

Thallus dark green, trichome constricted at cross - walls, 7 - 10μ broad, bent at ends and attenuated, cross cells granulated, end cell with slightly thickened convex outer membrane.

Habitat : From water body at Lothian Island, Sundarbans, 24 Parganas (S), West Bengal.

Specimen No. : N. S. 49. (Plate IV, Fig. 35).

36. *Oscillatoria tenuis* Ag. ex Gomont

Agardh, Alg. Dec. 2 : 25, 1813; Gomont, Monogr. Oscillariées, 220, pl. 7, figs. 2 - 3, 1892; Forti in De Toni, Sylloge Algarum, 5 : 166,1907; Frémy, Myxo. d' Afr. équat. franc., 217, fig. 186, 1929; Geitler, Kryptogamenflora, 959, fig. 611 f-g, 1932; Cyano, cotes d'Eur., 121, pl. 30, fig. 10, 1933. incl. *O. viridis* Vaucher, Hist. Conf. d'eau douce, 195 pl. 15, fig. 7, 1803; Desikachary, Cyanophyta, 208, pl. 38, fig. 8, 1959.

Thallus blue-green to olive-green, slimy, straight tirichomes, slightly constricted at cross-walls, apices slightly bent, not attenuated, 7-11μ broad, 1/3rd times as long as broad, cross-walls granulated, end cells hemispherical with thickened outer membrane.

Habitat : On river bed moist soil during low tide at Gosaba, Sundarbans, 24 Parganas (S), West Bengal.

Specimen No. : N. S. 9. (Plate IV, Fig. 36).

***Spirulina* Turpin em. Gardner**

Turpin in Dic. nat. de Levrault, 50 309,1827; em. Gardner, New Pacific Coast Marine Algae I, Univ. Calif Publ. Bot., 6. 379, 1917; Gomont Monogr. Oscillariées, 249, 1892.

Trichomes unicellular or multicellular, cylindrical, sheath absent, coiled into a more or less regular spiral, terminal cell rounded, without calyptra.

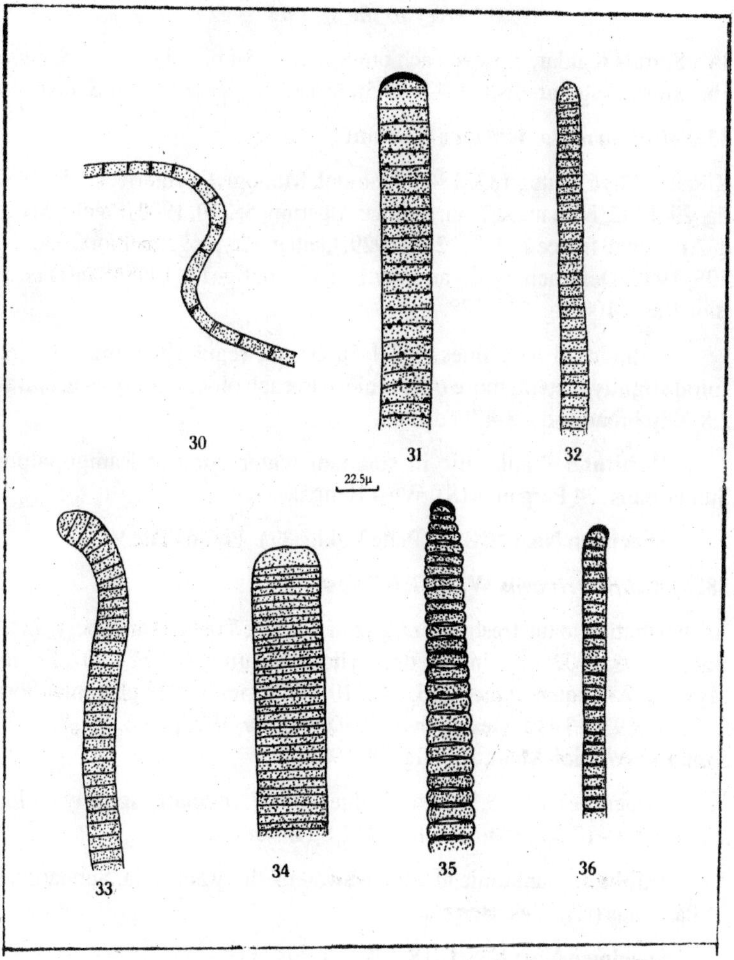

Plate IV

30. *Oscillatoria chlorina* Kütz. Ex Gomont; 31. *Oscillatoria limosa* Ag. Ex Gomont; 32. *Oscillatoria subbrevis* Schmidle; 33. *Oscillatoria curviceps* Ag. Ex Gomont; 34. *Oscillatoria princeps* Vaucher ex Gomont; 35. *Oscillatoria nigroviridis* Thwaites ex Gomont; 36. *Oscillatoria tenuis* Ag. Ex Gomont.

Key to the species

1a. Spirals regular, close to each other, 1.2 - 1.5μ broad - *S. major*
1b. Spirals regular, distant, 4.5 - 5.5μ broad - *S. princeps*

37. *Spirulina major* Kütz. ex Gomont

Kützing, Phyc. gene., 183, 1843; Gomont, Monogr. Oscillariées, 251, pl. 7, fig. 29, 1892; Forti in De Toni, Sylloge Algarum, 5; 210, 1907; Frémy, Myxo. d' Afr. équat. france 234, fig. 208, 1929; Geitler, Kryptpgamenflora, 930, fig. 595, 1932; Desikachary, Cyanophy 196, pl. 36, fig. 13, 1959; Santra *et. al.* Biol. Mem. 100, fig. 53, 1988.

Cylindrical trichomes, 1.2-1.5μ broad, regularly spirally coiled, spiral slightly distant, more or less close to each other, blue-green, spirals 2.8-3.2μ broad and 2.8-4.0μ distant.

Habitat : Planktonic in stagnant water body at Ramgopalpur, Sundarbans, 24 Parganas (S), West Bengal.

Specimen No. : N. S. 7. (Plate V, Fig. 37). Photo Plate VI.

38. *Spirulina princeps* W. et G. S. West

A contribution to the freshwater algae of Ceylon, Trans. Linn. Soc. (Lond.) Bot., 6 : 205, 1902; Forti in De Toni, Sylloge Algarum, 5 : 211, 1907; Frémy, Myxo, d' Afr. équat. franc., 236, fig. 210, 1929; Geitler, Kryptogamenflora, 931, fig. 593d, 1932; Desikachary, Cyanophyta, 197, pl. 36, fig. 7, 1959; Santra, *et. al.* Biol. Mem, 100, fig. 58, 1988.

Trichomes 4.5 - 5.5μ broad, blue - green, regularly spirally coiled, spirals 11.5 - 12.5μ broad and 10.5 - 11.8μ distant.

Habitat : Planktonic in brackishwater fishery at Haroa, Sundarbans, 24 Parganas (N), West Bengal.

Specimen No. : N. S. (P) 8. (Plate V, Fig. 38).

Arthrospira Stizenberger

Spirulina and *Artbrospira*, Hedwigia, 1 32,1852; Gomont, Monogr. Oscill., 246, 1892.

Trichomes multicellular, cylindrical, without sheath, loosely and regularly coiled, usually of relatively large diameter and large spirals, with comparitively short and fewer coils, cross-walls distinct, apices may or may not taper, terminal cell rounded, calyptra absent.

39. *Arthrospira platensis* (Nordst.) Gomont

Monogr. Oscillariées, 247, pl. 7, fig. 27, 1892; Desikachary, Cyanophyta, 190, pl. 35, fig. 2, 1959.

Thallus blue-green, trichomes constricted at cross walls, 6 - 8μ broad, not attenuated, regularly spirally coiled, spirals 26-32μ broad, distance between the spirals 45-50μ, cells nearly as long as broad, 2-7μ long, cross -walls granulated, end cells broadly rounded.

Habitat : In stagnant water at Ghushighata, Sundarbans, 24 Parganas (North), West Bengal.

Specimen No. : N. S. 19. (Plate V, Fig. 39).

Family - NOSTOCACEAE

Key to the Genera

1a. Heterocysts present -2
1b. Heterocysts absent, trichomes attenuated, end cells pointed - *Raphidiopsis*
2a. Intercalary heterocysts generally single -3
2b. Intercalary heterocysts in pairs - *Anabaenopsis*
3a. Filaments single or in a formless gelatinous mass - *Anabaena*
3b. Filaments in a definite colony - *Nostoc*

***Raphidiopsis* Fritsch *et* Rich**

Contributions to our knowledge of Freshwater Algae of Africa, Trans. roy. Soc. S. Afr., 18 : 91, 1929.

Trichomes short, more or less bent, without sheath, attenuated at one or both ends, pointed, heterocysts absent.

Key to the species

1a. Trichomes often curved - *R. curvatus*
1b. Trichomes generally straight - *R. indica*

40. *Raphidiopsis curvata* Fritsch et Rich

Contributions to our knowledge of Freshwater Algae of Africa, Trans. roy. Soc. S. Afr., 91, fig. 32, 1929; Geitler, Kryptogamenflora, 658, fig. 120, 1932; Desikachary, Cyanophy 422, pl. 65, figs. 5, 7 & pl. 79, fig. 4, 1959.

Trichome free floating, single, sigmoid or semi-circularly bent, not constricted at cross-walls, gas vacoules present, 9-11.5μ long, 4.5-5 μ broad in the middle, gradually tapering sharply, attenuated with pointed tips at both ends.

Habitat : Planktonic in a fishery at Haroa, Sundarbans, 24 Paıganas (North), West Bengal.

Specimen No. : N. S. 58. (Plate V, Fig. 40).

41. ***Raphidiopsis indica* Singh, R. N.**

J. Indian bot. Soc., 21 : 239 - 243, figs. 1 - 11, 1942; Desikachary, Cyanophyta, 22, pl. 79, figs. 1, 5, 1959; Santra *et. al.* Biol. Mem., 101, fig. 67, 1988.

Trichomes straight or slightly bent short, 1.8-2.5μ broad, attenuated, ends pointed, spores ellipsoidal, 3-3.5μ broad and 8.2-9μ long.

Habitat : On river bank of Bidyadhari at Minakhan, Sundarbans, 24 Parganas (North), West Bengal.

Specimen No. : N. S. 74. (Plate V, Fig. 41).

42. ***Anaboenopsis arnoldii* Aptekarj**

De nova Cyanophycearam species *Anabaenopsis arnoldii,* Nat. Syst. Inst. Cryypt. Horti Bot. Princi. U.S.S.R., 4: 54, figs, 1-8, 1926; Geitler, Kryptogamenflora, 809, fig. 519, 1932; Desikachary, Cyanophyta, 356, pl. 5, figs. 2-7, 1959,

Trichomes single, regularly spirally coiled with heterocysts at one end, other end with one or two heterocysts or spores, cells 7-9μ diameter, with gas vacoules, heterocysts intercalary also, in pairs 7.8-9.2μ broad and 8-10μ long.

Habitat : Planktonic from Haroa fishery at Sundarbans, 24 Parganas (North), West Bengal.

Specimen No. : N. S. 14. (Plate V, Fig. 42).

Anabaena Bory

Dict. class. d'hist. nat., 1: 307, 1822; Bory and Flahault, Revision des Nostacées, 224, 1888.

Trichomes uniformly broad throughout or apices may be somewhat attenuated, sheet absent if present then torn or forming a soft mucilaginous thallus, heterocysts generally intercalary, spores single or in long series, formed near the heterocysts or in between the heterocysts.

Key to the species

1a.	Akinetes spherical or subspherical	-2
1b.	Akinetes ellipsoidal with rounded ends	- *A. doliolum*
2a.	Trichomes 2 - 3.5μ broad, heterocysts narrower, akinetes 3.5 - 5.5μ broad	- *A. anomala*
2b.	Trichomes 6 - 7.5μ broad, heterocysts broader, akinetes 4μ broad	- *A. gelatinicola*

43. *Anabaena doliolum* Bharadwaia

The Myxophyceae of the United Provinces, I, Proc. Indian Acad. Sci., B, 2: 105, fig. 6A, D, 1935; Desikachary, Cyanophyta, 410, pl. 78, fig. 3, 1959.

Thallus mucilaginous, pale blue-green, single trichomes, tree swimming straight, curved or slightly coiled, upto 4μ broad, tapering at the end, with conical apical cell, almost pointed apex, cells barrel shaped, as long as broad or a little longer or shorter than broad, heterocysts barrel shaped, 5.5μ broad and upto 9μ long, spores ellipsoidal, upto 5μ broad.

Habitat : Planktonic in brackishwater fishery at Ramgopalpur, Sundarbans, 24 Parganas (South), West Bengal.

Specimen No. : N. S. 7. (Plate V, Fig. 43). Photo Plate VI.

44. *Anabaena anomala* Fritsch

Fritsch ex De, Proc. roy. Soc., London, B, 127 : 125, 1939 *(nomen nudum)* Desikachary, Cyanophyta, 398, 01.73, fig. 2, 1959; Santra *et. al.* Biol. Mem., 102, fig. 64, 1988.

Thallus gelatinous, blue green, trichomes densely aggregated, moniliform, cells 2-3μ diameter, spherical or nearly so, heterocysts intercalary and single, 3.5-5μ diameter, spherical.

Habitat : Free floating in water body at Lothian Island, Sundarbans, 24 Parganas (South), West Bengal.

Specimen No. : N. S. 50, (Plate V, Fig. 44).

45. *Anabaena gelatinicola* Ghose

A systematic and ecological account of blue-green algae from Lahore and Simla, J. Linn. Soc., Bot., London, 46 : 341, pl. 31, fig. 8, 1924; Desikachary,

Cyanophyta, 398, pl. 76, fig. 6, 1959; Santra *et. al.* Biol. Mem., 102, fig. 40, 1988.

Thallus thick, gelatinous, trichomes solitary, spirals densely ranged, cells sub spherical, 6-7.5μ broad, acute apex, heterocysts 7-8μ broad, spherical, spores in series, away from heterocysts, spherical.

Habitat : On *Porteresia coarctata* at Golabari, Sundarbans, 24 Parganas (South), West Bengal.

Specimen No.: N. S. 59. (Plate V, Fig. 45).

Nostoc Vaucher

Histoire des Conferveateaes d'eau douce, 203, 1803; Bonner and Flahault, Revision des Nostacae és hétérocystées, 181, 1888.

Thallus mucilaginous, gelatinous or conaccous, globose to oblong, foliose, filiform, solid or hollow, free or attached, filaments flexuous, curved or entangled, sheath sometimes distinct, generally diffluent, cells depressed, sphencal, barrel-shaped or cylindrical, heterocysts intercalary, spores spherical, or oblong in between heterocysts.

Key to the species

la. Thallus soft, formless, trichomes densely aggregated - *N. punctiforme*

1b. Thallus rigid, globose, trichomes less aggregated - *N. linckia*

46. *Nostoc punctiforme* (Kütz.) Hariot

J. de Bot., 5 : 31, 1891; Forti in De Toni, Sylloge Algarum, 388, 1907; Frémy, Myxo. d' Afr. équat. franc., 331, fig. 274, 1929; Geitler, Kryptogamenflora, 834, 1932; Desikachary, Cyanophyta, 374, pl. 69, fig. 1, 1959; Islam, pl. 1, fig. 2, 1981; Santra *et. al.* Biol. Mem., 102, fig. 65, 1988.

Thallus globose to sub globose, upto 25.3μ diam., deep bluish green coloured, numerous filaments densely entangled in hyaline, mucilaginous, delicate sheath, cells short, ellipsoidal, 3 - 3.8μ broad.

Habitat : Free floating among mass of Chaetomorpha sp. filaments in a brackish water fishery at Ramgopalpur, Sundarbans, 24 Parganas (South), West Bengal.

Specimen No. : N. S. 4. (Plate V, Fig. 46). Photo Plate VI.

47. ***Nostoc linckia*** **(Roth.) Bornet** ***ex*** **Born.** ***et*** **Flah.**

Bornet in Bornet and Thuret, Notes algologiques, 1.1, 86, pl. 18, figs, 1 - 12, 1880; Born. et Flah., Revision des Nostocacées hétérocystées, 192, 1888; Forti in De Toni, Sylloge Algarum, 5 : 391, 1907; Frémy, Myxo. d' Afr. équat. franc., 332, fig. 276, 1929; Geitler, Kryptogamenflora, 838, fig. 528b, 1932; Frémy, Cyano. Cotes d' eur., 175 pl. 58, fig. 1, 1933; Desikachary, Cyanophyta, 377, 1959; Santra *el. al.* Biol. Mem. 103, fig. 72, 1988.

Thallus irregularly expanding, tom, gelatinous, rigid, bluish green to blackish green, filaments densely entangled, sheath diffluent, distinct only in peripheral portions, trichome 3.5 - 4μ broad, cells short barrel shaped, heterocysts sub spherical.

Habitat : On *Avicennia* sp. pneumatophores along with *Rhizoclońium* filaments at Pathar Pratima, Sundarbans, 24 Parganas (South), West Bengal.

Specimen No. : N. S. 66. (Plate V, Fig. 47).

Family - SCYTONEMATACEAE

Scytonema **Ag.**

Systema Algarum, 26, 1824; Bornet and Flahault, Revision des Nostocacees hetercystees, 85, 1887.

Filaments false branched, formed laterally generally in between hetercysts; trichomes single in each sheath, straight.

48. ***Scytonema hofmanni*** **Ag.** ***ex*** **Born.** ***et*** **Flah.**

Agardh, Synopsis Algar. Sueciee, 117, 1817; Bornet and Flahault, Revision des Nostocacées, hétérocustées, 97, 1887; Forti in De Toni, Sylloge Algarum, 5 : 513, 1907; Frémy, Myxo. d' Afr. équat. franc., 313, fig. 266, 1929; Geitler, Kryptogamenflora, 772, fig. 495, 1932; Frémy, Cyano. cotes d' Eur., 170, pl. 53, fig. 5, 1933; Desikachary, Cyanophyta, 476, pl. 91, fig. 2,1959.

Thallus broadly expanded, cushion like, dark blue green, filamens 7 - 12μ broad, aggregated vertically, false branches aggregated, sheath firm, trichomes 5 - 10μ broad, cells unequal in length; heterocysts oblong.

Habitat : On stones and bricks at Ghushighata in stagnant water pool.

Specimen No. : N. S. 19. (Plate V, Fig 48).

Family - RIVULARIACEAE

***Calothrix* Ag.**

Syst. Alg., 24,1824; Bornet and Flahault, Revision des Nostocacées hétérocystées, 345, 1886.

Filaments single or in small bundles, arranged more or less in a parallel manner, mostly erect, unbranched, sheath firm, sometimes only at the base, heterocysts basal, seldom intercalary, spores single or in series next to the heterocyst.

49. *Calothrix contarenii* (Zanard.) Hornet et Flahault

Revision des Nostocacées hétérocystées, 355, 1886; Forti in De Toni. Sylloge Algarum, 5 : 610, 1907; Geitler, Kryptogamenflora, 600, 1932; Frémy, Cyano. cotes d' Eur., 142, pl. 3b, fig. 4,b 1933; Desikachary, Cyanophyta, 524, pl. 111, fig. 2, 5 - 8, 1959; Santra *et. al.* Biol. Mem. 105, fig. 26,1988.

Thallus firm, compact, dull green, trichomes swollen at the base, 4 - 7.8μ broad, ending in a long hair, 18-59μ long, heterocysts basal.

Habitat : Attached to free floating green algae in brackishwater fish pond at Ramgopalpur, Sundarbans, 24 Parganas, West Bengal.

Specimen No. : N. S. 4. (Plate V, Fig. 19)

Order - STIGONEMATALES

Family - NOSTOCHOPSIDACEAE

***Mastigocoleus* Lagerheim**

Note sur la Mastigocoleus genre des Algues de Pordre des Phycochromacées, Notarisia, 1 : 65, 1886; Bornet and Flahault, Revision des Nostocacées, hétérocysrées, 54, 1887.

Thallus free, variously bent, creeping on snail and mussel shells, trichomes made of one series of cells, irregularly branched, true branching, terminated by a heterocyst.

50. *Mastigocoleus testarum* Lagerh.

Note sur la *Mastigocoleus* genre del' ordre des Phycochromacées, Notarisia, 1 : 65, pl. 1, 1886; Bornet and Flahault, Revision des Nostocacées, hétérocystées, 5, 1887; Forti in De Toni, Sylloge Algarum, 5 : 564, 1907;

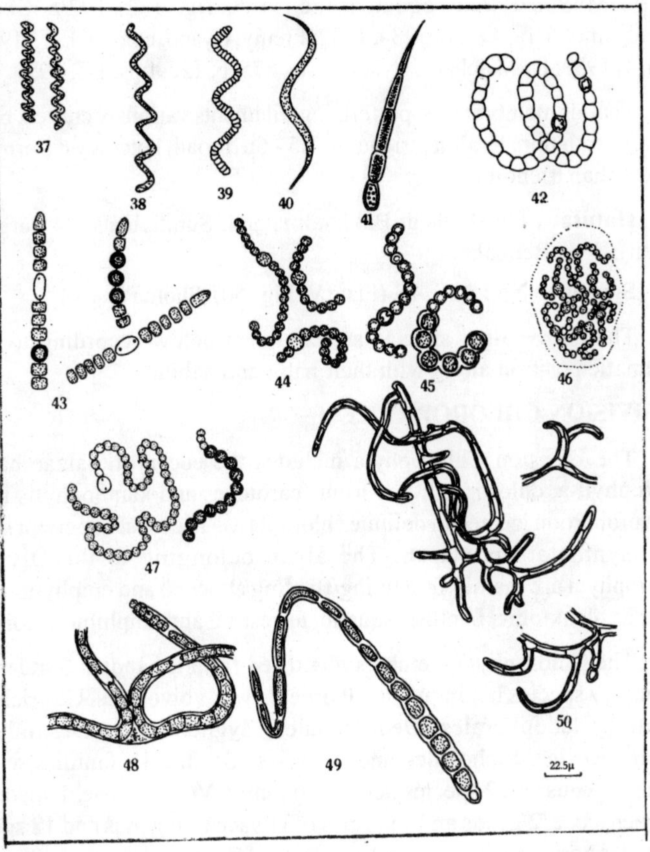

Plate V

37. *Spirulina major* Kütz. Ex Gomont; 38. *Spirulina princeps* W. et G. S. West; 39. *Arthrospira platensis* (Nordst.) Gomont; 40. *Raphidiopsis curvatus* Fritsch et Rich; 41. *Raphidiapsis indica* Singh, R. N.; 42. *Anabaenopsis arnoldii* Aptekarj; 43, *Anabaena doliolum* Bharadwaja; 44. *Anabaena anomala* Fritsch; 45, *Anabaena gelatinicola* Ghose; 46. *Nostoc punctiforme* (Kütz). Hariot; 47. *Nostoc linckia* (Roth.) Bornet ex Born *et* Flah; 48. *Scytonema hofmanni* Ag. Ex Born et Flah. 49. *Calothrix contarenii* (Zanard). Bornet et Flahault ; 50. *Mastigocoleus testarum* Lagerch.

Frémy, Myxo. d' Afr. équat. franc., 465, fig. 362, 1929; Geitler, Kryptogamenflora, 473, fig. 284, 1932; Frémy, Cyano. cotes d' Eur., 191, pl. 2, fig. 4, 1933; Desikachary, Cyanophyta, 575, p. 122, figs. 1-5, 1959.

Thallus membranous, perforating, filaments variously curved, 6-10μ broad, sheath thin, hyaline, trichome 3.5 - 6μ broad, heterocysts terminal, broader than trichome.

Habitat : On shells at Bakkhali island, Sundarbans, 24 Parganas (South), West Bengal.

Specimen No.: N.S. 46. (Plate V, Fig. 50). Photo Plate VI.

The above algal species are tabulated below according to their systematic position along with their habit and habitat.

6.3 DIVISION-CHLOROPHYTA

The Division Chlorophyta includes the eucaryotic algae having chlorophyll-a, chlorophyll-b, various carotenes and xanthophylls in the same proportion located in definite chloroplasts. The most important stored photosynthatate is starch. The algae belonging to the Division Chlorophyta are mainly free living/floating attached and epiphytic. They occur as planktonic, benthic, aquatic, terrestrial and amphibious forms.

The Chlorophyta members identified from the Indian Sundarbans include 39 species belonging to 10 orders, *viz.,* volvocales, Ulotrichales, Ulvales, Cladophorales, Oedogoniales, Zygnematales, Desmidiales, Chlorococcales, Siphonales and Charales. Of the 13 families and 21 genus, 2 genus and 2 species belongs to family Volvocaceae, 1 species of Ulotrichaceae, 2 genus and 7 species of Ulvaceae, 6 genus and 12 species of Cladophoraceae, 2 genus and 5 species of Zygnemataceae, 2 genus and 3 species of Desmidiaceae, 1 species each of Oedogoniaceae, Chlorellaceae, Radiococcaceae, 1 genus and 3 species of Hydrodictyaceae, 1 genus and 1 species of Codiaeae and 1 species of Characeae (Pie Charts-IIa, IIb, IIc). Of these 40 species, 14 species are exclusively planktonic, 1 species is epiphytic on other green algae, 6 species are periphytic, 1 species is benthic and the rest, *i.e.,* 17 are either free floating or attached to some hard substratum.

Table 16 : Habitat of Identified Cyanophyta Members From Indian Sundarbans

Order	*Family*	*Name of Species*	*Habit & Habitat*
Chroococcales	Chroococcaceae	*Microcystic litoralis*	Planktonic in brackishwater fisheries at high saline areas
		Microcystic bengalensis.	Planktonic in brackishwater fisheries at low saline areas.
		Gloeocapsa decortican	On wet soil at high saline area, near fresh water pond.
		Gloeocapsa punctata	On brick and cemented stairs of a fresh water pond.
		Gloeocapsa aeruginosa	As crusts on bricks and walls by a high saline canal
		Gloeocapsa rupestris	In stagnant water at mid saline area.
		Gloeocapsa kuetzingiana	On moist soil at mid saline area.
		Aphanocapsa littoralis	Planktonic/alongwith other algae in high saline brackishwater fisheries.
		Aphanocapsa pulchra	On wet soil as benthic flora
		Aphanothece stagnina	Free floating in a small fresh water pond.
		Merismopedia tenuissima	On wet soil and as plank ton from high saline areas.
	Entophy-salidaceae	*Johannesbaptistia pellucida*	Planktonic in low saline brackishwater fishery.

Contd...

Order	*Family*	*Name of Species*	*Habit & Habitat*
Chamaesi-phonales	Chamaesi-phonaceae	*Chamaesiphon curvatus*	Epiphytic on *Bostrychia*, *Polysiphonia* and *Caloglossa* in high saline areas.
	Dermocar-paceae	*Dermocarpa leibleiniae*	Epiphytic on *Lyngbya* sp. in high saline areas.
		Dermocarpa hemisphaerica	On *Cladophora* in mangrove forest floor.
		Stichosiphon sansibaricus	Epiphytic on *Lyngbya* sp. in high saline areas.
Pleuro-carpales	Hyellaceae	*Xenococcus chaetomorphae*	Epiphytic on *Chaetomorpha* filament.
		Xenococcus cladophorae	Epiphytic on *Cladophora* and *Enteromorpha*.
Nostocales	Oscillatoriaceae	*Microcoleus chthonoplastes*	On high saline muddy substratum
		Hydrocoleum lyngbyaceum	On high saline muddy substratum.
		Schizothrix lamyii	In stagnant water and on decomposed matter in high saline areas.
		Lyngbya birgei	Free floating and with other algae from paddy fields at high saline areas.
		Lyngbya hieronymusii	Planktonic and as free floating mass in high saline brackishwater fisheries.
		Lyngbya lutea	On wet soil adjacent to brackishwater fishery at high saline areas.

Contd...

Order	*Family*	*Name of Species*	*Habit & Habitat*
		Lyngbya majuscula	Bottom biota of brackishwater fishery at high saline areas.
		Lyngbya confervoides	Bottom biota of brackishwater fishery at high saline areas.
		Lyngbya semiplena	Among other algae in brackishwater fisheries at high saline areas.
		Phormidium stagnina	Planktonic & withother algae in high saline brack-ish water fisheries.
		Phormidium fragile	On twigs in a water body at high saline area.
		Oscillatoria nigroviridis	Planktonic/bottom biota in high saline areas
		Oscillatoria limosa	Periphytic on dead submerged substratum in high saline areas.
		Oscillatoria subbrevis	From dried up river bed alongwith *Porteresia coarctata* at high saline area.
		Oscillatoria curviceps	Planktonic in low saline bheries.
		Oscillatoria princeps	Planktonic in brackishwater fishery at low, mid and high saline areas.
		Oscillatoria chlorina	On river bed soil at high saline areas.
		Oscillatoria tenuis	On river bed soil at high saline areas.

Continued...

Order	*Family*	*Name of Species*	*Habit & Habitat*
		Spirulina princeps	Planktonic in brackishwater fisheries at low saline-mid saline areas.
		Spirulina major	Planktonic in stagnant water alongwith other BGA.
		Arthrospira platensis	In stagnant water body at mid saline areas.
	Nostocaceae Sub Family-Anabaenae	*Raphidiopsis indica*	Planktonic from high saline areas.
		Raphidiopsis curvata	Planktonic in fisheries at low saline areas.
		Anabaenopsis arnoldii	Planktonic in fisheries at low saline areas.
		Anabaena anomala	In water bodies at high saline areas.
		Anabaena doliolum	Planktonic in brackishwater fisheries.
		Anabaena gelatinicola	On *Porteresia coarctata* at high saline area.
		Nostoc punctiforme	Free floating among other green algae in brackishwater fisheries.
		Nostoc linckia	On wet riverbed in mid saline area.
	Sub-Family Rivulariaceae	*Calothrix contarenii*	Epiphytic on green algae at high saline areas.
	Sub-Family Scytone-mataceae	*Scytonema hofmanni*	On stones and bricks in stagnant pool at mid saline areas.
Stigonematales	Nosto-chopsidaceae	*Mastigocoleus testarum*	On shells collected from beach at high saline area.

Key to the Classes

Ia. Unicellular or multicellular filamentous, branched or unbranched green algae without ensheathing structures around the sex organs -Chlorophyceae

Ib. Multicellular, macroscopic plant body, erect, branched, differentiated into nodes and internodes with ensheathing structures around the sex organs -Charophyceae

CLASS - CHLOROPHYCEAE

Key to the Orders

1. Unicellular, multicellular or colonial thallus, cells or colonies motile either throughout vegetative phase or at least in some part of their life cycle; colonial members held together in gelatinous matrix. -Volvoccales
2. Thallus filamentous to foliaceous, simple or branched, without or with differentiation of attached base and free apex, single parietal chromatophore, pyrenoids present and uninucleate. -Ulotrichales
3. Thallus uniseriate or multiseriate, filamentous or parenchumatous, composed of cells united in more than one plane, forming foliose, ribbon shaped or tubular expanse. -Ulvales
4. Thallus filamentous, branched or unbranched, distinct basal hold-fast, uniseriate, multinucleate cytoplasm with greatly reticulate or fragmented chloroplast and one to many pyrenoids. -Cladophorales
5. Cylindrical, uninucleate cells, filamentous thallus branched or simple, cell division leading to 'apical cap' formation at the distal end of certain cells. - Oedogoniales
6. Unicellular or loose aggregate of cells or unbranched filaments with elaborated variously shaped chloroplasts, *viz.,* spiral, band-shaped, star-shaped or flat plates with many pyrenoids, formation of conjugation tubes. -Zygnematales
7. Thallus formed by, solitary cells or colonies of non - filamentous cells of definite sizes, and shapes, cells uninucleate or multinucleate. - Chlorococcales

8. Thallus variously branched, coenocytic, multinucleate, simple vesicular to well developed filamentous forms, chloroplasts numerous, small and discoid arranged in the periphery. -Siphonales

Order - VOLVOCALES

Family - VOLVOCACEAE

Key to the Genera

1a. All cells of a colony alike in size and function -*Pandorina*
1b. Cells of a colony of two different types - *Volvox*

***Pandorina* Bory 1824 em. Ehrb.**

Bory, Encycl. Math. Dict. hist. nat., 660, 1824; Pringsheim, Über Paarung der Schwarmsporon, 721 - 738, 1869; Dangeard, Botaniste, 7 : 192-211, 1900; Kurschikoff, Russk. Arkh. Protist. 2 : 179 - 194, 1923; Schreiber, Z. Bot. 17 : 337 - 376, 1925 Paşcher, Die Süsswasser-Flora, 4 : 423, 1927; Printz, Dienaturlichen Pflanzenfamilien, 3 : 56, fig. 17. 1927 ; Iyengar, J. Linn. Soc. Bot. 49: 323 - 375, 1933.

Colonies subspherical to ellipsoidal, hollow with 4, 8 or 16 cells embedded in a homogenous colonial envelope, chloroplast cup-shaped with one or more pyrenoids.

51. *Pandorina morum* (Müll.) Bory.

Bory de St. Vincent ; Encyl. Meth. p. 600, 1824; Pascher, Die Süsswasser - Flora, 4 : 427, figs. 387 - 389, 1927 ; Korschikoff, Vovocineae, 4 : 165, fig. 163, 1938; Iyengar and Desikachary, Volvocales, 418, fig. 243, 1981.

Colonies short ellipsoidal or nearly spherical 8 - 16 - 32 celled embedded in a common matrix, 105-120μ diameter, cells nearly obovate, 12 - 15μ diameter, obovate broadside towards outside, towards the inside angular by mutual compresseion when closely packed, chloroplast cup - shaped, massive, large eye - spot.

Habitat : Planktonic in brackishwater fishery at Haroa, Sundarbans, 24 Parganas (North), West Bengal.

Specimen No. : N. S. (P) 8. (Plate VI, Fig. 51).

Volvox L. 1758 era. Ehr. 1832

Linnaeus, System naturae, Regnum animale, 1758 ; Ehrenberg, Abh. Akad Berlin, 1830 : 0.68, 1832, Klein', L. Ber. dt. Ges. Preigburg, 5 : 29 - 92, 1891; Printz, Die natürlichen Pflanzenfamilien, 3 : 58, fig. 59, 1927 ; Pringsheim, Arch. Protistenk, 72 : 1-48, 1930; Iyengar, J. Linn. Soc. Bot. 49: 323-373, 1933; Smith, Trans. Am. microsc. Soc. 63 : 265-310, 1944.

Colonies globose-subglobose, numerous cells in a colourless gelatinous colonial matrix, cells round to spherical in surface view, chloroplast single, cup-shaped, protoplasmic connections may be present, gonidia variable per colony.

52. *Volvox* sp.

Linnaeus, Systerna naturae, Regnum animale, 1758 ; Ehrenberg, Abh. - Akad. Berlin, 1830: 0.68, 1832 ; Klcin, L. Ber. dt. Ges. Preiburg, 5 : 29 - 92, 1891; Printz, Die natürlichen Pflanzenfamilien, 3 : 58, fig. 59, 1927; Pringsheim, Arch. Protistenk, 72 : 1- 48, 1930 ; Iyengar, J. Linn. Soc. Bot. 49: 323 - 373, 1933 ; Smith, Trans. Am. microsc. Soc. 63 : 265 - 310, 1944.

Colonies globose - subglobose, 650 - 730μ diam., numerous cells in a colourless gelatinous colonial matrix, cells round in surface view, 5.3 - 6μ diam, chloroplast single, cup -shaped, protoplasmic connections cannot be seen, gonidia 8-10 per colony, each gonidia 50-63μ diam.

Habitat: Planktonic in brackishwater fishery at Haroa, Sundarbans, 24 Paraganas (North), West Bengal, during monsoon.

Specimen No.: N. S (P) 2 (Plate VI, Fig. 52).

Order - ULOTRICHALES

Key to the Families

1a. Plants filamentous, unbranched - Ulotrichaceae
1b. Plants membranous - Ulvaceae

Family - *ULOTRICHACEAE*

Uronema Lagerheim 1887

Malphigia 1: 517, 1887b

Simple, unbranched, filamentous algae, mostly attached by a narrow base sometimes expanded to form expanding disc, grow attached to or as epiphytes on other substratum or algae, terminal cell of the filament tapers to an acuminate tip, terminal cell varies in size, is straight or bent or slightly swollen, cells generally cylindrical, often elongated, thin walled and each contains a parietal plate like chloroplast with one or more pyrenoids, chloroplast extend throughout the whole length of the cell or may cover only a portion, nucleus centrally located in the hyaline protoplasm enclosed by chloroplast.

53. *Uronema confervicolum* Lagerheim

Malphigia 1 : 518, pl. 1, 1887b; Heering, in Pascher's Süsswasserfl, 6: 36, fig. 37, 1914; Palik, Index Horti Botan., 19; Ramanathan, Ulotrichales, p. 50, pl. 13, figs. A - G, 1964.

Filaments multicellular, straight or curved, upto 1.2 mm long, attached by a slightly large basal cell, cells 11.7 - 13.6μ long, 5.8 - 6.8μ broad, terminal cell pointed, cells cylindrical, chloroplast parietal, extending nearly the full length of the cells.

Habitat : Attached to other Chlorophyceae algae in a brackishwater fishery at Ramgopalpur, Sundarbans, 24 Paraganas (South), West Bengal.

Specimen No.: N. S. 5. (Plate VI, Fig. 53). Photo Plate XIII.

Family - ULVACEAE

Key to the Genera

1a. Adult plants at least in part tubular -*Enteromorpha*
1b. Adult plants foliaceous, blades 2 cell thick -*Ulva*

***Enteromorpha* Link 1820**

Thallus capillary like to large sized, simple or chiefly alternately branched, or with the branches showing uniseriate filamentous tips, at first attached, sometimes later free floating, adult holdfasts formed by downgrowths from cells of the stalk portion, cells very closely packed, single chloroplast, generally with one pyrenoid.

Key to the species

1a. Plants tubular throughout except near the base and in filamentous proliferations - 2
1b. Plants simple or nearly simple, strands or blades without conspicuously thickened cell walls, usually large, cells not in distinct longitudinal series *E. intestinalis*

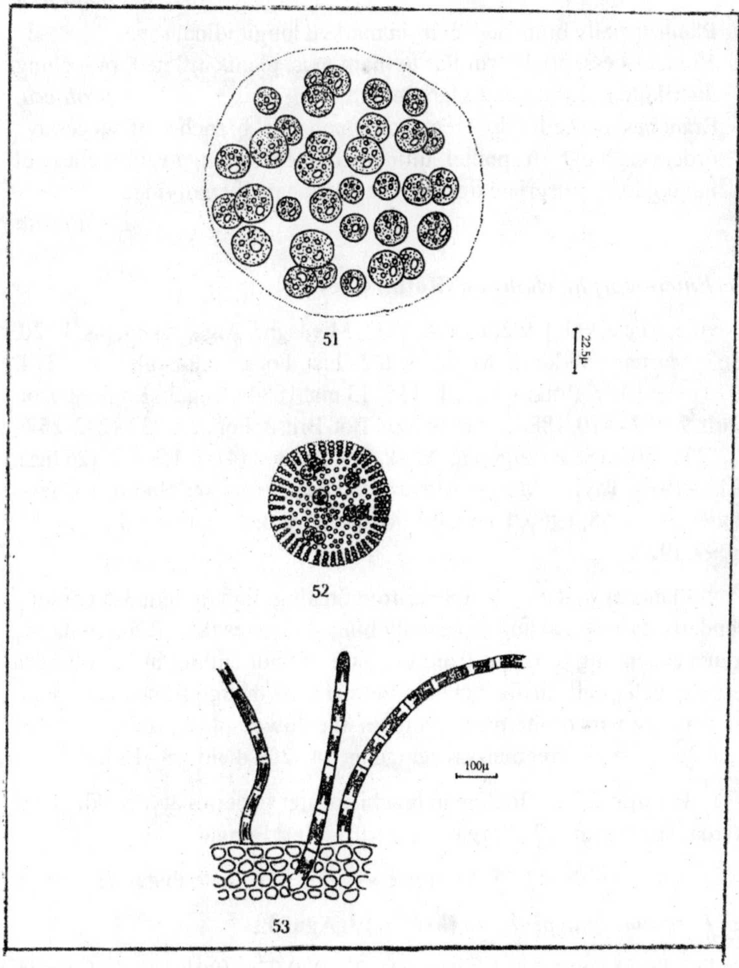

Plate VI

51. *Pandorina morum* (Müll.) Bory ; 52. *Volvox* sp.; 53. *Uronema confervicolum* Lagerheim.

2a. Plants usually branched below, cells not in longitudinal rows -*E. compressa*
2b. Plants usually branched, cells in marked longitudinal rows -3
3a. Branches essentially similar to main axis, plants inflated, branching distributed throughout plant length, -*E. prolifera*
3b. Branches markedly different from main axis, branches of successive orders each in turn smaller, ultimate branches elongate, branches not having long uniseriate tips, main axis repeatedly redivided -*E.clathrata*

54. *Enteromorpha clathrata* (Roth.) Greville

Greville, Algae, vol. 1 & 2, 1 - 456, 1833; Montagne, Algae, 1- 64, pls. 1 - 20, 1863; Martens, Vidensk. Medd. Natur - hist. Foren. Kjøbenh. 297 - 314, 1870 ; Hemsley, Botany I, 2 : 1- 135, 13 pls. 1884 ; Hauck, Englers' Bot. Jahrb. 9: 457 - 470, 1888 ; Murray Jour. Bot. Brit. & For., 27: 237 - 242, 257 - 262, 298 - 305, 1889; Børgesen, Dansk, Bot. Arkiv. 1 (4) : 1-158 + 2, 126 figs., 1913 - 1920; Taylor, Marine Algae of the East. Trop. and Subtrop. Coasts of the Am., p. 58, 1960; Islam, Bibliotheca Phyco. pl. 3, figs. 26-27, pl. 13, fig. 99, 1976.

Plants at first attached, later free floating, light green, 3-5 cm long slenderly cylindrical and repeatedly branched, axes 0.5 - 2.5 mm diam., branches tending to taper from the base but not ending in a prolonged series of cells, cells in distinct longitudinal rows though somewhat irregular arranged in older plants and in the lower plants of your cells subrectangular or irregularly rectangular, 14 - 20μ diam, 18 - 28 long.

Habitat : Free floating in brackishwater fisheries at Minakhan and Haroa, Sundarbans, 24 Parganas (North), West Bengal.

Specimen No.: N. S. 58. (Plate VII, Fig. 54). Photo Plate VII.

55. *Enteromorpha prolifera* (Müller) J. Agardh

Collins, Proc. Amer. Acad. Arts and Si. 37 : 229-270, 1901; Vikers, Liste des algues marines de la Barbade, Am. Sci. Nat. Bot. ix, 1 : 45-66, 1905; Collins and Hervey, Proc. Amer. Acad. Arts and Sci. 53 : 1-195, pls. 1-6, 1917; Hoyt, Bull. Bur. Fisheries (U.S.) 36 : 367-556, pls. 84-109, 1920; Taylor, Paper Michigan Acad. 26 : 69-701, 1941; Islam. pl. 4, figs. 7-10, Bangladesh J. Botany, 1973; Islam, Biblio. Phyco., 11, 1976; Santra & Pal. Indian Biol. pl. 11, figs 20-22a, 1988.

Plants solitary or tufted, much branched, generally remaining attached, cylindrical, main axis upto 2.5mm broad, more or less abundantly proliferously branched, cells rounded, subangular, irregular, 10-19μ diam., in younger parts always in longitudinal series, pyrenoids prominent.

Habitat : Abundant in brackishwater fisheries at Haroa, Minakhan, Malancha, Sarberia of Sundarbans, 24 Parganas (North), West Bengal.

Specimen No. : N.S. 58. (Plate VII, Fig. 55). Photo Plate VII.

56. ***Enteromorpha compressa*** **(Linnaeus) Greville**

Montagne, Ann. Sci. Nat., Bot., iii, 14 : 283-309, 1850; Martens, Videnske Medd. Natur-hist. Foren. Kjøbenh., iii, 2 (18-20) : 297-314, 1870; Mazé & Schramm, Essai de classification des alg/ues de la Guadeloupe, xix + 283 + iii pp, 1870-77; Dickie, Jour. Linn., Soc. Bnot., 14 : 146-152, pl. 11, 1874 ; Hamsley, Journ. Linn. Soc. Bot., 3 : 1-229, 53 pls., 1884; Journ. Bot., Brit. & For. 26 : 193-196, 1889; Mobius, Hedwigia, 28:309-347, pls. 10, 11, 1890; Taylor, Amer. Journ. Bot., 17 : 635, 1 fig., 1930b ; Srinivasan, Phycoligica Indica, pl. Li, 1973; Islam, Biblio. Phyco., 10, 1976.

Plants of medium height, proliferous growth, attached to other substratum which remain submerged in water, bright green fronds tubular, crispate, compressed or collapsed, expanded or dilated above, varying in dimension, 2-12 mm. wide, lower portion long, tapering and characteristically branched with several branches arising from the gradually contracted stalk like base which are similar to them ain axis, cells irregular, 10-15 μ diameter, cells rounded to polygonal.

Habitat: Periphytic on twigs and other plants submerged in brackishwater fisheries at Minakhan and Haroa, Sundarbans, 24 Parganas North), West Bengal.

Specimen No.: N. S. 58, (Plate VII, Fig. 56). Photo Plate VII.

57. ***Enteromorpha intestinalis*** **(Linnaeus) Link.**

Martens, Videnske Medd. Natur - hist. Foren. Kjøbenh, iii, 2 (18 - 20): 297 - 317, 1870 ; Mazé & Schramm, Essai de classification des algøes de la Guadelopue, xix + 283 + iii pp, 1870 - 77; Zeller, Vidensk. Medd. Naturh. Foren. Københann, 1875 : 426 - 432; Hauck, Engler's Bot. Jahrb., 9: 457 - 470, 1888; Murray, Jour. Bot. Brit. & For., 27: 237-242, 257 - 262, 298 - 305, 1889; Collins, Proc. Amer. Acad. Arts and Sci. 37 : 229 - 270, 1901 ; Hoyt, Bull. Bur. Fisheries (U.S.) 36 : 367 - 556, pls. 84 - 109, 1920 ; Srinivasan,

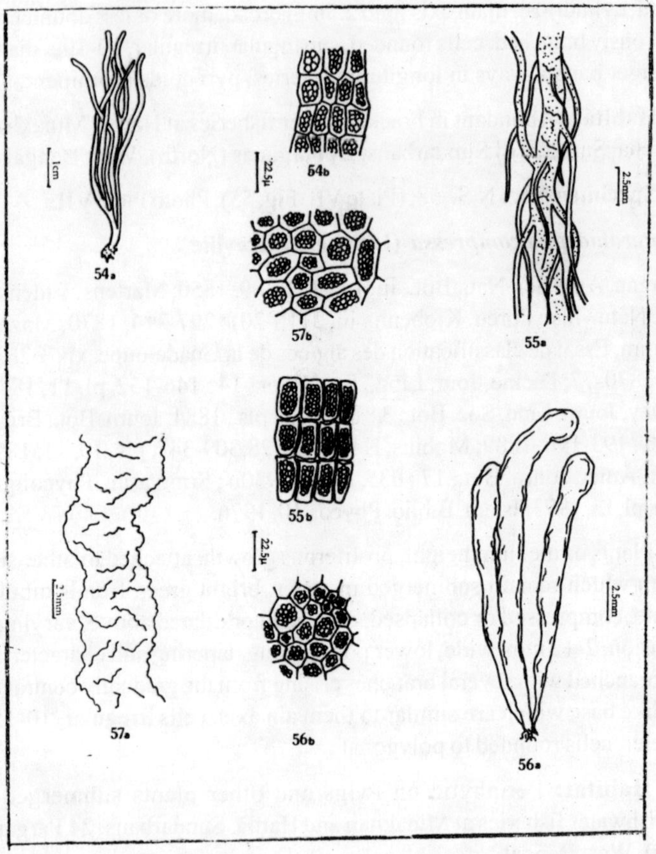

Plate VII

54a Habit sketch of *Enteromorpha clathrata* (Roth.) Greville; 54b, Cells of *Enteromorpha clathrata* (Roth.) Greville; 55a. Part of the thallus of *Enteromorpha prolifera* (Müller) J. Agardh; 55b. Cells of *Enteromorpha prolifera* (Müller) J. Agardh; 56a. Habit sketch of *Enteromorpha compressa* (Linnaeus) Greville; 56b. Cells of *Enteromorpha compressa* (Linnaeus) Greville; 57a. Part of the thallus of *Enteromorpha intestinalis* (Linnaeus) Link.; 57b. Cells of *Enteromorpha intestinalis* (Linnaeus) Link.

Phyco. Indica, pl. L, 1973 ; Islam, Bangladesh J. Bot. pl. 4, figs. 1-4, 1973; Islam, Bibliotheca Phyco. 10, 1976; Santra & Pal, Indian Biol. pl. II, figs. 18, 19, 1988.

Gregarious growth, at first attached to substratum by a holdfast, free floating later, sub-intestiniform tubular, with contortions, light green to yellowish green, membranous, fronds tapering below, in the upper portions elongate to attenuate, tubular, cylindrical, clavate or generally inflated and bullate, simple or rarely sparingly branched from the base, or proliferous cells not arranged in any order in the inflated portions, round- polyhedral, in surface view 6-15μ diam, the cells in section radially rounded, oblong, lateral walls thin but rather thick on the inner side, chloroplast parietal.

Habitat : Abundant in brackishwater fisheries at Haroa, Minakhan, Malancha, Sarberia of Sundarbans, 24 Parganas (North), West Bengal.

Specimen No.: N. S. 58. (Plate VII, Fig. 57), Photo Plate VII.

Ulva Linnaeus, 1753

Plants at first irregularly sub-filiform, soon expanded, marginally attached plane or crispate, orbicular, lobed, two layers of cells, cells showing single chromatophore, usually on the external faces, with one or two pyrenoids.

Key to the species

1a. Plants normally divided into narrow segments, in transverse sections cells along the central portion of the mature thallis much taller than those near the margins -*U. fasciata*

1b. Plants simple or with thick broad lobes -2

2a. Blades not naturally perforate, cells in transverse sections nearly square or longer than broad -*U. lactuca*

2b. Small sized thallus, lobed, blades taper to form a holdfast, cells nearly similar, irregularly arranged - *U. patengensis*

58. *Ulva lactuca* Linnaeus

Harvey, Smithsonian Contrib. To Knowledge, 10 : ii + 1 -140, 1858; Montagne, Algae. Vol. I, pp. 7-69, 1853 ; Martens, Didenske Medd. Natur-hist. Foren. Kjøbenh., iii, 2 (18-20) : 297 - 314, 1870 ; Mazé & Schramm, Essai de classification des alg\ues de la Guadeloupe, xix + 283 + iii pp.,

1870-77 ; Hemsley, Journ. Linn. Soc. Bot., 3 : 1 - 229, 53 pls., 1884; Journ. Bot., Bnit. & For. 26: 193-196, 1889; Möbius, Hedwigia, 28: 309-347, pls. 10, 11, 1890; Børgesen, Dansk. Bot. Arkiv. 1 (4) : 1-158 + 2, 126 figs., 1913-1920 ; Srinivasan, Phyco. Indica, pl. XLIX, p. 49, figs. 1-2, 1973; Islam, Bibliotheca Phyco. p. 11, 1976.

Plants attached by a small and inconspicuous holdfast, foliaceous, bright green, blades lanceolate to broadly ovate to rounded, often irregularly and deeply incised to form lobes, margins of lobes ruffled and wavy or undulate, membrane thick near the base, marginal portions somewhat thin, cells usually as tall as broad, closely packed, in cross-section sub-quadrate with rounded corners.

Habitat : Attached to bricks and other hard substratum on the river bank of Matla at Purandar, Sundarbans, 24 Parganas (South), West Bengal.

Specimen No.: N. S. 8. (Plate VIII, Fig. 58). Photo Plate VIII.

59. *Ulva fasciata* Delile

Harvey, Smithsonian Contrib. To Knowledge, 10 : ii 1+ 1 -140, 1858; Mazé & Schramm, Essai de classification des alg\ues de la Guadeloupe, xix + 283 + iii pp., 1870 - 77; Dickie, Jour. Linn. Soc. Bnot., 14: 146 - 152, pl. 11, 1874 ; Murray, Jour. Bot. Brit. & For., 27 : 237-242, 257 - 262, 298 - 305, 1889; Collins, Prod. Amer. Acad. Arts and Sci. 37:229-270, 1901; Grieve, Trans. Bot. Soc. Edinburgh, 24(1) : 7-12; Børgesen, Dansk. Bot. Arkiv. 1(4) : 1 - 158 + 2, 126 figs., 1913 - 1920 ; Hoyt, Bull. Bur. Fisheries (U.S.) 36: 367-556, pls. 84 -109, 1920 ; Hamel & Hamel - Joukov, Algues des Antilles Fracaises, 1- 3, 1931 ; Questel, Marine Algae, vii + 224 pp., 1942; Mottos, Arquivos do Museu Paranaertse, 9 ; 245-260, 1952.

Plants large, form a very small hapteron and stalk, blades cuneate in the lower portions, expanding above, irregularly lobed, generally irregularly or pinnately divided into linear lobes, which attain massive lengths, 0.5 - 2.5 cm broad, margins entire to irregularly ruffled, central portion paler than the margins.

Habitat : Attached to bricks and other hard substratum on the river bank of Matla at Purandar, Sundarbans, 24 Parganas (South), West Bengal.

Specimen No.: N. S. (Plate VIII, Fig. 59), Photo Plate VIII.

60b. T. S. thallus of *Ulva patengensis* Salam & Khan; 60c. Superficial cells of *Ulva patengensis* Salam & Khan.

60. *Ulva patengensis* Salam and Khan

Salam & Khan, Phykos, 129 - 131, fig. 1-4,1980; Santra & Pal, Indian Biologist, pl.II, figs. 14 - 17, 1988.

A small sized thallus upto 2 cm tall, 0.5 - 1 cm broad, light green, blades ovate to lanceolate, lobed, blades tapering gradually to a narrow stipe forming holdfast, cells irregularly arranged, massive chloroplast.

Habitat: Abundant ; attached to bricks, pillars and bamboo poles on the river at Purandar, Sundarbans, 24 Parganas (South), West Begnal.

Specimen No. : N. S. 8. (Plate VIII, Fig. 60).

Order - OEDOGONIALES

Family - OEDOGONIACEAE

***Oedogonium* Link, 1820**

Epistola de Algis aquaticis in genera disponendis in Nees von Esenback, C.G.H Horae Physicae barolinensis, pp. 1- 8, pl. 1.

Filaments single, unbranched, vegetative cells cylindrical sometimes undulate or nodulose, basal cell with hold fast, terminal cell obtuse, apiculate or hyaline, oogorua and antheridia produced by direct division of vegetative cells.

61. *Oedogonium undulatum* (Brebisson) Al. Braun

In De Barym Abhandl. D. Senck bergischen Gesellsch p. 94, 1854; Borge, Svenska Vet.-Akad. P. 4, pl. 1, fig. 2, 1899 ; Tiffany, The Oedogoniaceae, p. 118, pl. XLII, fig. 407, 1930.

Dioecious, nannadrous, cylindrical cells in unbranched filaments in whcih certain cells have transversely striate walls at the, distal end, filaments, undulate, basal cell modified to form a holdfast, apical cell broadly rounded or acuminate, cell slightly enlarged at the upper ends, cells uninucleate, single reticulate chloroplast, strands of reticulum parallel to the long axis, many pyrenoids at the intersection of reticulum, species collected is a nannandrous form, dwarf antheridial filaments present, elongate, rarely near oogonium, oogonium subglobose or ellipsoid-globose, vegetative cells 33-110μ long and 15-30μ broad.

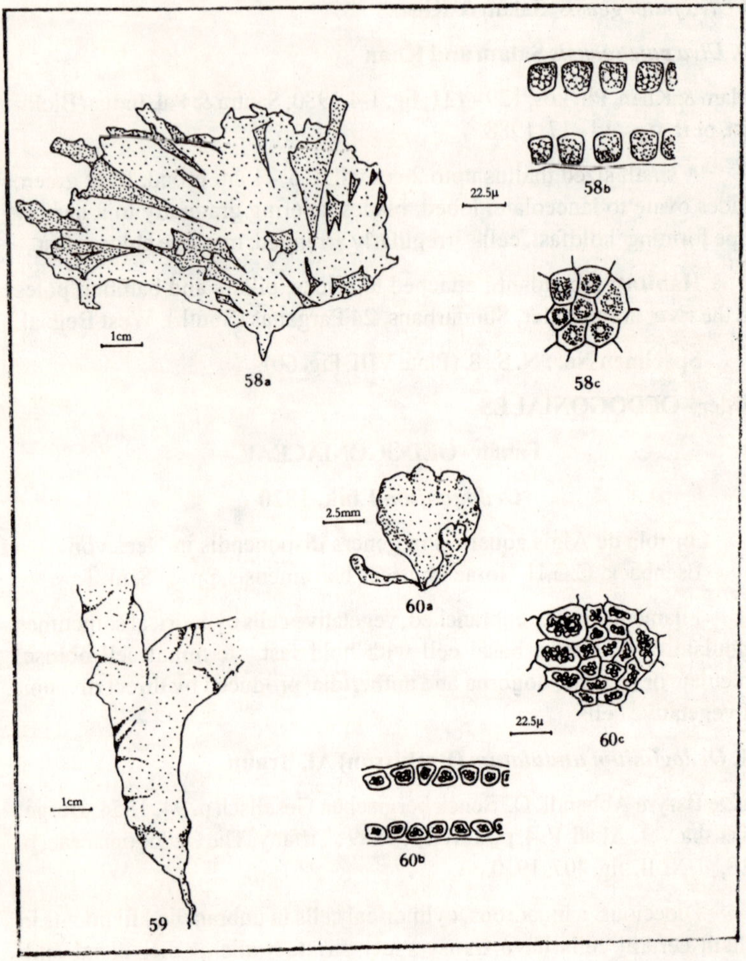

Plate VIII

58a. Thallus of *Ulva lāctuca* Linnaeus; 58b. T.S. thallus of *Ulva lactuca* Linnaeus; 58c. Superficial cells of *Ulva lactuca* Linnaeus; 59. *Ulva fasciata* Delile; 60a. Thallus of *Ulva patengensis* Salam & Khan;

Habitat : Attached to submerged substratum in a water body at Ramgopalpur, Sundarbans, 24 Parganas (South), West Bengal.

Specimen No.: N. S. 6. (Plate IX, Figs. 61a, 61b, 61c). Photo Plate IX.

Order - CLADOPHORALES

Family - CLADOPHORACEAE

Key to the Genera

1a. Filaments unbranched or with few short simple branchlets -2
1b. Filaments progressively, often abundantly branched -3
2a. Filaments may be attached by a basal end, cells very large, symmetrical, filaments coarse, a mass not completely collapsing on removal from water -*Chaetomorpha*
2b. Flaments much slender, straight or contorted, usually without rhizoidal structures, cells of varying sizes, collapse on removal from water -*Lola*
2c. Filaments slender or massive, irregular contour, periphytic, with few lateral or spur branches -*Rhizoclonium*
3a. Thallus profusely branched, irregularly dichotomous branching, cell's broader at upper part near septa. - *Cladophora*
3b. Well branched creeping thallus, long aseptate rhizoidal structure -*Cladophorella*

***Cladophora* Kützing, 1843**

Plants filamentous, sparingly to repeatedly branched, attached by rhizonial extension, prostrate system often give rise to erect shoots, chromatophores reticulate, many pyrenoids.

62. *Cladophora echinus* (Bias.) Küetz.

De Toni, Sylloge Algarum, 1: 343, 1889 - 1924 ; Möbius, Hedwigia, 28: 309-347, pls. 10, 11, 1890; Hamel, Ann. Crypt. Exotique, 2: 53 - 59, 9 figs, 1929; Taylor, Marine algae of the eastern tropical & sub tropical coasts of the Americas, p. 94, 1960; Islam, Bibliotheca Phyco. pl. 3, figs. 23 - 24, 1976.

Plants about 0.6 - 1.5cm tall, profusely branched, irregularly dichotomous branching, dark olive green coloured, cells broader at upper part near septa, cells of axis 90 - 120μ broad, 310 - 590μ long, cells of terminal branches 15 - 30μ broad, 23-68μ long.

Habitat: Periphytic on submerged twigs in brackishwater fisheries at Kumirmari, Sundarbans, 24 Parganas (South), West Bengal.

Specimen No.: N. S. 5(S). (Plate IX, Fig. 62). Photo Plate X.

Cladophorella Islam

Plants well branched, creeping thallus, long aseptate rhizoidal stuctures present.

63. *Cladophorella sundarbanensis* Islam

Islam, Rev. Alg. 7(4): 276 - 289, 1964 ; Islam, Bangladesh J. Botany, pl. 11, fig. 8-14, 1973; Islam, Bibliotheca Phyco. p. 16, 1976.

Plants light green to yellowish green, prostrate system creeping on substratum, well branched thallus, attached to the soil or pneumatophores with long aseptate rhizoidal structures, 2-7mm long, cells nearly cylindrical, 30 - 60μ diam., 78 - 280 long μ long, scattered akinetes present.

Habitat : Benthic on forest floor at Dhanchi Island, Sundarbans, 24 Parganas (S), West Bengal.

Specimen No. : N. S. 80. (Plate IX, Fig. 63), Photo Plate X.

Lola Hamel

Plants filamentous, unbranched, slender, usually without rhizoidal structures, entangled, free floating, straight filaments, cells of varying sizes.

Key to the species

1a. Filaments not straight	-*L. implexa*
1b. Filaments straight	-2
2a. Filaments cylindrical	-*L. capillaris*
2b. Filaments contorted	-*L. tortuosa*

64. *Lola capillaris* (Kützing) Hamel

Islam, Bangladesh J. Bot. pl. 1, figs. 4, 5, 1973; Islam, Bibliotheca Phyco. pl. 2, figs. 11-14, 1976 ; Santra & Pal, Indian Biol., pl. III, figs. 30, 31, 1988.

Filaments cylindrical, not straight, vegetative cells, 45-54.5μ diameter and 45-150μ long, cells cylindrical, some long, some short, slightly swollen at septum, basal rhizoid like growth present.

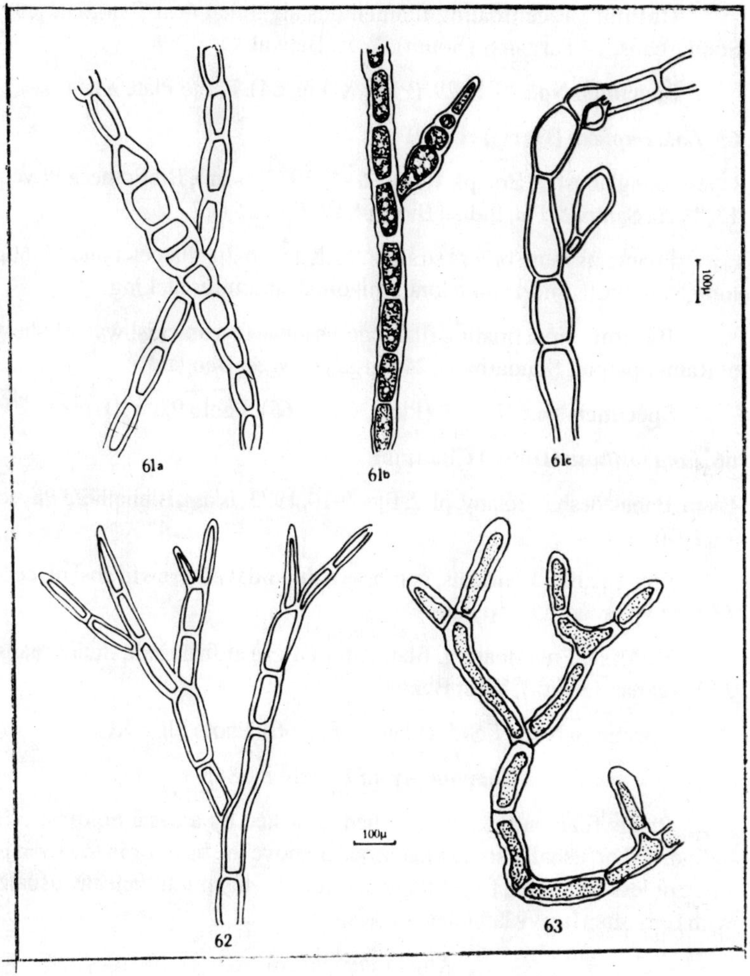

Plate IX

61a. *Oedogonium undulatum* (Brebisson) Al. Braun.; 61b. Dwarf antheridial filament of *Oedøgonium undulatum* (Brebisson) Al. Braun.; 61c. Dwarf antheridial filament of *Oedgonium undulatum* (Brebisson) Al. Braun.; 62. *Cladophora echinus* (Bias.) Kütz.; 63. *Cladophorella sundarbanensis* Islam.

Habitat : Free floating filamentous algal mat near Jharkhali jetty, Sundarbans, 24 Parganas (South), West Bengal.

Specimen No.: N. S. 29. (Plate X, Fig. 64). Photo Plate XI.

65. *Lola implexa* (Harv.) Hamel

Islam, Bangladesh J. Bot. pl. 1, figs. 6 - 8, 1973, Islam, Bibliotheca Phyco. 13, 1976: Santra & Pal, Indian Biol., pl. III, figs. 25 - 27.

Filaments more or less straight, cells 15-26.2μ diameter and 45-60μ long, some cells short some long, rhizoidal structures lacking.

Habitat: Free floating filamentous algae in a brackishwater fishery at Ramgopalpur, Sundarbans, 24 Parganas, West Bengal.

Specimen No.: N. S. 7. (Plate X, Fig. 65). Photo Plate XI.

66. *Lola tortuosa* (Dülw.) Chapman

Islam, Bangladesh J. Botany, pl. 2, figs. 9-10, 1973; Islam, Bibliotheca Phyco. 13, 1976.

Free floating filaments, contorted, rhizoidal outgrowth absent, cells 18.5-22.5μ broad, 35-45μ long.

Habitat: Free floating, filamentous algae at Jharkhali, Sundarbans, 24 Parganas (South), West Bengal.

Specimen No.: N. S. 5. (Plate X, Fig. 66). Photo Plate XI.

Chaetomorpha Kützing, 1845

Plants filamentous, unbranched, attached by a basal holdfast cell, cylindrical or usually somewhat broader above the base, or in the form of more or less entangled free Filaments, cells highly multinucleate, usually with firm often heavy lamellated walls.

Key to the species

la. Plants erect, filaments clearly basally attached, filaments of nearly uniform diameter except very near the holdfast, filaments 100μ diam. or more, basal cells about 2.5-4.2 times as long as, suprabasal cell C. aerea

1b. Plants of flexuous entangled filaments without specific basal attachment -2

2a. Filaments under 100μ diameter -*C. gracilis*
2b. Filaments over 100μ diameter, filaments 80 - 180μ diam.-*C. brachygona*

67. *Chaetomorpha brachygona* Harvey

Harvey, Smithsonian contrib. to Knowledge, 10, ii + 1 - 140, pls. 37-50, 1858; Murray, Journ. Bot. Brit. & For., 27 -242, 1889; Collins, Proc. Amer. Acad. Arts & Sci., 37 : 229 - 270, 1901; Bøergesen, Dansk. Bot. Arkiv. 1(4) : 1-158, 1913: Collins & Harvey, Proc. Amer. Acad. Arts, and Sci., 53 : 1-195, 1917; Howe.Algae, ix + 585 pl. 1918 ; Howe, Algae, vii + 695 pp., 1920; Howe, Journ. Washington Acad. Sci., 18(7) : 186 - 194, 1928: Hoyt, Bull. Bur. Fisheries [v.s.] 36 : 367-356, 1920; Taylor, Carnegie Inst. Wash, Publ. 461 : 115-124, 1935, Taylor, Contri. U.S. Nat. Herb., 28 : 549 562, pl. 20, 1940; Taylor, Rep. Allen Hancock Atlantic Exped., 2, 193 pp., 20, pl. 1942 ; Islam, Bibliotheca Phyco. pl. 1976 ; Santra & Pal, Indian Biol. pl. III, fig. 24, 1988.

Thallus fialmentous, filamént cylindrical, generally soft, flexuous - entangled cell wall lamellated, hyaline and thick, diameter of filaments 90-225μ , cells 75 - 190μ long.

Habitat: Free floating and attached to submerged in a brackishwater fishery at Ramgopalpur, Sundarbans, 24 Parganas (South), West Bengal.

Specimen No.: N. S. 6. (Plate X, Fig. 67). Photo Plate XII.

68. *Chaetomorpha gracilis* Kützing

Maze & Schramm, Essai de classification des algues de la Guadeloupe, xix + 283 + iii pp, 1870 ; Murray, Journ. Bot. Brit. & For., 27 : 237- 242, 1889; Möbius, Hedwigia, 28 : 309-347, pls. 10, 11, 1890; Bøergesen, Dansk. Bot. Arkiv. 1(4) : 1 - 158, 1913; Collins & Hervey, Proc. Amer. Acad. Arts. and Sci., 53: 1 - 195, 1917 ; Taylor, Carnegie Inst, Wash., Publ. 379, 25: 219, 1928; Taylor, Amer. Journ. Bot., 16 : 621 - 630, 1929; Taylor. Rev. Algol, 5 : 279-313, 1931; Taylor, Mich. Acad. Sci., Arts & lett. 17 : 395 - 407, pl. 26, 1933 ; Islam, Bangladesh J. Bot. pl. 3, figs. 1 - 5, 1973; Islam, Bibliotheca Phyco. pl. 2, figs. 15-17, 1976; Santra & Pal, Indian Biol., pl. III, figs. 28, 29, 1988.

Plants entangled, thallus filamentous, filments not straight, cells cylinderical short and ltong, wall thin, cross walls slightly inflated, constricted at septum, cells 90-97.5μ diameter, 57.5 - 142.5μ long.

Habitat: Alongwith *Ulva* sp. thallus on brick walls at Purandar, Sundarbans, 24 Parganas (South), West Bengal.

Specimen No.: N.S. S. (Plate X, Fig. 68). Photo Plate XII.

69. ***Chaetomorpha aerea*** **(Dillwyn) Kützing**

Dickie, Journ. Linn. Soc. Bot., 14 : 146 - 152, pl. 11, 1874; Hemsley, Algae, pp. 104-128, 1884; Murray, Journ. Bot. Brit. & For., 26: 193 - 196, 1889; Collins, Proc. Amer Acad. Arts & Sci., 37- 229 - 270, 1901; Bøergesen, Dansk. Bot. Arkiv. 1(4) : 1-158, 1913; Hoyt, Bull. Bur. Fisheries [v.s.] 36 : 367 - 356, 1920, Taylor, Paper Michigan Acad. Sci., Arts & Lett. 39 : 85 - 109, 1954 ; Islam Bangladesh J. Bot. pl. 3, figs. 6 -7, 1973, Islam, Bibliotheca Phyco. pl. 2, fig. 18, 1976, Santra & Pal, Indian Biol. pl. III, fig, 23, 1988.

Plants gregarious, bright green, thallus filamentous, unbranched, cell wall thick, lamellate, hyaline, basal cell long, sub clavate, rhizoidal, with disc like base lobed or fimbricate at the margins, filaments slender towards base, 37.5 - 52.5μ broad, 60 - 120μ long, stiff and straight, little constricted at septa.

Habitat : Free floating mass of algal filaments in a brackishwater fishery at Ramgopalpur, Sundarbans, 24 Parganas, West Bengal.

Specimen No.: N. S. 4. (Plate X, Fig. 69). Photo Plate XII.

Rhizoclonium Kützing, 1843

Plants filamentous, simple or sparingly branched, initially basally attached, but eventually free or with lateral rhizoidal branches, tapering and spreading, of few cells, main filaments usually entangled, somewhat long cells, unevenly articulated, cells with moderate to quite thick walls, chromatophores with numerous pyrenoids, nuclei one to a few in each cell.

Key to the species

1a.	Cells 50μ - 150μ diam., branches few to many	–2
1b.	Filaments 20-30μ diam., branches few, scattered, rhizoids numerous	*-R.riparium*
2a.	Filaments 60-80μ diam., branches few to many at right angles with main axis, sometimes pluricellular, rhizoids present	*-R. hookeri*
2b.	Filaments 95 - 150μ diam., knee like processes present, rhizoids plenty	*-R. grande*

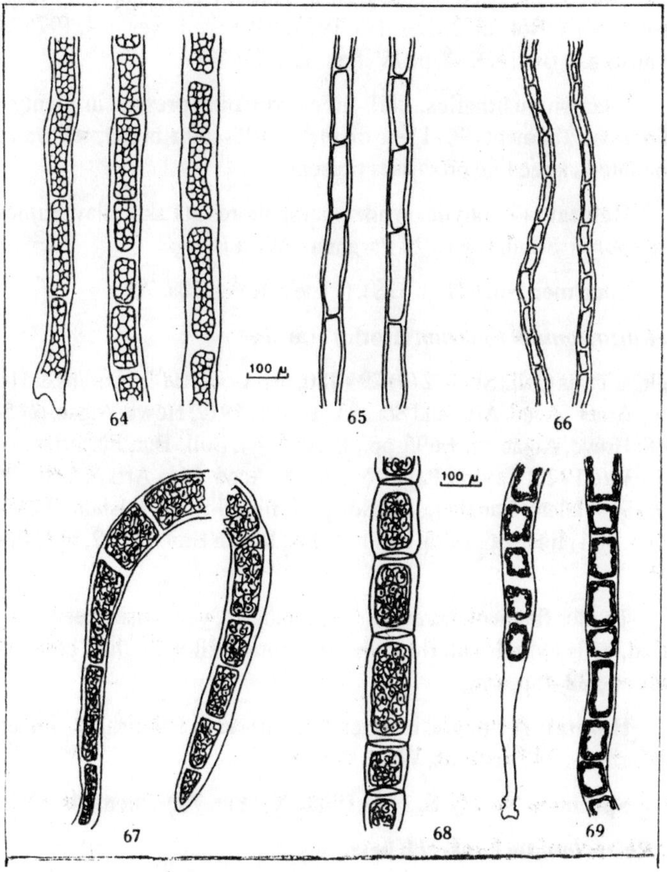

Plate X

64. *Lola capillaris* (Kützing) Hamel; 65. *Lola implexa* (Harv.) Hamel; 66, *Lola tortuosa* (Dillw.) Chapman; 67. *Chaetomorpha brachygona* Harvey; 68, *Chaetomorpha gracilis* Kützing; 69. *Chaetanorpha aerea* (Dillwyn) Kürzing.

70. ***Rhizoclonium grande*** **Boergesen**

Boergesen, Det. Kgl. Danske vidensk. Selskab. Biol. Med. 14, figs. 5-6, 1935; Boergesen, Biol. Medd. Dansk. vidensk. Selskab., I, 1946; Islam, Bangladesh J. Bot., pl. 2, figs. 1-5, 1973; Islam, pl.1, figs. 1-4, 1976; Santra & Pal, Indian Biol. XX : 2, pl. IV, figs. 32, 33, 1988.

Filamentous thallus, dark green, rhizoids present in plenty, cells cylindrical, filaments 96 -158μ diameter, 105-150μ broad, cell wall thick, lamellated, knee like processes present.

Habitat : Periphytic on pneumatophhores at Pakhiralay, Sajnekhali, Jamespur in Sundarbans, 24 Parganas, West Bengal.

Specimen No. : N. S. 3 (S). (Plate X1, Fig. 70a, 70b).

71. ***Rhizoclonium riparium*** **(Roth.) Harvey**

Collins, Tufts Coll. Stud., 2 (3) 79 - 480, pls. 1- 18, 1909 ; Collins & Hervey, Proc. Amer. Aced. Arts. and Sci., 53 : 1 - 195, 1917; Howe, Algae, ix+585 pl. 1918; Howe, Algae, vii + 695 pp., 1920; Hoyt, Bull. Bur, Fisheries [vs.] 36: 367 - 356, 1920; Taylor, Paper Michigan Acad. Sci., Arts & Lett. 39 : 85-109, 1954; Islam, Bangladesh J. Bot. pl. 2, figs. 6-8, 1973; Islam, Bibliotheca Phyco. pl. 1, figs. 5-6, 1976; Santra & Pal, Indian Biol. XX : 2, pl.1, figs.5-6, 1988.

Thallus filamentous, expanded, entangled, yellowish-green, filaments curled, cells cylindrical, rhizoids numerous, cell wall thin, cells 26-30μ diam. and 32-45μ long.

Habitat: Periphytic on pneumatophores at Pakhiralay, Sajnekhali, Sundarbans, 24 Parganas, West Bengal.

Specimen No. : N. S. 3 (S). (Plate X1, Fig. 7 1). Photo Plate XIII.

72. ***Rhizoclonium hookeri*** **Küetz.**

Collins, Tufts, Coll. Stud., 3(2) : 91-109, pls. 1-2, 1909; Collins & Harvey, Proc. Amer. Acad. Arts. and Sci., 53 : 1-195, 1917; Howe, Algae, ix + 585 pl. 1918; Howe, Algae, vii + 695 pp., 1920; Taylor, Carnegie Inst. Wash., Publ. 379, 25 : 219, 1928 ; Taylor, Amer, Journ. Bot., 16 : 621-630, 1929; Feldmann & Lami, Comptes Rendus Acad. Sci. 203 (17) : 883-885, 1937; Islam, pl. 7, figs. 2-5, 1973; Islam, pl. 1, figs. 7-10, 1976; Santrat Pal, Indian Biol. XX : 2, pl. IV, fig. 36, 1988.

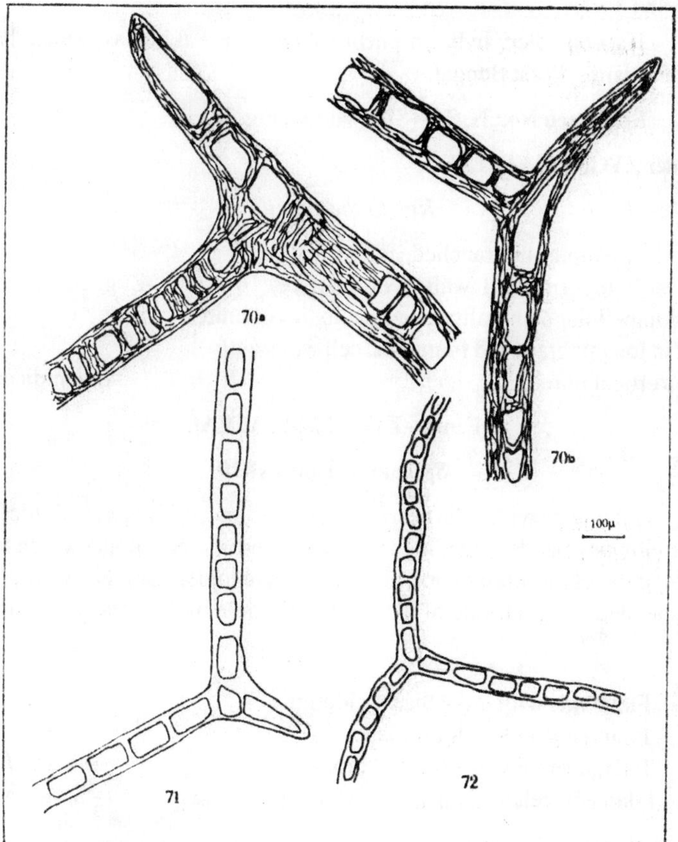

Plate - XI

70a & 70b. *Rhizoclonium grande* Boergesen; 71. *Rhizoclonium riparium* (Roth.) Harvey; 72. *Rhizoclonium hookeri* Kützing.

Branched, filamentous thallus, filaments rather stiff, entangled, short, often pluricellular, sometimes rebranched, branches usually at right angles with main axis, main and branches of nearly same dimensions, rhizoids present, cell wall lamellated of irregular thickness, cells 65-764μ diam., 70- 85μ long.

Habitat : Periphytic on pneurnatophores at Pakhiralay, Sundarbans, 24 Parganas, West Bengal.

Specimen No.: N. S. 3 (S). (Plate Xl, Fig. 72).

Order-ZYGNEMATALES

Key to the Families

1a. Filamentous, unbranched, cylindrical cells with unsegmented walls without pores -Zygnematitaceae
1b. Unicellular cells solitary or in colonies or united in long unbranched filaments, cell walls with vertical pores -Desmidiaceae

Family-ZYGNEMATACEAE

Spirogyra Link, 1820

Filaments with cells nearly as broad as long or longer. Chloroplasts are elongate bands extending from end to end of a cell, coiled in the form of spirals, chloroplasts contain several pyrenoids, uninucleate, nucleus suspended in the middle of the central vacoule by cytoplasmic strands.

Key to the species

1a. Filaments with more than 4 chloroplasts -2
1b. Filaments with 3 chloroplasts -*S. ternata*
2a. Filaments slender, 60 - 75μ broad -*S. dubia*
2b. Filaments relatively massive, 120-150μ broad - *S. setiformis*

73 *Spirogyra dubia* Kg.

Randhawa, Zygnemaceae, p. 329, fig. 317, 1959; Transeau, The Zygnemataceae, p. 178, pl. 28, fig. 8, 1951; Islam, Bangladesh J. Bot., p. 198, pl. 5, figs. 52 - 54, 1984.

Filaments with cells longer than broad, 60 - 72.5μ broard, 75-105μ long, end cells plane, chloroplasts 5-6 running from comer to comer of the cell forming a reticulate appearance.

Habitat: Free floating in low saline water bodies at Haroa, Sundarbans, 24 Parganas North, West Bengal.

Specimen No.: N.S 58 (Plate XII, Fig. 73). Photo Plate XIV.

74. ***Spirogyra setiformis*** **(Roth.) Kg.**

Randhawa, Zygnemaceae, p. 321, fig. 301, 1959; Transeau, The Zygnemataceae, p.173, pl. 27, figs. 4 - 5, 1951; Islam, Bangladesh J. Botany, p. 208, pl. 2, fig. 24 - 25, pl. 4, figs. 42 -43, 1984.

Thallus nearly brown, cells of the filaments regularly cylindrical, usually of an size throughout the filament, massive filaments, cells 120 - 142 μ broad, 126 - 250μ long, chloroplasts 4 - 6 running parallel in angular strips.

Habitat: Free floating in brackishwater fishery at Minakhan, Sundarbans, 24 Parganas (N) West Bengal.

Specimen No.: N. S. 73 (Plate XII, Fig. 74).

75. ***Spirogyra ternata*** **Ripart**

Yamagishi, Sci. Rep. Tokyo, Kyoiku Daigaku, p. 97, pl. 6, figs. 4 - 6, 1966; Islam, Bangladesh J. Botany, pl. 212, pl. 4. fig. 44 - 46, 1984.

Filaments slender, blackish brown masses, cells 62 - 65μ broad, 138-150μ long, sometimes inflated, chloroplasts - 3, arranged in a reticulate fashion.

Habitat : Free floating in brackish water fishery at Minakhan, Sundarbans, 24 Parganas, North, West Bengal.

Specimen No. : N. S. 73 (Plate XII, Fig. 75).

Family - DESMIDIACEAE

Key to the Genera

1a.	Cells with a median constriction	-2
1b.	Cells without a median constriction	-*Closterium*
2a.	Cells with spines	-3
2b.	Cells without spines	-*Cosmarium*
3a.	Cells with transverse ring of spines on lateral walls	-*Triplastrum*
3b.	Cells with spines at the apices	- *Triploceras*

Closterium Nitzsch, 1817

Cells elongate, without a median constriction, attenuated at the poles, 2 chloroplasts one in each semicell, chloroplasts may be entire or with longitudinal ridges, pyrenoids arranged in axial series.

76. *Closterium acutum* (Lyngb.) Bréb. ex Ralfs

Irene - Marie, Flore Desmidiale de la region de Montreal, 81, pl. 3, figs. 25-26, 1938; Smith, Bull. Wisc. Geol. Nat. Hist. Survey, 57. I: 1 - 243, 1920; Agarkar, Phykos, p.1, fig. 7, 1969; Suxena, Biblio, Phyco 66, p. 66, pl. II, fig. 19, 1983.

Cells, elongate, distinctly lunate, markedly attenuated at the poles, brown to yellowish brown in colour, semicells not too distinct, 3.2 - 4.5μ broad, 50 -51μ long, pyrenoids few, arranged regularly in axial series.

Habitat: Planktonic in brackish water fisheries at Minakhan and Golabari, Sundarbans, West Bengal.

Specimen No.: N. S. 60. (Plate XII, Fig. 76).

Cosmarium Corda, 1834

Unicellular with compressed cells, length nearly equal to or slightly greater than breadth with a deep median constriction, certain species with nearly circular outline with a very narrow and shallow sinus, walls without spines, each semicell contains a single axial chloroplast.

Key to the species

1a. Cells 13-18μ long, 12-13μ diameter -*C. regenlli*
1b. Cells 33-35μ long, 25-28μ diameter -*C. paucighaulatum*

77. *Cosmarium regenlli* Wille fa.

Sestt & Prescott, Hydrobiologia, p. 68 pl. 23, fig. 12, 1961; Agarkar, Phykos, p.63, fig. 62, 1971; Islam & Zaman, J. Asiatic Soc. Bangladesh (Sc.), 1:45-65, 1975 ; Islam & Haroon, Int. Rev. ges. Hydrobiol. 65 (4) : 551- 604, 1980; Islam, *et. al.* Bangladesh J. Bot., p. 49, pl. 1, figs. 19 -21, 1992.

Cells 13.5-18μ long, 12-13.2μ broad, isthmus 4-4.5μ semi-cells distinctly depressed or flat at the apices, walls nearly smooth.

Habitat: Planktonic in brackishwater fisheries at Minakhan and Golabari, Sundarbans, West Bengal.

Specimen No.: N.S. 60. (Plate XII, Fig. 77).

78. *Cosmarium paucigranulatum* Borge fa.

Islam, *et. al.* Bangladesh J. Botany, p. 49, pl. 2, fig. 24, 1992.

Cells 33-35μ long, 25-28μ broad, isthmus 7.5-9μ apices slightly flat, walls smooth.

Habitat: Planktonic in brackish water, fishery at Minakhan, Sundarbans, 24 Parganas North, West Bengal.

Specimen No.: N.S. 60. (Plate XII Fig. 78).

Triploceras Bailey, 1851

Elongate cells with the sides being undulate with transverse whorls of mamilate protuberances.

79. *Triploceras gracile* Bailey

Smith, Bull. Wisconsin Geol. Nat. Hist. Survey, 57 (2) :1-227, 38 pl., 1924; Smith, The Fresh Water Algae of the United States, p. 321, fig. 231A 1950.

Solitary cells, elongate, with flattened apices, with truncated processes bearing 2-3 short spines, sides of the semi cells undulate with transverse whorls of mamillate protuberances, cells 25 - 30μ long, 2.5 - 5μ broad.

Habitat: Planktonic at Minakhan, Sundarbans, 24 Parganas North, West Bengal.

Specimen No.: N. S. 60. (Plate XII, Fig. 79).

Triplastrum Iyengar & Ramanathan, 1942

Cells solitary linear, slightly inflated and truncated at poles, inflated portion with 3-4 shallow lobes terminating in small spines constricted at the central portion.

80. *Triplastrum abbreviatum* (Turner) Iyengar & Ramanathan

Turner, Kongl. Sevenska Vet. Akad. Handl., pl. 4, fig. 17, 1892; Iyengar & Ramanathan, J. Indian Bot. Soc., 21(3&4) : 228, 1942; Islam, *et. al.* Bangladesh J. Bot. pl. 1, fig.1-8, pl.2, figs. 19-29, pl. 3, figs. 36 - 41, 1980.

Solitary, linear, straight cells, 6-10 times longer than broad, conspicuous and distinct median constriction forming a shallow angular notch, apices trucated, inflated with 3- 4 lobes, each lobe bearing spines, chloroplast 2 - 3 in each semi cell arranged in median series.

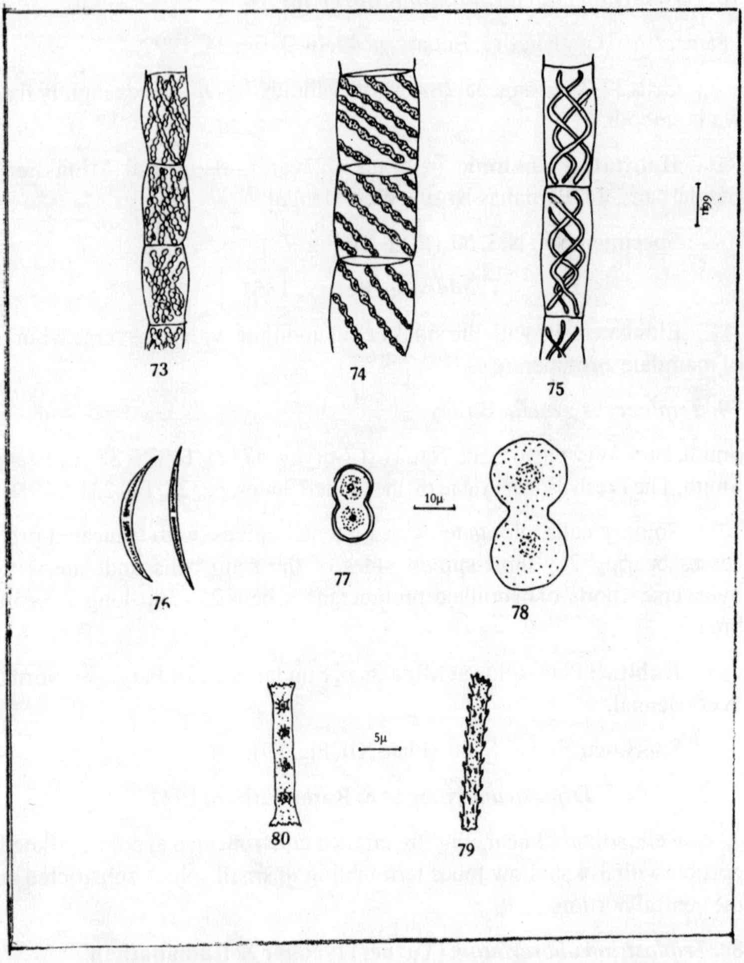

Plate XII

73. *Spirogyra dubia* Kg. ; 74. *Spirogyra setiformis* (Roth.) Kg; 75. *Spirogyra ternata* Ripart ; 76. *Closterium acutum* (Lyngb.) Breb, Ex. Ralfs. 77. *Cosmarium regenlli* Wille fa.; 78. *Cosmarium paucigranulatum* Boerge fa. ; 79. *Triploceras gracile* Bailey; 80. *Triplastrum abbreviatum* (Turner) Iyengar & Ramanathan.

Habitat: Planktonic in brackish water fishery at Haroa, Sundarbans, West Bengal.

Specimen No.: N. S. 59. (Plate XII, Fig. 80).

Order- CHLOROCOCCALES

Key to the Families

la.	Cells always in a colony with definite orientation and number of cells	-2
1b.	Solitary cells or coenobium, no definite orientation of cells	- **Oocystaceae**
2a.	Number of cells in colonies or coenobium ranges from 2 - 256	- **Hydrodictyaceae**
2b.	Number of cells in coenobium always a multiple of 2, cells with smooth or ornamented walls	-**Scenedesmaceae**

Family - HYDRODICTYACEAE

***Pediastrum* Meyen, 1829**

Colonies free floating with 2-128 polygonal cells arranged in a stellate plate one cell in thickness, coenobium entire or perforate, more than 16 cells then tendency to be in concentric rings, peripheral cells of colonies differ in shape from the interior cells and may have 1, 2 or 3 processes, not found in interior cells, cell walls may be smooth.

Key to the species

1a.	Coenobium entire	- 2
1b.	Coenobium perforate	- *P. duplex*
2a.	Cells 8.5 - 15μ diam., colonies large	- *P. boryanum*
2b.	Colonies 20-30μ diam.	- *P. tetras*

81. *Pediastrum boryanum* (Turp.) Menegh.

Brunnthaler, 1915; Smith, The Fresh Water Algae, of the United States, p. 243, fig 15, 1950; Ahmed, *et. al.* Biblio, Phyco. p. 396, pl. 1, fig. E. 1983; Pandey, *et. al.,* Biblio, Phyco., pl. 116, pl. 1, fig. 7, 1983.

Free floating colonies with polygonal cells arranged in a stellate plate of one cell thickness, coenobium entire, peripheral cells of the colony with two angular processes blunt at the tip, walls smooth, cells 9-15μ broad, upto 18μ long.

Habitat : Planktonic in brackish water fisheries at Haroa and Minakhan, Sundarbans, West Bengal.

Specimen No.: N. S. 13. (Plate XIII, Fig. 81).

82. *Pediastrum tetras* (Ehr.) Ralfs

Pandey, *et. al.* Biblio. Phyco., p. 116, pl. 1, fig. 5, 1983.

Coenobium entire, 20 - 35μ in diameter, cells 8 - 15μ in diameter.

Habitat : Planktonic in brackish water fishery at Golabari.

Specimen No. : N. S. 60 (Plate XII, Fig. 82).

83. *Pediastrum duplex* (Meyan)

Pandey *et. al.* Biblio. Phyco. p. 116, pl. 1, fig. 3, 1983; Ahmed, *et. al.* Biblio. Phyco., p. 393, pl. I, fig. C., 1983.

Coenobium perforate, cells 7-8.2μ in diameter, cells quadrate with concave sides, perforation lens shaped, walls smooth, peripheral cells with 2 processes.

Habitat: Planktonic in brackish water fishery at Haroa.

Specimen No.: N. S. 8(P) (Plate XIII, Fig. 83). Photo Plate XIV.

Family - OOCYSTACEAE

Key to the Genera

1a. Cells solitary, spherical, with smooth walls, single chloroplast - *Chlorella*

1b. Cells in colonies, in a gelatinous matrix, without spines, spherical - *Radiococcus*

***Chlorella* Beijerinck, 1890**

Unicellular, planktonic, single or aggregated forms, cells spherical to broadly ellipsoidal, single prominent cup-shaped, parietal chloroplast.

84. *Chlorella vulgaris* Beijerinck

Ahmed, *et. al.* Biblio. Phyco., p. 398, pl. II, fig. C., pl. XII, Fig. A, 1983.

Unicellular, firee floating planktonic, single or aggregated form, cells small spherical, single prominent, cup-shaped, parietal chloroplast, nucleus in a transparent cavity towards one side of the chloroplast, cells 4.8-5.5μ diam., 4 daughter cells within the mother cell wall also found.

Habitat : Planktonic in a fishery at Kharibari, 24 Parganas (North), West Bengal.

Specimen No. : N. S. 8 (P) (Plate XIII, Fig. 84).

***Radiococcus* Schmidle, 1902**

Brunnthaler, Oesterr. Bot. Zeitschr. 63 : 1-8, 3 figs. 1913 : Islam & Aziz, Bangladesh J. Bot., 16(1) : 89 - 92, 1987.

Colonial, free floating alga has globose cells, pyramidately grouped in fours, within a broad radially fibrillar gelatinous matrix.

85. *Radiococcus* sp.

Colonial, more or less, round, sub spherical to amorphous colonies formed by adhering daughter colonies 105-111μ diam., cells globose to broadly ovoid, 12-15μ diam. each with parietal chloroplast and a pyrenoid.

Habitat : Planktonic in brackishwater fishery at Haroa, 24 Parganas (N), West Bengal.

Specimen No. : N. S. 8(P). (Plate XIII, Fig. 85).

Family - SCENEDESMACEAE

***Scenedesmus* Meyen, 1829**

Flat, rarely curved, plate of ellipsoidal to spindle shaped cells arranged in alternating or double series with their long axes paralles to one another, cell wall smooth, corrugated, granulate or spicate, with or without marginal or lateral teeth or spines young cells have a single longitudinal laminate chloroplast containing one pyrenoid, cells uninucleate.

Key to the species

1a. Spines absent - *S. bijuga*
1b. Spines present - *S. quadricauda*

86. *Scenedesmus bijuga* (Turp.) Lagerheim

Ahmed, *et. al.* Biblio. Phyco. p. 409, pl. IV, fig. G, 1983.

Cells 2 - 4, arranged in a single row, cells nearly oblong, similar in shape, apices rounded, closely appressed to each other, wall smooth, cells 5 - 6μ diam., 7-14μ long.

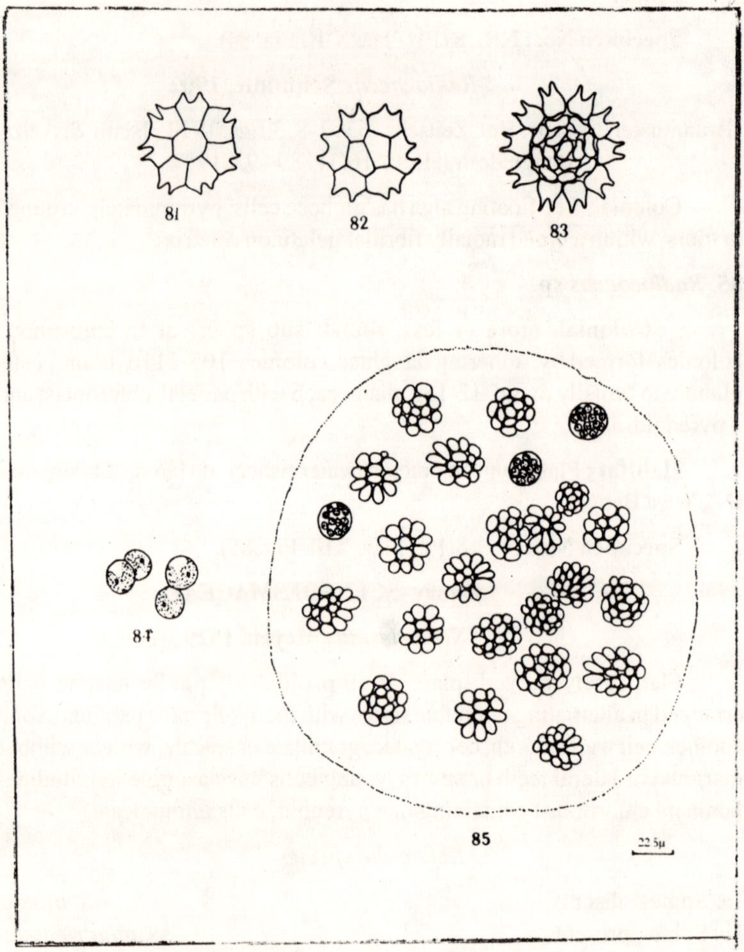

Plate XIII

81. *Pediastrum boryanum* (Turp.) Menegh.; 82. *Pediastrum tetras* (Ehr.) Ralfs.; 83. *Pediastrum duplex* (Meyan) ; 84. *Chlorella vulgaris* Beijerinck; 85. *Radiococcus* sp.

Habitat: Planktonic in brackish water fishery at Haroa, Sundarbans, 24 Parganas, West Bengal.

Specimen No.: N. S. 8(P). (Plate XIV, Fig. 86).

87. ***Scenedesmus quadricauda*** **(Turp.) Brébisson**

Ahmed, *et. al.* Biblio. Phyco. p. 412, pl. VI, fig. A, 1983; Pandey, *et. al.* Biblio Phyco. p. 124, pl. 3, fig. 2, 1983.

Colonies of upto 8 cells, usually 2-4 celled, arranged in a single row, cells oval to oblong, outer cells with single spines on each corner, spines quite long and curved, cells 4-8μ in diam., 15 - 18μ long.

Habitat: Planktonic in brackish water fishery at Haroa, Sundarbans, 24 Parganas, West Bengal.

Specimen No.: N. S. 8(P). (Plate XIV, fig. 87).

Order- SIPHONALES

Family - CODIACEAE

Boodleopsis **A. & E. S. Gepp. 1911**

Plants filamentous, forming uncalcified green cushions, filaments loosely interwoven or marginally free, ditrichotomously or irregularly widely and repeatedly branched, lower filaments thicker-walled and larger, occassionally emitting rhizomoids, the upper filaments more closely branched, with branches sharply constricted at base.

88. ***Boodleopsis sundarbanensis*** **Islam**

Islam, Bangladesh J. Botany, 2(1), pl. I - IV, figs. 1-18, 1973.

Thallus much branched, light green, densely congested, coenocytic filaments forming loose vaucheroid mat on muddy substratum, branching dichotomous in early phase, irregular later, main axis 45-112.5μ broad, primary branches 15-90μ broad, secondary branches 9-38μ broad.

Habitat : On muddy substratum at upper littoral zone at Ramgopalpur degraded forest area, Sundarbans, 24 Parganas (South), West Bengal.

Specimen No.: N. S. 6. (Plate XIV, Fig. 88). Photo Plate XV.

Reported earlier by Islam (1973) for the first time from the mangrove forests of Sundarbans , Khulna and Chittagong Districts, Bangladesh.

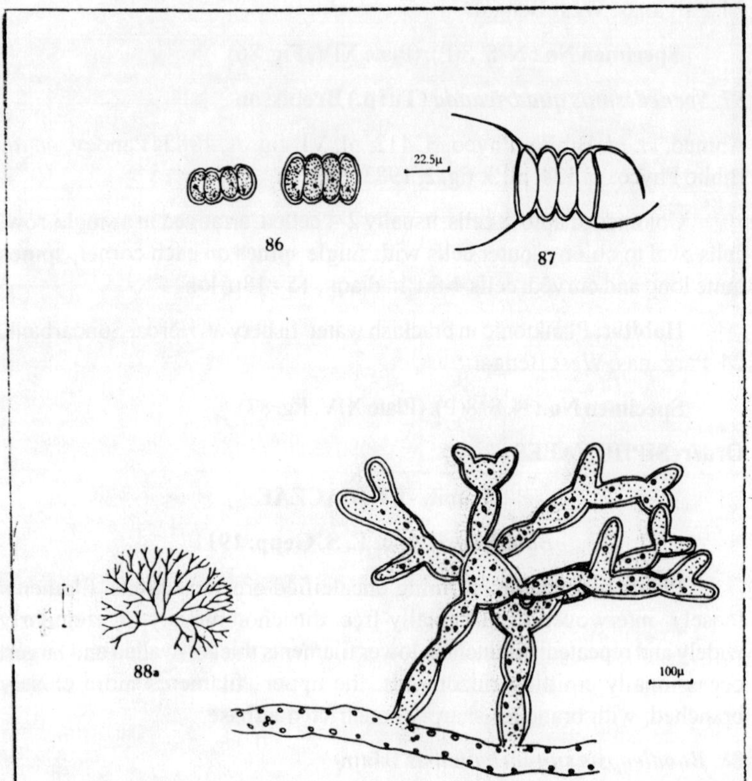

Plate XIV

86. *Scenedesmus bijuga* (Turp.) Lagerheim; 87. *Scenedesmus quadricanda* (Turp.) Brébisson; 88a. Habit sketch of *Boodleopsis sundarbanensis* Islam; 88b. Cellular details of *Boodleopsis sundarbanensis* Islam.

Class-CHAROPHYCEAE

Order - CHARALES

Family - CHARLEAE Von Leonh.

***Chara* Linnaeus**

Gen. Plant. ed.5, pl. 491, 1754.

Stems and branches corticate or ecorticate, stipulodes present, though sometimes rudimentary, branchlets of 4 or more segments, bract cells, 4 or more at a node, oogonia and antheridia produced one above the other from periphery of branchlet nodes, antheridium below the oogonium, oospores terete.

89. *Chara zeylanica* Willdenow

Mem. Ac. Roy. Berlin. p. 86, pl. 2, fig. 1, 1805. (=*C. vericillata* Roxburgh; *C. gymnopus* Br. var. *macilenata* Br.; *C. polyphylla* var. *ceylonica* Br.); Pal *et. al.* Charophyta p. 105, figs. 253-255, 1962.

Monoecious, stem stout, firm, 550-780μ thick, internodes 1-3 times the length of the branchlets, stipulodes in two series, well developed, acute, whorls of 11-12 rather stout spreading branchlets, branchlets of 9-11 segments, the lowest segment very short and ecorticate, terminal segment short, hardly exceeding the bracts, nucules 580-620μ long, 180-275μ broad, spiral cells showing 14-15 convolutions, coronula slightly spreading, globules 270-300μ in diameter.

Habitat : In a fish pond at Jharkhali, Sundarbans, 24 Parganas (South), West Bengal.

Specimen No. : N.S. 61. (Plate XV, Figs. 89a, 89b, 89c, 89d & 89e). Photo Plate XIV.

The above algal species are tabulated below according to their systematic position along with their habit and habitat.

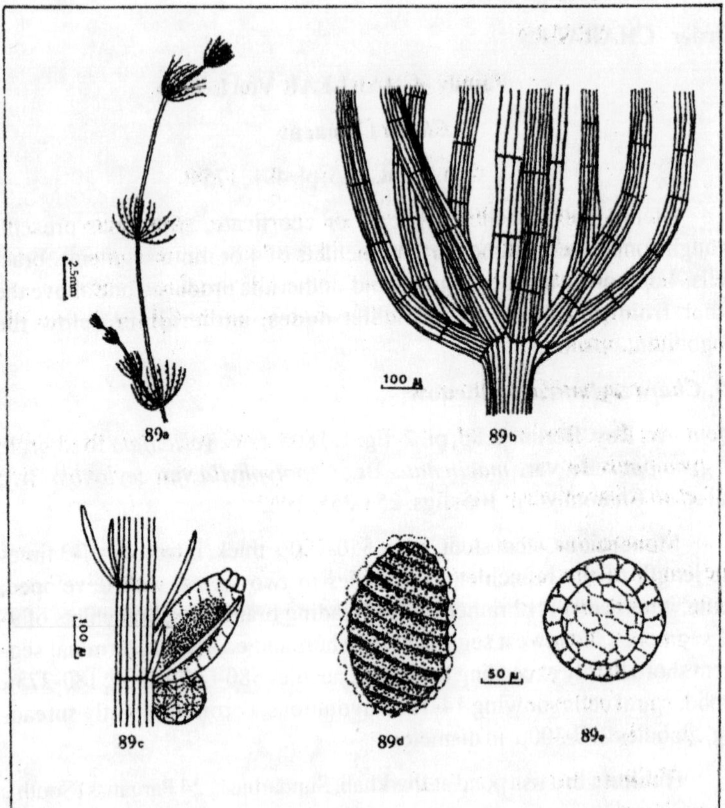

Plate XV

89a. Vegetative Thallus of *Chara zeylanica* Willdenow; 89b. Vegetative thallus of *Chara zeylanica* Willdenow; 89c. An internode of *Chara zeylanica* Willdenow showing nucule & globule; 89d. Nucule of *Chara zeylanica* Willdenow; 89e. Globule of *Chara zeylanica* Willdenow.

Table 17 : Habitat of Identified Chlorophyta Members From Indian Sundarbans

Class Order & Sub Order	*Family*	*Name of Species*	*Habit & Habitat*
CHLOROPHYCEAE			
		Volvox sp.	Planktonic in brackish-water fishery at low saline area.
Volvocales	Volvocaceae	*Pandorina morum*	Planktonic in brackish-water fishery at low saline area.
Ulotrichales	Ulotrichaceae	*Uronema confervicola*	Epiphytic on other green algae in brackishwater fisheries in high saline areas.
Ulvales	Ulvaceae	*Entermorpha clathrata*	Alongwith *Ulva* spp. attached to bricks and other hard substratum at high saline areas.
		Enteromorpha compressa	Periphytic on twigs/sub merged plant parts in brackishwater fisheries at mid saline areas.
		Enteromorpha intestinalis	Abundant in brackish-water fisheries at low to mid saline areas
		Enteromorpha prolifera	Abundant in brackish-water fisheries at low to mid saline areas.
		Ulva lactuca	Attached to bricks on the high saline river bank. *Ulva patengensis* Attached to bricks on the high saline river bank.
		Ulva fasciata	Attached to bricks on the high saline river bank.

Contd...

Class Order & Sub Order	*Family*	*Name of Species*	*Habit & Habitat*
Cladophorales	Cladophoraceae	*Chaetomorpha aerea*	Free floating mass in high saline brackish water fisheries.
		Chaetomorpha gracilis	Aongwith *Ulva* sp. thallus on bricks.
		Chaetomorpha brachygona	Free floating/attached to submerged substratum in high saline brackishwater fisheries.
		Lola capillaris	Free floating algal mass in brackishwater fisheries at mid to high saline areas.
		Lola implexa	Free floating in brackishwater fisheries at mid to high saline areas.
		Lola tortuosa	Floating in brackishwater fisheries at high saline areas.
		Rhizoclonium grande	Periphytic on pneumatophores at high saline areas.
		Rhizoclonium hookeri	Periphytic on pneumatophores at high saline areas.
		Rhizoclonium riparium	Periphytic on pneumato phores at high saline areas.
		Cladophora echinus	Epiphytic on submerged substratum in brackishwater fisheries at high saline areas.
		Cladophorella sundarbanensis	Benthic on forest floor in high saline areas.

Contd...

Class Order & Sub Order	Family	Name of Species	Habit & Habitat
Oedogoniales	Oedogoniaceae	*Oedogonium smeyndulatum*	On submerged substratum/ free floating in brackishwater fisheries at high saline areas.
	Zygnemataceae	*Spirogyra dubia*	Free floating in water bodies at low saline areas.
		Spirogyra setiformis	Floating in water bodies at low to mid saline areas.
		Spirogya ternata	Floating in water bodies at low to mid saline areas.
Desmidiales	Desmidiaceae	*Triplocerus gracile*	Planktonic in mid saline brackishwater fisheries
		Cosmarium paucigranulatum	Plankton in mid saline brackishwater fisheris.
		Cosmarium regenlli	Planktonic in mid saline brackishwater fisheries.
		Closterium acutum	Planktonic in brackishwater fisheries at mid to high saline areas.
Chlorococcales	Chlorellaceae	*Chlorella vulgaris*	Planktonic in low saline brackishwater fisheries.
	Radiococcaceae	*Radiococcus* sp.	Planktonic in brackishwater fishery at low saline areas.
		Pediastrum boryanum	Planktonic in brackishwater fishery at low to mid saline areas.
	Hydrodictyaceae	*Pediastrum duplex*	Planktonic in low to mid saline brackishwater fisheries

Contd...

Class Order & Sub Order	Family	Name of Species	Habit & Habitat
		Pediastrum tetras	Planktonic in low saline brackishwater fisheries.
		Scenedesmus bijuga	Planktonic in low saline brackishwater fisheries.
	Scenedesmaceae	*Scenedesmus quadricauda*	Planktonic in low saline brackishwater fisheries.
Siphonales	Codiaceae	*Boodleopsis* sundarbanensis	On muddy substratum at upper littoral zone at high saline zone.
Charophyceae			
Charales	Characeae	*Chara zeylanica*	In brackishwater fisheries at high saline areas.

6.4. DIVISIONS – (a) CHRYSOPHYTA (b) PHAEOPHYTA

Popularly known as the golden brown algae (Chrysophyta) and brown algae (Phaeophyta) they have chlorophyll-a, chlorophyll-c, β-Carotene and fucoxanthin as their important pigments.These algae areusually found as benthic or planktonic forms in the IndianSundarbans. Table-18 gives an account of the identified Chrysophyta and Phaeophyta members fromthe Indian Sundarbans. Only two Phaeophyta member, *viz., Dictyota* sp. and *Coplomenia sinuosa* have been identified fromthe Indian Sundarbans. *dictyota* sp. is found growing as periphytic algae on pneumatophores, other aerial root systems and lower trunk regionsofvarious mangroves and also as benthic algae along with other red algae in the upper littoral zonesofhigh saline areas in the sea ward fringes. It belongs to theorder Dictyotales and family Dictyotaceae. *Calpomenia sinuosa* occurs as a benthic form embedded in littoral muddy shores and forest floor. It belongs to the order dictyosiphonales and family Punctariaceae. The 44 species of Chrysophyta identified from the IndianSundarbans belong to2 classes, *viz.,* Xanthopyceae and Bacillariophyeae. Only two species belonging to genus *Vaucheria,* family Heterosiphonales have been identified from the Indian Sundarbans; both these occur as dense mat on the forest floor in the littoral regions or as part of the bottom biota as benthos in medium to high salineareas of Sundarbans. The algal flora representing the Bacillariphyceae were found mainly as planktonic forms and only a few (3 species belonging to family Achnanthaceae), *viz., Achnanthes microcephala, A. minutissima* and *Cocconeis placentula* as epiphytic species on the red algae *Pterosiphonia pinnata*. The planktonic forms (39 species) are represented in 5 orders, *viz.*, Coscinodiscineae, Biddulphineae, Rhizosoleniineae, Fragilariineae and Naviculineae. The 13 families include 4 genus and 7 species belonging to Coscinodiscineae, 1 species of Hemidiascaceae, 1 genus and 2 species of Biddulphiaceae, 2 genus and 6 species belonging to Chaetoceraceae, 1 genus and 3 species of Bacteriastraceae, 1 species of Corethronaceae, 1 genus and 2 species of Rhizosoleniaceae, 3 genus and 3 species of Fragilariaceae, 1 species of Diatomaceae, 6 genus and 7 species of Naviculaceae, 1 species of Gomphonemaceae, 2 genus and 2 species of Cymbellaceae and 3 species and 3 species of Bacillariaceae (Pie Charts– IIa, IIb, IIc). Out of these planktonic forms 32 species are exclusively high saline loving and are mostly found in the river water nearer to the Bay and 7 species are found to inhabit the brackish water fisheries.

(a) CHRYSOPHYTA

Key to the Classes

1a. Unicellular or multicellular, non silcified cell walls **-Xanthophyceae**
1b. Unicellular or colonial, highly silcified cell wall, with 2 overlapping, halves that fit together **-Bacillariophyceae**

Class- XANTHOPHYCEAE
Order - HETEROSIPHONALES

Family - VAUCHERIACEAE

Filamentous thallus filaments long, variously branched, unseptate, coenocytic are attached to the substratum by means of branched haptera or rhizoids.

***Vaucheria* De Candolle, 1805**

Thallus filamentous, inarticulate, laterally and irregularly branched, ocassionally assuming a dichotomous appearance. Filaments form tufted masses, usually branched rhizoids. The filaments are more or less cylindrical or rarely numerous small chloroplasts towards the exterior mid minute nuclei towards the inaterior chloroplasts without pyrenoids.

90. *Vaucheria prescottii* Islam

Islam, Pak. Acad. Sci. 2(1) : pl. 4, figs. 1 - 7, 1965; Islam, Bangladesh J. Bot. pl. 10, fig.9. 10, 1973; Islam, Bibliotheca Phyco. P. 26, 1976.

Coenocytic thallus forming dense mat on decomposed matter, filaments branched, 60-75μ diameter, septation scarsely present, oogonia solitary with short broadly elliptical with sub globular swelling at the apex, 105-120μ diameter, 202-206μ long.

Habitat: Dense mat on decomposed matter at Ramgopalpur, Sundarban, 24 Parganas, West Bengal.

Specimen No.: N. S. 10. (Plate XVI, Fig 90). Photo Plate XVI.

91. *Vaucheria* sp.

Coenocytic filaments forming a dense mat, 30 - 65μ broad, dichotomously branching.

Habitat : Dense brown mat on regularly inundated forest floor at Ramgopalpur, Sundarbans, 24 Parganas (S), West Bengal.

Specimen No. : N. S. 12. (Plate XVI, Fig. 91). Photo Plate XVI.

Class - BACILLARIOPHYCEAE

Key to the Orders

1a. Valve striae arranged basically with respect to a point, an annulus or a central areola **- Biddulphiales**

1b. Valve striae arranged basically with respect to a line **- Bacillariales**

Order - BIDDULPHIALES

Key to the Suborders

1a. Valves without marginal ring of processes -2

1b Valves generally with a marginal ring of processes **- Coscinodiscineae**

2a. Valves primarily unipolar, no marginal ring of processes present **- Rhizosoleniineae**

2b. Valves primarily bipolar, no marginal ring **- Biddulphiinea**

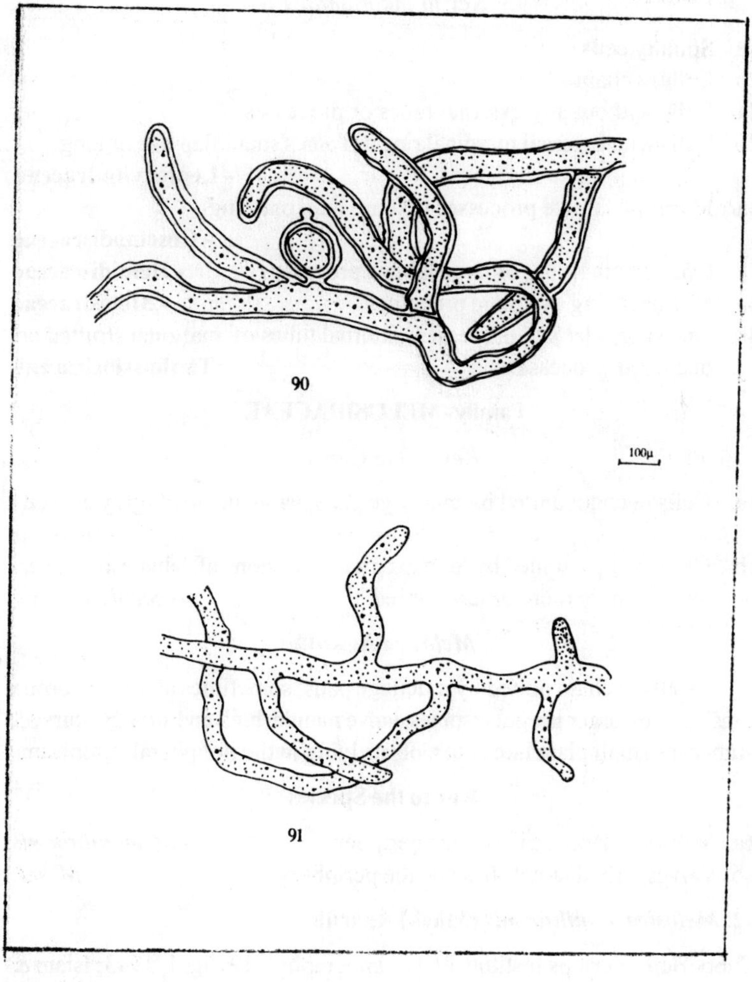

Plate XVI

90. *Vaucheria prescottii* Islam; 91. *Vaucheria* sxp

Suborder - COSCINODISCINEAE

Key to the Families

1a. Solitary cells -2
1b. Cells in chain -4
2a. Cells without any external tubes or processes -3
2b. Cells with external marginal ring of spines small flap like or long **- Leptocylindraceae**
3a. Marginal labiate processes in more than one ring **- Coscinodiscaceae**
3b. One marginal ring of large labiate processes **- Hemidiscaceae**
4a. Marginal ring of libiate processes **- Melosiraceae**
4b. One or few labiate processes, external tubes of marginal strutted or occluded processes **- Thallassiosiraceae**

Family - MELOSIRACEAE

Key to the Genera

1a. Cells in chain united by mucilage pads, valve mantle highly curved - *Melosira*
1b. Cells in chain united by long external extensions of labiate processes, valve mantle more or less curved - *Stephanopyxis*

***Melosira* Agardh**

Cells in chain united by mucilage pads, sometimes also by a corona consisting of larger irregular spines, valve mantle, high and strongly curved, numerous small plate like chloroplasts lying in the peripheral cytoplasm.

Key to the Species

1a. Valves without striae on the periphery - *M. moniliformis*
1b. Valves with distinct striae on the periphery - *M. sol.*

92. *Melosira moniliformis* (Muel.) Agardh

Cupp, Bull. 'Scripps Institute of Oceanography', 39, fig. 1, 1943; Islam & Aziz, Bangladesh J. Botany, pl. 1, fig. 4, 1975.

Frustules short, cylindrical, united in chains, cells short and cylindrical, valves circular, convex at walls, neck present, cell diameter 30µ - 33µ.

Habitat : Plankton from Haldibari, Sundarbans, 24 Parganas (S), West Bengal.

Specimen No. : N. S. 77. (Plate XVII, Fig. 92). Photo Plate XVII.

93. *Melosira sol* (Ehr.) Kuetz.

Frenguelli, Rev. Algologique, pl. 5, fig. 4, 1960; Islam & Aziz, Bangladesh J. Bot. pl. 10, fig. 12, 1975.

Valves with distinct striae on the periphery, seen in valve view only, frustules not observed, valve diameter 64μ - 68μ, central field large and very thick.

Habitat : Plankton from Patharpratima, Sundarbans, 24 Parganas (S), West Bengal.

Specimen No. : N. S. 65. (Plate XVII, Fig, 93); very rare specimen. Photo Plate XVII.

***Stephanopyxis* Ehrenberg**

Cells in chains united by long external extensions of labiate processes, valve mantle high, more or less curved, valve wall with large hexagonal areolae with large external foramina, resting spores present (semiendogenous).

94. *Stephanopyxis palmeriana* (Grev.) Grun.

Wildeman, Les algues de la flora de Buitengorg, 324, 1900; Cupp, Bull. 'Scripp's Institute of Oceanography', 40, fig. 4, 1943; Venkataraman, Proc. Nat. Sci. India, 310, fig. 1, 1958; Yamaji, The Plankton of the Japanese Coastal Waters, 28, pl. 5, fig. 1, 1968; Islam & Aziz, Bangladesh J. Bot., pls. 1 & 10, figs. 11, 12, 122, 1975.

Oblong cells in short chains, united by 8 to 10 spines arising from the margins, frustules 80μ - 96μ long, 60μ - 65μ diameter, spines 6μ - 8μ long.

Habitat : Plankton from Haldibari, Sundarbans, 24 Personas (S), West Bengal.

Specimen No. : N. S. 77. (Plate XVII, Fig. 94); only in very few number, Photo Plate XVII.

Family - THALASSIOSIRACEAE

Cyclotella (Kutzing) Brebisson

Cells usually solitary, valves tangentially undulated, valve wall aveolate, central field distinctly different from the rest of the valve, one to many strutted processes within the central field, central field reticulate, rugose or with warts or granules, one marginal ring of strutted processes, one marginal labiate process.

95. *Cylotella glomerata* Bachmann

Hustedt, Osterreichs und der Schweiz, p. 105, fig. 81 c, d, 1930; Venkateswarlu, Bibliotheca Phycologia, p. 15, pl. 1, fig. 2, 1983.

Valves 7μ - 7.6μ in diameter, girdle view not observed, 12-13 striae in 10μ, central field finely punctuate.

Habitat : Plankton from Pirkhali, Sundarbans, 24 Parganas (S), West Bengal.

Specimen No. : N. S. 76. (Plate XVII, Fig. 95); found in low number. Photo Plate XVII.

Family - COSCINODISCACEAE

Coscinodiscus Ehrenberg

Solitary cells, no external tubes or processes, marginal labiate processes in more than one ring, valves circular, pervalver axis not very high and diameter smaller than 2 mm.

Key to the species

1a. Frustules disc shaped - *C. eccentricus*
1b. Frustules coin shaped - *C. gigas*
1c. Frustules wedge shaped - *C. granii*

96. *Coscinodiscus eccentricus* Ehr.

Cleve, Bull. Des Resultats adult pendant les courses periodiques, p. 216, 1904; Fritsch, The Structure and Reproduction of the Algae, fig. 187 A, 1935; Cupp, Bull. 'Scripp's Institute of Oceanography', 52, fig. 14, pl. 1, fig. 1, 1943; Islam & Aziz, Bangladesh J. Botany, pl. 1, figs. 7 - 8, 1975; Hasle & Syvertson, In : Identifying Marine Phytoplankton, p. 62. pl. 6, 1997.

Valves flat, cells, disc shaped, areolae arranged in slightly curved, nearly parallel rows, central field large, merging with distinct outer divisions to give an impression of seven divisions, valves 38μ - 45μ diameter.

Islam and Aziz, 1975 have described this species as *C, excentricus* Ehr. Based on valve view, Hasle & Syvertson, 1997 have described this species as *Thalassiosia eccentrica* (Ehrenberg) Cleve, based on valve view and girdle view and kept *Coscinodiscus eccentricus* Ehr. ss basionym. Since in the samples only the valve view was seen, the present species has been described as *Coscinodiscus eccentricus* Ehr.

Habitat : Plankton from Golabari and Canning, Sundarbans, 24 Parganas (S), West Bengal.

Specimen No. : N. S. 50A, N.S. - 50B. (Plate XVII, Fig. 96) Photo Plate XVIII.

97. *Coscinodiscus gigas* Ehr.

Yamaji, The Plankton of the Japanese Coastal Waters, 21, 98, 1968. Hendey, Nova Hedwigia, 374,1971; Islam & Aziz, Bangladesh J. Botany, pl. 10, fig. 123, 1975.

Valves large and thin walled, 76μ - 80μ diameter, central field not distinguishable from the peripheral region, areolae continuous from centre to periphery.

Habitat : Plankton from Jharkhali and Pirkhali, Sundarbans, 24 Parganas (S), West Bengal.

Specimen No. : N. S. 76. (Plate XVII, Fig. 97).

98. *Coscinodiscus granii* Gough

Cupp, Bull. 'Scripp's Institute of Oceanography', 56, fig. 21, 1943; Yamaji. The Plankton of the Japanese Coastal Waters, 12, 1968; Islam & Aziz, Bangladesh J. Botany, pl. 10, fig. 121, 1975.

Valves medium sized with coarse hexagonal areolae, 45μ - 50μ diameter, areolae forming definite rosettes in the centre, as well as, in the periphery, radial areolation.

Habitat : Plankton from Harwood Point, Sundarbans, 24 Parganas (S), West Bengal.

Specimen No. : N. S. 64. (Plate XVII, Fig. 98a, 98b). Photo Plate XVIII.

Family - HEMIDISCACEAE

Hemidiscus Wallich, 1860

Cells low cylindrical to discoid, few hyaline unperforate bands present, valve areolation radial often fasciculate, one marginal ring or large labiate processes similar in shape and generally in size, valves semi-circular.

99. *Hemidiscus cuneiformis* Wallich

Wallich, Transaction of the Microscopical Society, New Ser., p. 42, pl. 2, figs. 3 & 4, 1860; Hustedt, Osterreichs und der Schweiz, p. 904, fig. 542, 1930; Cupp, Bull. 'Scripp's Institute of Oceanography', 170, fig. 121, 1943; Hendey, Fisheries Investigations Ser. IV., p. 94, pl. 22, fig. 9, 1964; Yamaji, The Plankton of the Japanese Coastal Waters, 15, 1968; Hendey, Nova Hedwigia, 376, 1971; Islam & Aziz, Bangladesh J. Botany, pl. 6, fig. 79 - 80, 1975; Hasle & Syvertsen, In : Identifying Marine Phytoplankton, p. 128, pl. 22, 1997.

Frustules nearly spherical, 460µ - 480µ diameter, each valve lies compressed to the other at the centre.

Habitat : Plankton from Harwood Point, Sundarbans, 24 Parganas (S), West Bengal.

Specimen No. : N. S. 64. (Plate XVII, Fig. 99).

Family - LEPTOCYLINDRACEAE

Corethiron Castracane

Cells cylindrical with more or less dome shaped valves, girdle composed of many bands, valves with marginal long, (barbed) and short (hooked) spines, chloroplasts rounded or oval shaped, numerous.

100. *Corethron hystrix* Hensen

Cupp, Bull. 'Scripp's Institute of Oceanography', 70, fig. 36, 1943; Subrahmanyan, Proc. Ind. Acad. Sci. 24B, figs. 99, 101 & 103, 1946; Yamaji, The Plankton of the Japanese Coastal Waters, 138, 1968; Islam & Aziz, Bangladesh J. Botany, pl. 1, fig. 3, 1975.

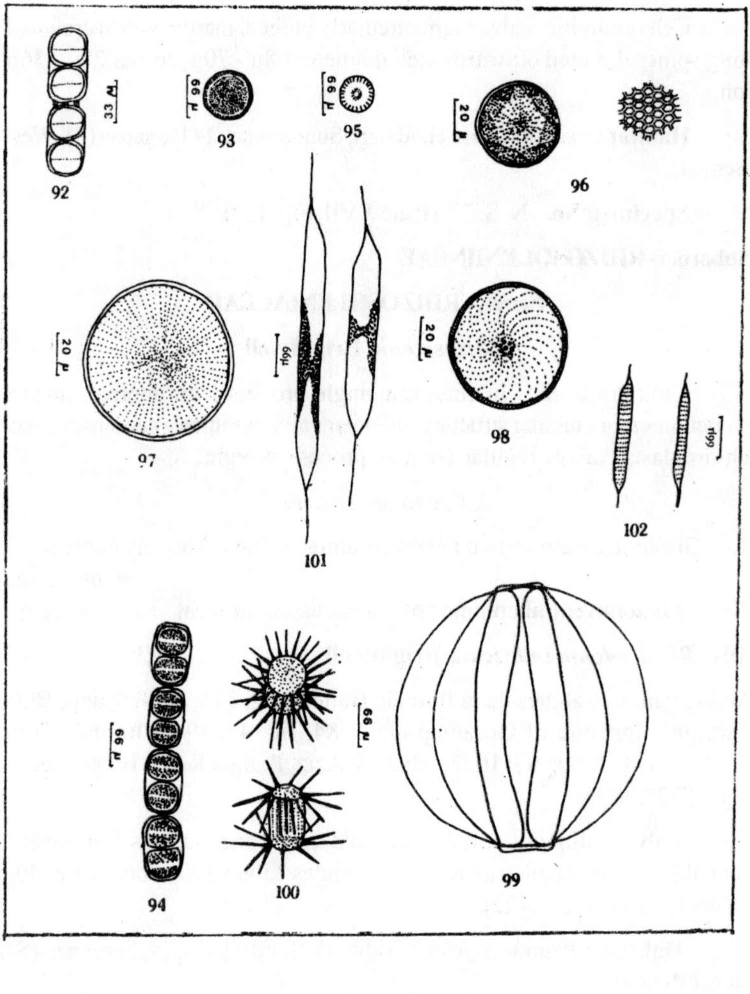

Plate XVII

92. *Melosira moniliformis* (Muel.) Agardh; 93. *Melosira sol* (Ehr.) Kuetz.; 94. *Stephanopyxis palmeriana* (Grev.) Grun.; 95. *Cyclotella glomerata* Bachmann; 96. *Coscinodiscus eccentricus* Ehr.; 97. *Coscinodiscus gigas* Ehr.; 98. *Coscinodiscus granii* Gough; 99. *Hemidiscus cuneiformis* Wallich; 100. *Corethron hystrix* Hensen; 101. *Rhizosolenia imbricata* Brightwell; 102. *Rhizosolenia setigera* Brightwell.

Cells cylindric, valves semicircularly bulged, margin with a crown of long spines directed outwards, cell diameter 68μ - 70μ, spines 28μ - 36μ long.

Habitat : Plankton from Haldibari, Sundarbans, 24 Parganas (S), West Bengal.

Specimen No. : N. S. 77. (Plate XVII, Fig. 100).

Suborder- RHIZOSOLENIINEAE

Family - RHIZOSOLENIACEAE

***Rhizosolènia* Brightwell**

Cylindrical cells in chains, a single process with internal labiate, sometimes more tubular structure, unipolar valve symmetry numerous small chloroplasts, valves regular conical, process straight.

Key to the species

1a. Girdle segments in two lateral columns, valves obliquely conical - *R. imbricata*

1b. Two dorsiventral columns of bands, valves conical - *R. setigera*

101. *Rhizosolenia imbricata* Brightwell

Wildeman, Les algues de la flora de Buitengorg, 313, 1900; Cupp, Bull. 'Scripp's Institure of Oceanography', 84, fig. 47, 1943; Round, Nova Hedwigia, 463, fig. 4G, 1972; Islam & Aziz, Bangladesh J. Botany, pl. 2, figs. 27-29, 1975.

Valve obliquely conical, pointed ends, and possess long apical spinal processes swollen at the base, frustules 250μ - 320μ long, 25μ - 40μ diameter, spines, 20μ - 22μ.

Habitat : Plankton from Haldibari, Sundarbans, 24 Parganas (S), West Bengal.

Specimen No. : N. S. 77. (Plate XVII, Fig. 101).

102. *Rhizosolenia setigera* Brightwell

Cupp, Bull. 'Scripp's Institute of Oceanography, 88, fig. 49, 1943; Yamaji, The Plankton of the Japanese Coastal Waters, 109, 1968; Islam & Aziz, Bangladesh J. Botany, pl. 1, fig. 17, 1975.

Valves oblique to conical with apical thin cone like processes, frustules 320μ - 325μ long, 60μ - 62μ diameter, spines 70μ - 75μ long.

Habitat : Plankton from Haldibari, Sundarbans, 24 Parganas (S), West Bengal

Specimen No. : N. S. 77. (Plate XVII, Fig. 102).

SUBORDER - Biddulphiineae

Key to the Families

1a. Cells with bipolar elevations -2
1b. Cells with long setae - *Chaetocerotaceae*
2a. Cells in straight chains, valves elliptica - *Hemiaulaceae*
2b. Cells in zig zag chains, valves elliptical to lanceolate - *Biddulphiaceae*

Family - HEMIAUIACEAE

***Climacodium* Grunow, 1868**

Cells straight but usually forming somewhat twisted chains, valves elliptical, apertures between cells in chains large.

103. *Climacodium frauenfeldianum* Grunow

Hustedt, Osterriechs und der Schweiz, p. 776, figs. 453 & 454, 1930; Cupp. Bull. 'Scripp's Institute of Oceanography', 147, fig. 105, 1943; Yamaji, The Plankton of the Japanesh Coastal Waters, 25, 1968; Islam & Aziz, Bangladesh J. Botany, pl. 7, fig. 91, 1975; Hasle & Syvertsen, In : Identifying Marine Phytoplankton, p. 172, pl. 33, 1975,

Frustules united to form chains, without chaete, valve surface concave, 120μ - 128μ diameter, apertures large.

Habitat : Plankton from Harwood Point, Sundarbans, 24 Parganas (S), West Bengal.

Specimen No. : N. S. 64. (Plate XVIII, Fig. 103). Photo Plate XIX.

Family - CHAETOCEROTACEAE

Key to the Genera

1a. Generally two setae per valve, one at each end of the apical axis - *Chaetoceros*
1b. More than two setae per valve, arranged regularly around its margin - *Bacteriastrum*

Chaetoceros Ehrenberg

Cells more or less rectangular in girdle view, cells elliptical to almost or rarely circular in valve view, opposite setae of adjacent cells touch one another near their origin.

Key to the species

1a. Chloroplasts one or a few plates, no chloroplast in setae - 2
1b. Chloroplasts numerous throughout the cell including setae - *C. flexuosus*
2a. Cells with, two, chloroplasts - 3
2b. Cells with two chloroplasts - *C. laciniosus*
3a. Cells in curved or helical chains, intercalary setae bent in one direction - *C. curvisetus*
3b. Unicellular species - *C. tenuissimus*
3c. Valves of adjacent cells do not touch - *C. subsecundus*

104. *Chaetoceros curvisetus* Cleve

Cupp, Bull. 'Scripp's Institute of Oceanography', 137, fig. 93, 1943; Yamaji, The Plankton of the Japanese Coastal Waters, 135, 1968; Islam & Aziz, Bangladesh J. Botany, pl. 6, figs. 75A, 75B, 1975.

Chains coiled, frustules bent to join from end to end, chaete arising from corners of cells, all bent towards outside of curved axis, frustules 8μ - 12μ long, 10μ - 10.5μ diameter chaete 23μ - 30μ long.

Habitat : Plankton from Harwood Point, Sundarbans, 24 Parganas (S), West Bengal.

Specimen No. : N. S. 64. (Plate XVIII, Fig. 104). Photo Plate XIX.

105. *Chaltoceros flexuosus* Mangin

Islam & Aziz, Bangladesh, J. Botany, pl. 4, fig. 58 - 59, 1975; Hasle & Syvertsen, In : Identifying Marine Phytoplankton, p. 203, pl. 41, 1997.

Valve surface concave, leaving lens - shaped apertures, chaste arising from both margins, short, frustules 7μ - 15μ long, 30μ - 40μ diameter, chaste 25μ long.

Habitat : Plankton from Pirkhali, Sundarbans, 24 Parganas (S), West Bengal.

Specimen No. : N. S. 76. (Plate XVIII, Fig. 105); present in few numbers. Photo Plate XIX.

106. *Chaetoceros laciniosus* Schütt

Schütt, Beirichte der deutschen botanischen Gesellchaft, p. 38, pl. 4, fig. 5, 1895; Hustedt, Osterreichs und der Schweiz, p. 701, fig. 401, 1930; Cupp, Bull. 'Scripp's Institute of Oceanography', p. 128, fig. 80, 1943; Hendey, Nova Hedwigia, p. 127, pl. 13, fig. 2, 1964; Evensen & Hasle, Beiheft zur Nova Hedwigia, p. 160, figs. 42 - 45, 1975; Rines & Hargraves, Biblio. Phyco. p. 83, figs. 167 - 169, 1988; Hasle & Syvertsen, In : Identifying Marine Phytoplankton, p. 209, pl. 43, 1997.

Cells in long loose chains, thin setae, basal part at first parallel later perpendicular to chain axis, outer most part bent towards chain, apertures high, elliptical and square to rectangular.

Habitat : Planktonic at Jhilla, Sundarbans, 24 Parganas (S), West Bengal.

Specimen No. : N. S. 81. (Plate XVIII, Fig. 106).

107. *Chaetoceros tenuissimus* Meunier

Hustedt, Osterreichs und der Schweiz, p. 756, fig. 438, 1930; Rines & Hargraves, Biblio Phyco. p. 97, 1988; Hasle & Syvertsen, In : Identifying Maine Phytoplankton, p. 222, pl. 47, 1997.

Cells small with nearly equal apical and pervalver axis 18µ - 20µ, setae straight to slightly curved emerging from poles of valve at 45° to both apical and pervalver axes.

Habitat : Plankton from brackishwater fishery at Ramgopalpur, Sundarbans, 24 Parganas (S), West Bengal.

Specimen No. : N. S. 7. (Plate XVIII, Fig. 107).

108. *Chaetoceros subsecundus* (Grunow) Hustedt

Hustedt, Osterreichs und der Schweiz, p. 709, fig. 404. 1930; Cupp, Bull. 'Scripp's Institute of Oceanography', p. 130, fig. 83, 1943; Hendey, Nova Hedwigia, p. 128, pl. 10, fig. 1, 1964; Islam & Aziz, Bangladesh J. Botany, pl. 4, figs. 60 - 61, 1975; Yamaji, The Plankton of the Japanese Coastal Waters, 24, 100, 1968; Rines & Hargraves, Biblio. Phyco. p. 76, figs. 150, 151 & 153, 1988; Hasle & Syvertsen, In : Identifying Marine Phytoplankton, p. 220, pl. 46, 1997.

Convex valves, setae from valve margin, terminal chaete diverging at an acute angle, frustules 18μ - 22μ long, 60μ - 75μ diameter.

Habitat : Plankton from Harwood Point, Sundarbans, 24 Parganas (S), West Bengal.

Specimen No : N. S. 64. (Plate XVIII, Fig. 108). Photo Plate XIX.

***Becterisarrum* Shadbolt**

Cells cylindrical, numerous small lobed chloroplasts, setae of two adjacent cells fused for a certain distance beyond point of origin and branched again, terminal setae unbranched and curved.

Key to the species

1a. Terminal setae similar on both ends of chains - 2

1b. Terminal setae on either end different, inner setae very long - *B. cosmosum*

2a. Terminal setae of both ends directed towards the chain - *B. delicatulum*

2b. Terminal setae first transverse to chain axis, then abruptly obliquely curved towards the chain in their outer part

B. varians

109. *Bacteriastrum varians* Lauder

Hustedt, Osterreichs und der Schweiz, p. 610 - 624, fig. 353, 1930; Cupp, Bull. 'Scripp's Institute of Oceanography', p. 95, figs. 55-58, 1943; Subrahmanyan, Proc. Ind, Acad. Sci., 127, figs. 170-172 & 175, 1946; Islam & Aziz, Bangladesh J. Botany, pl. 1, fig. 13, 1980; Hasle & Syvertsen, In : Identifying Marine Phytoplankton, p. 186, pl. 37, 1997.

Valves with 10 chaete, tip of chaete slightly bent, valves 20μ diameter, chaete 55μ long.

Habitat : Plankton from Harwood Point, Sundarbans, 24 Parganas (S), West Bengal.

Specimen No. : N. S. 64. (Plate XVIII, Fig. 109). Photo Plate XX.

110. *Bacteriastrum cosmosum* Pav.

Cupp, Bull. 'Scripp's Institute of Oceanography', 99, fig. 58, 1943; Subrahmanyan, Proc. Ind. Acad. Sci., 128, fig. 177, 1946; Islam & Aziz, Bangladesh J. Botany, pl. 1, fig. 14, 1980.

Valves with 9 chaete, 13μ - 14μ diameter, chaete bent slightly, smooth 40μ - 45μ long.

Habitat : Plankton from Harwood Point, Sundarbans, 24 Parganas (S), West Bengal.

Specimen No. : N. S. 644 (Plate XVIII, Fig. 110).

111. *Bacteriastrum delicatulum* Cleve

Cupp, Bull. 'Scripp's Institute of Oceanography', 96, fig. 55, 1943; Davis, Mich, St. Univ. Press, fig. 77, 1955; Islam & Aziz, Bangladesh J. Botany, pl. 3, fig. 49 - 51, 1975.

Valves with 12 solid long chaete, tips of each chaete bifurcated into forks, frustules 12μ - 14μ diameter, chaete 55μ - 70μ long.

Habitat : Plankton from, Harwood Point, Sundarbans, 24, Parganas (S), West Bengal.

Specimen No. : N.S. 64. (Plate XVIII, Fig. 111). Photo Plate XX.

Family - BIDDULPHIACEAE

Biddulphia Gray

Valves bipolar, elliptical or lanceolate, bipolar elevations present, cells in straight or zig - zag chains, two or more labiate processes per valve, numerous small chloroplasts lying against valve wall.

Key to the species

1a. Processes away from the elevations, face Valve flat or slightly concave in the in the middle - *B. mobiliensis*

1b. Processes close to slender elevations, valve face flat or concave throughout between processes - *B. sinensis*

112. *Biddulphia mobiliensis* Bailey

Fritsch, The Structure & Reproduction of Algae, figs. 213 A - C, 1935; Wildeman, Les algues de la flora de Buitengong, 342, fig. 342, 1900; Islam & Aziz, Bangladesh J. Botany, pl. 7, fig. 92, 1975; Hasle & Syvertsen, In : Identifying Marine Phytoplankton, p. 216, pl. 49, 1997.

Frustules 138μ - 140μ long, 99μ - 10μ diameter, horns 15μ long, spines 45.5μ - 48μ long, valve end concave, horns short but spines long.

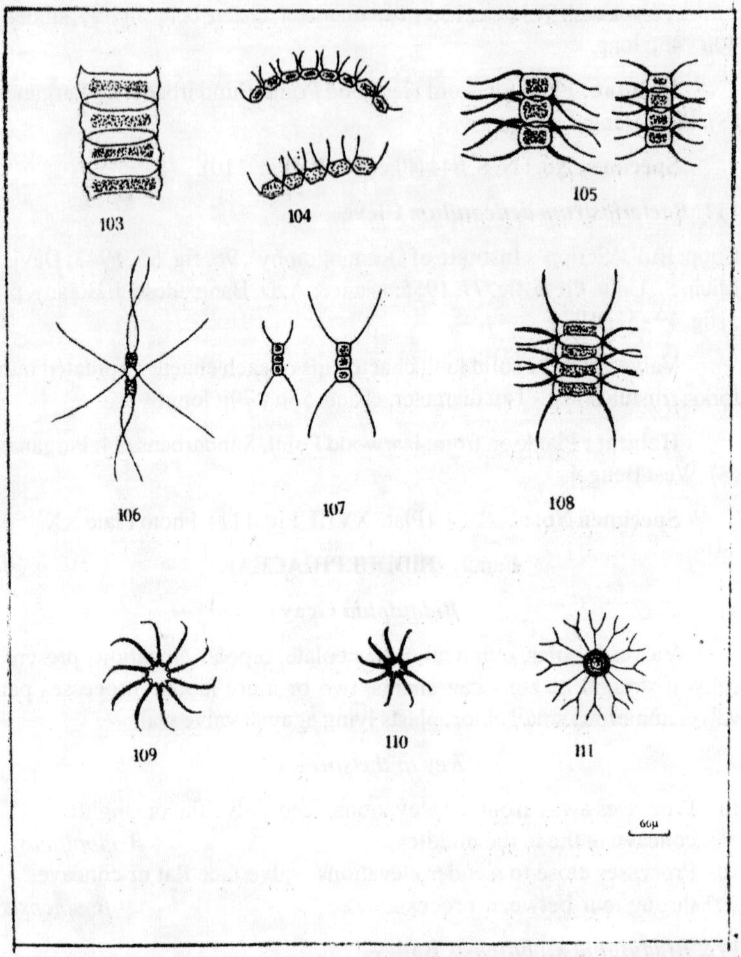

Plate XVIII

103. *Climacodium frauenfeldianum* Grunow; 104. *Chaetoceros curvisetus* Cleve; 105. *Chaetoceros flexuosus* Mangin; 106. *Chaetoceros laciniosus* Schütt; 107. *Chaetoceros tenuissimus* Meunier; 108. *Chaetoceros subsecundus* (Grunow) Hustedt; 109. *Bacteriastrum varians* Lauder; 110. *Bacteriastrum cosmosum* Pav.; 111. *Bacteriastrum delicatulum* Cleve.

Habitat : Plankton from Harwood Point, Sundarbans, 24 Parganas (S), West Bengal.

Specimen No. : N. S. 64. (Plate XIX, Fig. 112). Photo Plate XXI.

113. *Biddulphia sinensis* Greville

Yamaji, The Plankton, of the Japanese Coastal Waters, pl. 5, fig. 15, pp. 38, 105, 1968; Islam & Aziz, Bangladesh J. Botany, pl. 7, fig. 93, 1975; Hasle & Syvertsen, In : Identifying Marine Phytoplankton, p. 216, pl. 49, 1997.

Valve and either concave or convex, short horns and long spines, frustitles 150μ - 180μ long, 160μ diameter, horns 15μ long, spines 43μ - 55μ long.

Habitat : Plankton from Harwood Point, Sundarbans, 24 Parganas (S), West Bengal.

Specimen No. : N. S. 64. (Plate XIX, Fig. 113). Photo Plate XXI.

Order-BACILLARIALES

Key to the Sub orders

1a. Araphid pennate diatoms, sternum present, raphe absent **- Fragilariineae**

1b. Raphid pennate diatoms, sternum and raphe present **- Bacillariineae**

Family - FRAGILARIACEAE

Key to the Genera

1a. Cells with similar ends in valve as well as girdle view -2

1b. Cells with dissimilar ends in star like or spiral chains - *Asterionella*

2a. Rectangular cells closely appressed together in ribbons - *Fragilaria*

2b. Cells narrowly linear in stellate colonies - *Synedra*

***Asterionella* Hassall**

Cells with dissimilar ends in valve as well as girdle views, cells joined by valve faces of expanded foot poles, in star like spiral chains, foot poles angular in girdle view and more or less rounded in valve.

114. ***Asterionella japonica*** **Cleve**

Cleve & Möller, In : Diatoms, p. 3, no. 307, 1882; Cupp, Bull. 'Scripp's Institute of Oceanography', pp. 188 - 190, figs. 138, 139, 1943; Round *et. al.,* In : The Diatoms, Biology and 'Morphology of the Genera', pp. 392, 393, 1990; Hasle & Syvertsen, In : Identifying Marine Phytoplankton, p. 241, pl. 50, 1997; Islam & Morshed, Bangladesh J. Botany, pl. 2, figs. 19 - 22, 1985.

Cells arranged in flat, ring like colonies, basal part of frustules broader, nearly oval, gradually narrow, conical flask shaped in girdle view, frustules 55μ - 58μ long, basal part 6μ - 8μ broad.

Habitat : Plankton from Patharpratima, Sundarbans, 24 Parganas (S), West Bengal.

Specimen No. : N. S. 66. (Plate XIX, Fig. 114). Photo Plate XXII.

Fragilaria **Lyngbya**

Cells with numerous narrow bands, chloroplasts large, two per cell, valves broadly to narrowly elliptical, apices rounded to slightly attenuated to almost capitate, narrow sternum, striae present.

115. ***Fragilaria vaucheriae*** **(Kutz.) Peterson, A. Cl.**

Cleve-Euler, K. S. vet. Akad. Handle. Fjarde Ser. II, p. 42, fig. 353a. c, k, l, 1953; Suxena, Biblio. Phyco., pl. V, fig. 79 a - b, 1983.

Length of frustules 130μ - 135μ, width of individual cells 3μ, closely apressed to each other.

Habitat : Plankton from Pirkhali, Sundarbans, 24 Parganas (S), West Bengal.

Specimen No. : N. S. 76. (Plate XIX, Fig. 115). Photo Plate XXII.

Synedra **Ehrenberg**

Cells narrowly linear in stellate colonies, longer valves rostrate, smaller valves lanceolate, narrow sternum, transapical striae present.

116. ***Synedra ulna*** **(Nitzsch) Ehr.**

Hustedt, Osterreichs und der Schweiz, p. 151, f. 159, 1930; Venkateswarlu. Biblio. Phyco. pl. 1; figs. 6a, b, 1983; Sarode & Kamat, Biblio. Phyco. pl. 1, fig. 12, 1983.

Valves, 112μ - 130μ long, 7-8μ broad, linear to linear lanceotare, tapering at poles, ends rounded, axial area linear, striae 9 - 11 in 10μ.

Habitat : Plankton from Haroa and Golabari brackishwater fisheries, Sundarbans, Parganas (S), West Bengal.

Specimen No. : N. S. 77. (Plate XIX, Fig. 116).

Family - DIATOMACEAE

Diatoma De Candolle

Araphid diatoms with labiate processes at one end of each valve, processes are at diagonally opposite ends.

117. *Diatoma vulgare* Bory

Hustedt, Osterreichs und der Schweiz, figs. 103a, 108, 1930; Islam & Aziz, Bangladesh J. Botany, pl. 8, figs. 95 - 97, 1975.

Valves lanceolate, broad in relation to length, transverse costae run across, frustules 30μ - 32μ long, 8μ - 11μ, broad.

Habitat : Plankton from Haldibari, Sundarbans, 24 Parganas (S), West Bengal.

Specimen No. : N. S. 77. (Plate XIX, Fig. 117).

Suborder-BACILLARIINEAE

Key to the Families

1a. Isovalvar cells, both valves of a cell have 'naviculoid' raphe -2

1b. Hetervalvar cells, one valve has raphe with two longitudinal slits, other valve has no raphe - *Achnanthaceae*

2a. Valves elongate with variable outline, raphe eccentric, running along margin of one valve -3

2b. Valves linear, lanceolate or elliptical, raphe straight or sigmoid - *Naviculaceae*

3a. Cells rectangular, or spindle shaped valves elongate, raphe strongly eccentric - *Bacillariaceae*

3b. Cells oblong, inflated in the middle, raphe slightly eccentric - *Cymbellaceae*

Family - ACHNANTHACEAE

Key to the Genera

1a. Valve linear with rounded, apices - *Achnanthes*
1b. Valves broad to elliptical, central area more or less rounded - *Cocconeis*

Achnanthes Bory

Heterovalvar cells, one valve with two longitudinal slits, other valve without raphe, cells in ribbons, valve linear with rounded apices, raphe straight, narrow sternum.

118. ***Achnanthes microcephala*** **(Kütz) Grunow**

Venkateswarlu, Biblio, Phyco., pl. 1, figs. 12 a, b, 1983; Sarode & Kamat, Biblio. Phyco., pl. IV, fig. 42, 1983.

Valves 18μ - 22μ long, 3μ - 3.5μ broad, narrow, linear with broadly rounded ends, frustules slightly bent at the centre, valves with or without raphe, striae 30 - 32 in 10μ in both.

Habitat : Epiphytic on *Polysiphonia* sp. collected from Panchmukhani, Sundarbans, 24 Parganas (S), West Bengal.

Specimen No. : N. S. 78. (Plate XIX, Fig. 118). Photo Plate XVIII.

119. ***Achnanthes minutissima*** **(Kütz) var.** ***cryptocephala*** **Grunow**

Venkateswarlu, Biblio. Phyco., pl. I, figs. 13 a, b, 1983; Sarode & Kamat, Biblio. Phyco., pl. IV, fig. 44, 1983.

Valves 16μ - 18μ long, 2μ - 2.2μ broad, linear, raphe valve with thin raphe, central area slightly widened, striae 30 - 32 in 10μ.

Habitat : Epiphytic on *Pterosiphonia* sp. collected from Burirdhapi, Sundarbans, 24 Parganas, West Bengal.

Specimen No. : N. S. 72. (Plate XIX, Fig. 119).

Cocconeis Ehrenberg

Heterovalvar cells, one valve has raphe other valve without raphe, cells in amorphous, stellate, radiating colonies, broad to elliptical, central area roundish.

120. *Cocconeis placentula* Ehr.

Hustedt, Osterreichs und der Schweiz, p. 190, f. 261, 19,30; Cleve - Euler, K. S. vet. Akad. Handle. Fjarde Ser. II, Bd., p. 8, 1953; Venkateswarlu, Biblio. Phyco., pl. 1, figs. 9 a, b, 1983; Sarode & Kamat, Biblio. Phyco., pl. III, fig. 38, 1983.

Valves 22µ - 28.6µ long, 14.8µ broad, elliptical, central area roundish, raphe valve with thin straight raphe, finely punctnaate.

Habitat : Epiphytic on *Polysiphonia* sp. at Pirkhali, Sundarbans, 24 Parganas (S), West Bengal.

Specimen No. : N. S. 76. (Plate XIX, Fig. 120).

Family - NAVICUILACEAE

Key to the Genera

1a.	Valves linear, lanceolate or elliptical	- 2
1b.	Valves flattened, sigmoid	- 3
1c.	Valves oblong to lanceolate with capitate rounded apices	- 4
2a.	Raphe straight, stauros absent	- *Navicula*
2b.	Raphe straight, stauros present	- *Stauroneis*
2c.	Raphe straight, transapical striae lineate	- *Pinnularia*
3a.	Valves gently sigmoid to lanceolate, raphe more or less straight	- *Pleurosigma*
3b.	Valves and raphe strongly sigmoid	- *Gyrosigma*
4a.	Transverse striae, radially arranged throughout the valves	- *Gomphonema*
4b.	Longitudinal finely punctate striae present	- *Anomoeneis*

***Navicula* Bory**

Valves linear, lanceolate or elliptical, raphe generally straight, raphe not raised on a ridge.

Key to the species

1a.	Valves lanceolate with acutely rounded ends	- *N. cryptocephala*
1b.	Valves narrowly lanceolate with acute ends	- *N. radiosa*

121. *Navicula cryptocephala* Kütz.

Hustedt, Osterreichs und der Schweiz, p. 295, fig. 496, 1930; Cleve - Euler, K. S. vet. Akad. Handle. Fjarde Ser., p.154, fig. 813 a-b, 1953; Suxena, Biblio. Phyco., pl. VI, fig. 106 a-b, 1983.

Valves 16μ - 29.6μ long, 4μ - 5μ broad, lanceolate with acutely rounded ends, striae 14 - 15 in 10μ, radial in the middle and convergent at the ends.

Habitat : Attached to submerged substratum at Jharkhali, Sundarbans, 24 Parganas, West Bengal.

Specimen No. : N. S. 76. (Plate XIX, Fig. 121). Photo Plate XXIII.

122. *Navicula radiosa* Kütz.

Hustedt, Osterreichs und der Schweiz, p. 299, f. 513, 1930; Cleve -Euler, K. S. vet. Akad. Handle. Fjarde Ser., 1953, III, p. 155; Venkateswarlu, Biblio. Phyco., pl. III, fig. 33, 1983; Sarode & Kamat, Biblio. Phyco., pl. IX, fig. 105, 1983.

Valves 58μ-75.5μ long, 9.2μ -11.6μ broad, striae 11-13 in 10μ narrowly lanceolate with acute ends, axial area narrow, striae radial in the middle and convergent at the end.

Habitat : Pktonic at Golabari, Sundarbans, 24 Parganas (S), West Bengal.

Specimen No. : N. S. 50A. (Plate XIX, Fig. 122).

***Stauroneis* Ehrenberg**

Cells in tight ribbons, valves linear, lanceolate or elliptical, raphe straight, one chloroplast per cell, stauros present.

123. *Stauroneis phoenienteron* (Nitzsch.) Ehr.

Hustedt, Osterreichs und der Schweiz, p. 255, fig. 404, 1930; Sarode & Kamat, Biblio. Phyco., pl. V1, fig. 68, 1983; Pandey *et. al.*, Biblio. Phyco., pl. 1, fig. 1, 1983.

Valves 88.4μ - 96μ long, 19 - 21.6μ broad, lanceolate, with slightly constricted and capitate broadly rounded, ends, central area stauroid, striae 18-20 in 10μ , radial throughout,

Habitat : Plankton from Burirdhapri, Sundarbans, 24 Parganas (S), West Bengal.

Specimen No. : N. S. 70. (Plate XX, Fig. 123).

***Pinnularia* Ehrenberg**

Cells rectangular in girdle view, valves lanceolate, sometimes with slightly produced ends, transapical striae lineate.

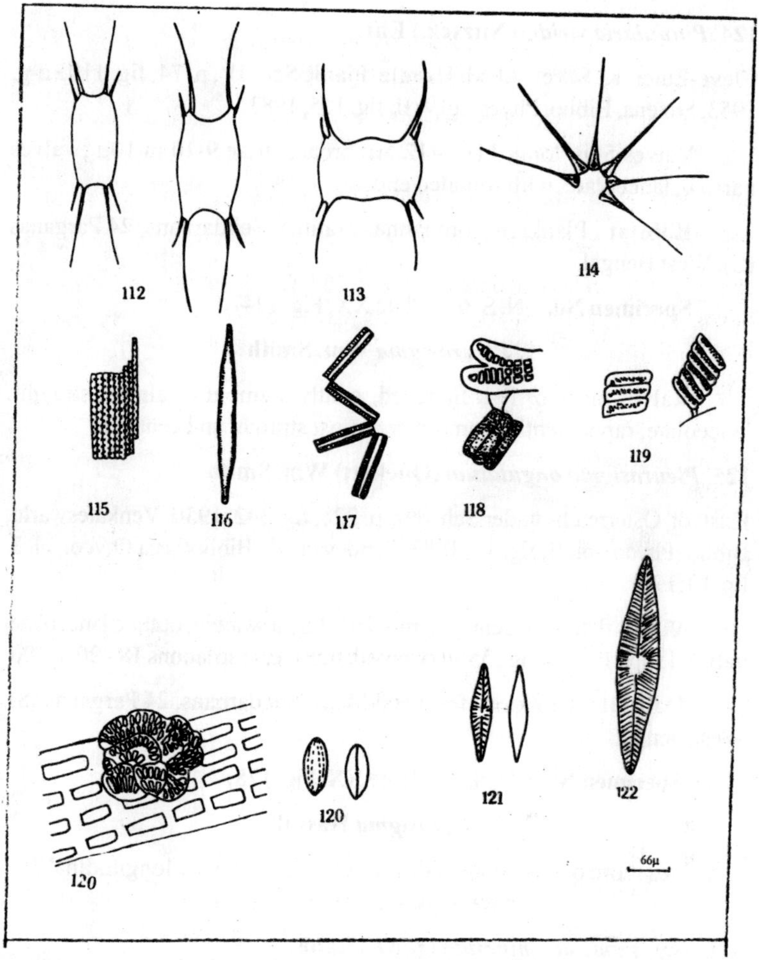

Plate XIX

112. *Biddulphia mobiliensis* Bailey; 113. *Biddulphia sinensis* Greville; 114. *Asterionella japonica* Cleve; 115. *Fragilaria vaucheriae* (Kütz.) Peterson, A. Cl; 116. *Synedra ulna* (Nitzsch.) Ehr.; 117. *Diatoma vulgare* Bory; 118. *Achnanthes microcephala* (Kütz.) Grunow; 119. *Achnanthes minutissima* (Kütz.) var. *cryptocephala* Grunow; 120. *Cocconeis placentula* Ehr.; 121. *Navicula cryptocephala* Kütz.; 122. *Navicula radiosa* Kütz.

124. ***Pinnularia viridis*** **(Nitzsch.) Ehr.**

Cleve-Euler, K. S. vet. Akad. Handle. Fjarde Ser., IV, p. 74, fig. 1103 e-g, 1953; Suxena, Biblio. Phyco., pl. VII, fig. 125, 1983.

Valves 50μ long, 11μ - 12.5μ broad, striae 9-10 in 10μ, valves narrow, lanceolate, with rounded ends.

Habitat : Plankton from Pathar Pratima, Sundarbans, 24 Parganas (S), West Bengal.

Specimen No. : N. S. 65. (Plate XX, Fig. 124).

Pleurosigma **Wm. Smith**

Valves more or less flattened, gently sigmoid or almost straight, lanceolate, raphe gently sigmoid or almost straight and central.

125. ***Pleurosigma angulatum*** **(Quekett) Wm. Smith**

Hustedt, Osterreichs under Schweiz, p. 228, fig. 342, 1930; Venkateswarlu, Biblio. Phyco., pl. II, fig. 17, 1983; Pandey *et. al.*, Bibliotheca Phyco., pl. 3, fig. 10, 1983.

Valves flattened, gently sigmoid or almost straight, outline lanceolate, valves 150μ -165μ long, 35-48μ broad, transverse striations 18 - 20 in 10μ.

Habitat : Plankton -from Haldibari, Sundarbans, 24 Parganas (S), West Bengal.

Specimen No. : N. S. 77. (Plate XX, Fig. 125).

Gyrosigma **Hassall**

Valves more or less flattened, strongly sigmoid with longitudinal & transverse & siae, radically striae.

126. ***Gyrosigma acuminatum*** **(Kütz.) Rabh.**

Hustedt, Osterreichs under Schweiz, p. 222, fig. 329, 1930; Venkateswarlu, Biblio, Phyco., pl. I, fig. 16 a, b, 1983 ; Sarode & Kamat, Biblio. Phyco., pl. IV, fig. 48, 1983.

Valves 110.6μ -162.8μ long, 20 - 22μ broad, sigmoid, gradually tapering from the middle towards end, ends broadly rounded, raphe sigmoid, axial area narrow, central area small.

Habitat : Plankton from Haldibari, Sundarbans, 24 Parganas (S), West Bengal.

Specimen No.: N. S. 77. (Plate XX, Fig. 126). Photo Plate XXIII.

***Gomphonema* Hustedt**

Valves oblong to lanceolate with capitate rounded apices, transverse & sriae, radically arranged.

127. *Gomphonema sphaerophorum* Ehr.

Hustedt, Osterreichs und der Schweiz, 372, fig. 695, 1930 ; Venkateswarlu, Biblio. Phyco., pl. IV, fig. 60,1983; Pandey *et. al.,* Biblio. pl. 2, fig. 18, 1983; Suxena, Biblio. Phyco., pl. VII, fig. 140, 1983.

Valves 19µ - 30µ long, 9µ - 11µ broad, transverse striations 15 - 17 in 10µ , radially arranged.

Habitat : Plankton from Sajnekhali, Sundarbans, 24 Parganas (S), West Bengal.

Specimen No. : N. S. 15. (Plate - XX, Fig. 127).

***Anomoeneis* Pfitzer**

Valves broad or narrow to lanceolate with a capitate rounded ends, radially finely punctate striae.

128. *Anomoeneis exilis* (Grun.) Cleve

Hustedt, Osterreichs und der Schweiz, Kieselalg, II (6), p. 752, fig. 114 d, 1959; Sarode & Karmat, Biblio. Phyco., pl. VI, fig. 73, 1983; Venkateswarlu, Biblio. Phyco., pl. II, fig.22, 1983.

Valves 30µ - 35µ long, 3.6-4.2µ broad, narrow, lanceolate with capitate rounded ends, striae 28 in 10µ radial, finely punctate.

Habitat : Planktonic at Minakhan, Sundarbans, 24 Parganas (N), West Bengal.

Specimen No.: N. S. 59 A. (Plate XX, Fig. 128).

Family - CYMBELLACEAE

Key to the Genera

1a. Valves oblong to lanceolate - *Cymbella*
1b. Valves oblong to lanceolate with, slightly rounded capitate apices - *Amphora*

Cymbella Agardh

Valves linear oblong to lanceolate with slightly eccentric raphe.

129. *Cymbella ehrenbergii* Kütz

Cleve-Euler, K.S. vet. Akad. Handle. Fjarde Ser., IV, p. 148, fig. 121 c, e, 1955; Suxena, Biblio. Phyco., pl. VII, fig. 130, 1983; Venkateswarlu, Biblio. Phyco., pl. III, fig. 48, 1983.

Valve 98μ - 128μ long, 30μ - 35μ broad, striations more towards the ends, less in the middle.

Habitat : Plankton from Basanti, Sundarbans, 24 Parganas (S), West Bengal.

Specimen No. : N. S. 18, (Plate XX, Fig. 129).

Amphora Ehrenberg

Valves oblong to lanceolate with slightly rounded capitate apices, raphe slightly eccentric.

130. *Amphora* veneta Kütz.

Hustedt. Osterreichs und der Schweiz, Bacill. p. 345, fig. 631, 1930; Venkateswarlu, Biblio. Phyco., pl. III, fig. 42 a, b,1983.

Valve 20μ - 22μ long, 10μ broad, striae 20 - 23 in 10μ.

Habitat : Plankton from Haldibari, Sundarbans, 24 Parganas (S), West Bengal.

Specimen No. : N. S. 77. (Plate XX, Fig. 130); very rare presence in the sample collected.

Family - BACILLARIACEAE

Nitzschia Hassall

Cells usually in chains, united by overlapping valve ends to form a stepped chain, raphe eccentric, frustules straight.

Key to the species

1a. Keel punctate 7-8 in 10μ -*N. obtusa*
1b. Keel punctate 9-10 in 10μ - *N. sublinearis*
1c. Keel punctate 18-19 in 10μ - *N. acicularis*

1

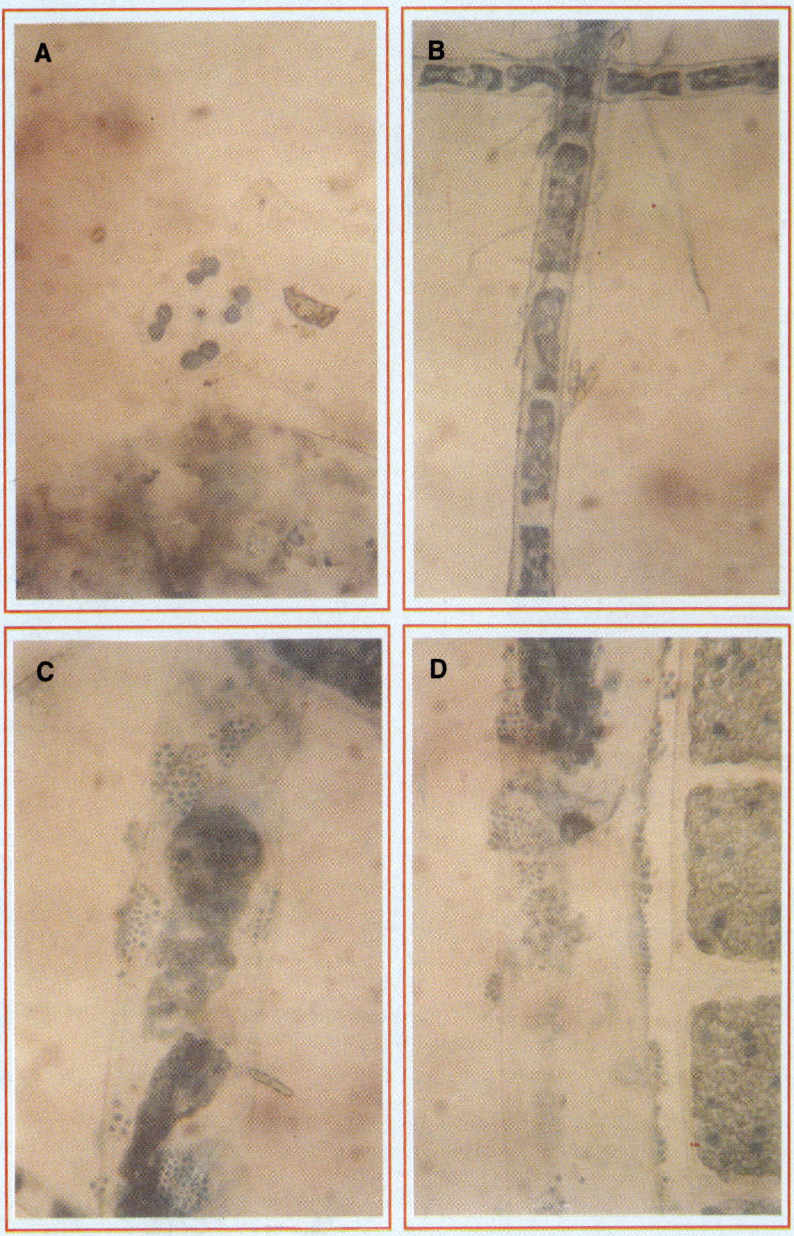

Plate - I : A. *Gloeocapsa Kuetzingiana* Nag. (X 200);B. *Stichosiphon sansibaricus* (Hieron.) Drouet et Daily (X 200); C. & D. *Xenococcus cladophorae* (Tilden) Setchell et Gardner (x 200).

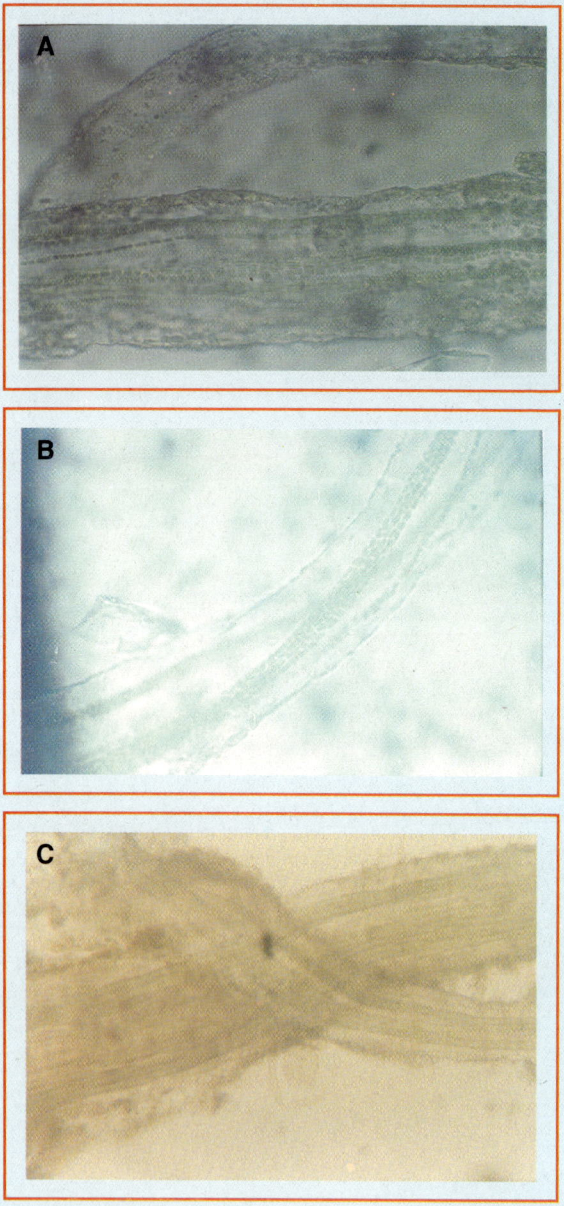

Plate - II : A. & B. *Microcoleus chthonoplastes* Thuret ex Gomont (X 200); C. *Schizothrix lamyii* Gomont (X 200).

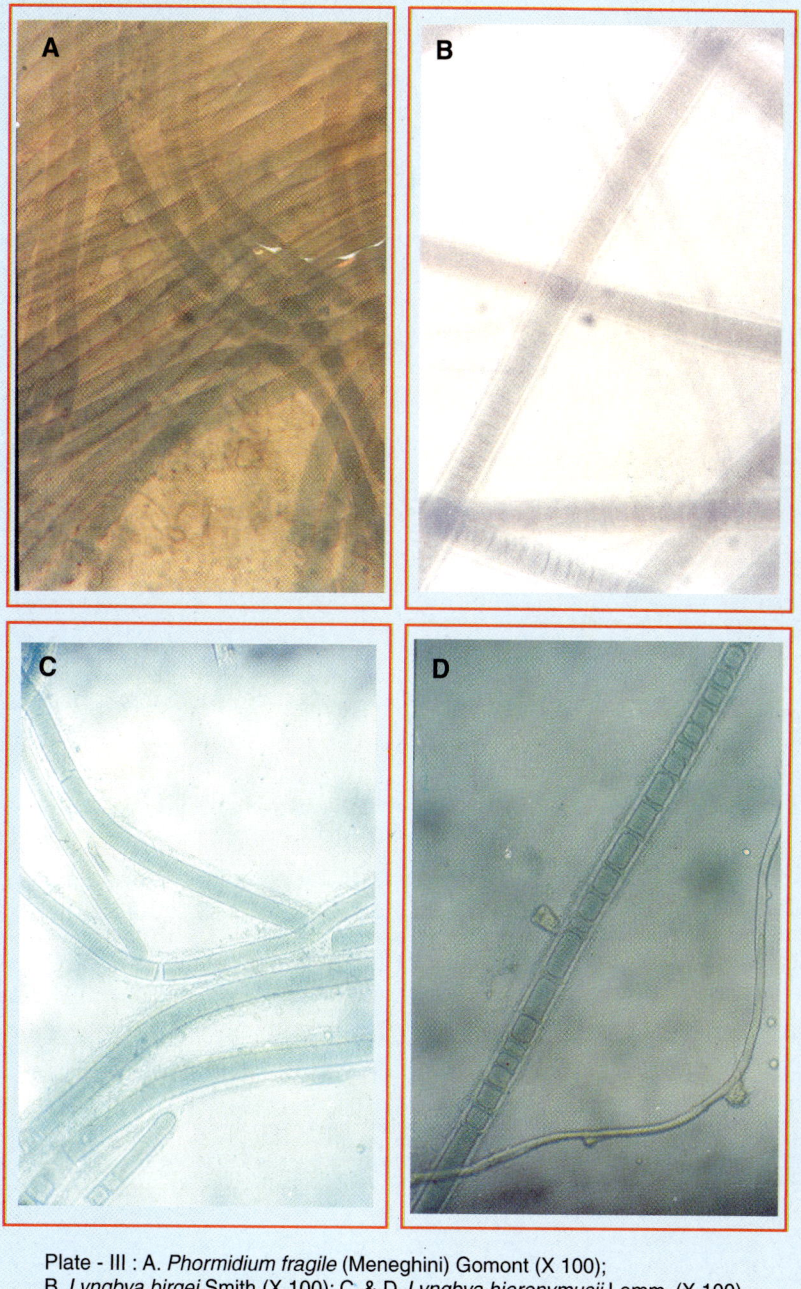

Plate - III : A. *Phormidium fragile* (Meneghini) Gomont (X 100); B. *Lyngbya birgei* Smith (X 100); C. & D. *Lyngbya hieronymusii* Lemm. (X 100).

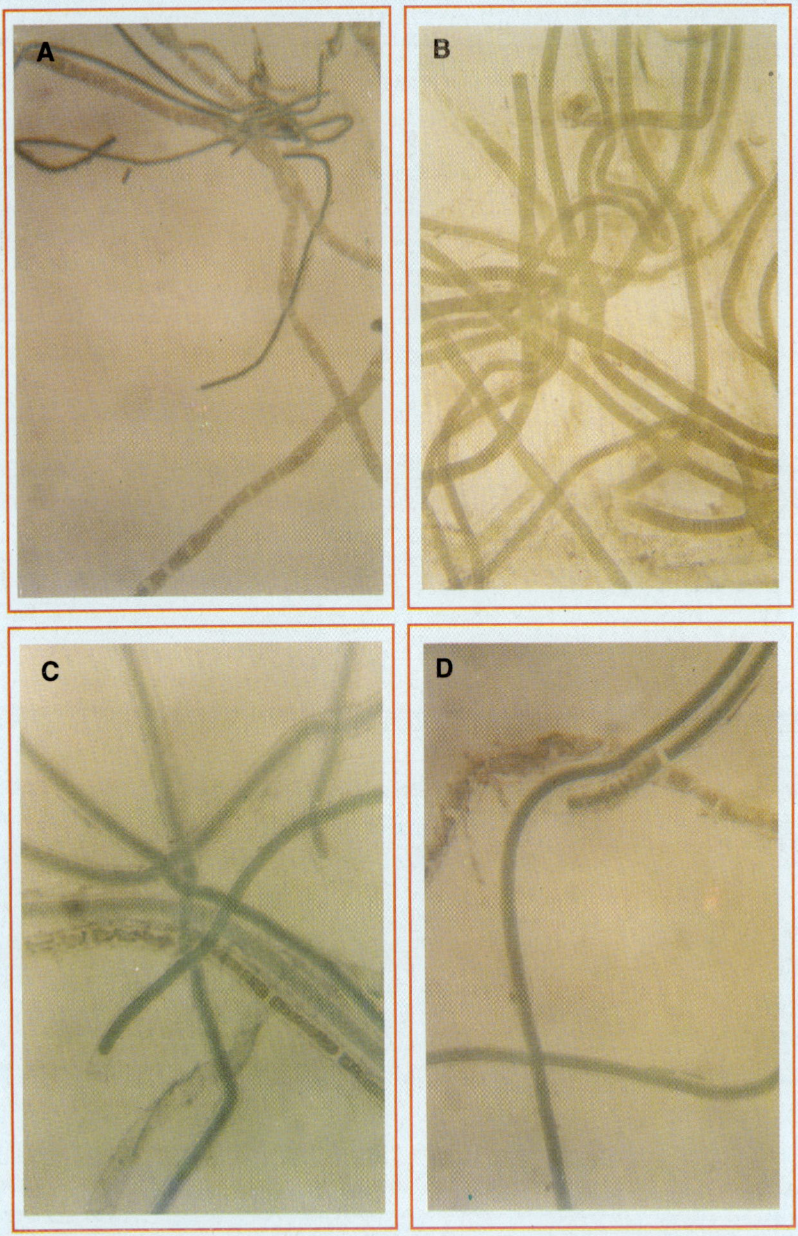

Plate - IV : A. *Lyngbya lutea* (Ag.) Gomont (X 100); B. *Lyngbya majuscula* Hearvey ex Gomont (X 100); C. & D. *Lyngbya confervoides* C. Ag. Ex Gomont (X 100).

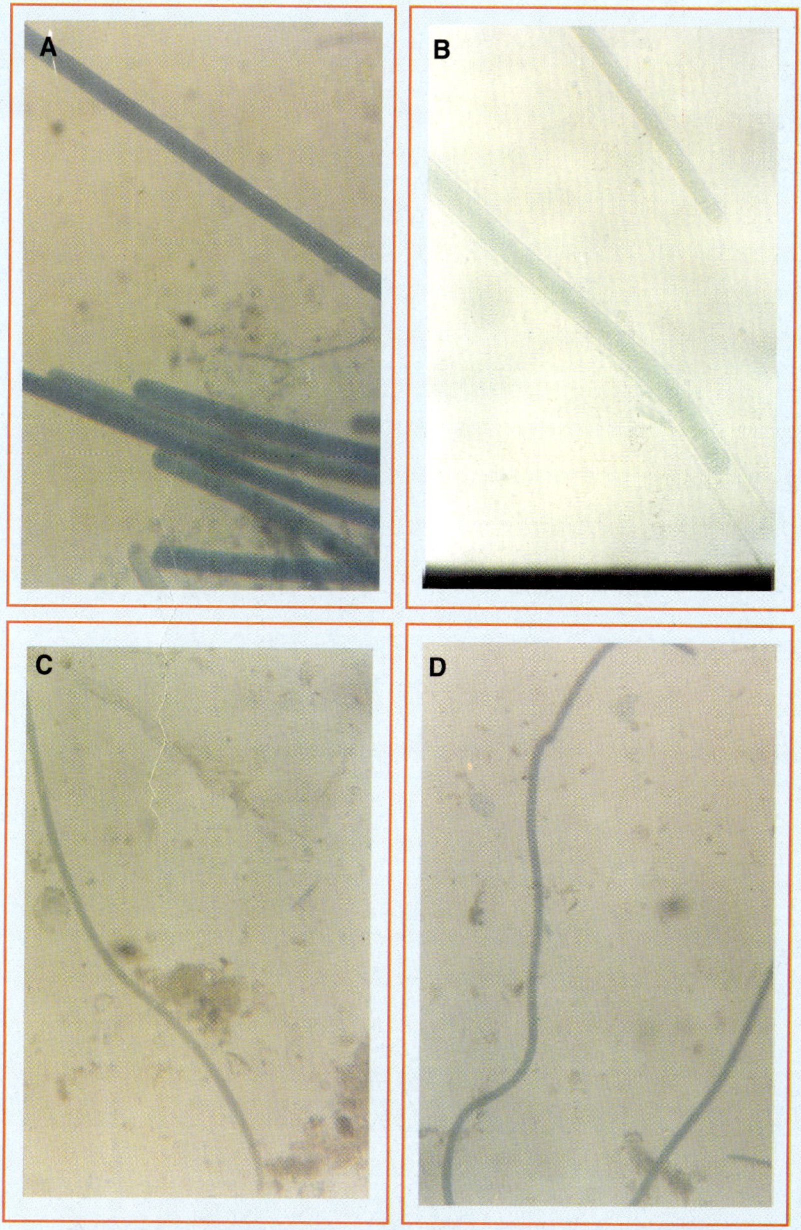

Plate - V : A. *Oscillatoria limosa* Ag. Ex Gomont (X 100); B. (a) *Oscillatoria subbrevis* Schmidle &(b) *Lyngbya birgei* Smith (X 100); C. & D. *Oscillatoria chlorina* Kutz. Ex Gomont (X 100).

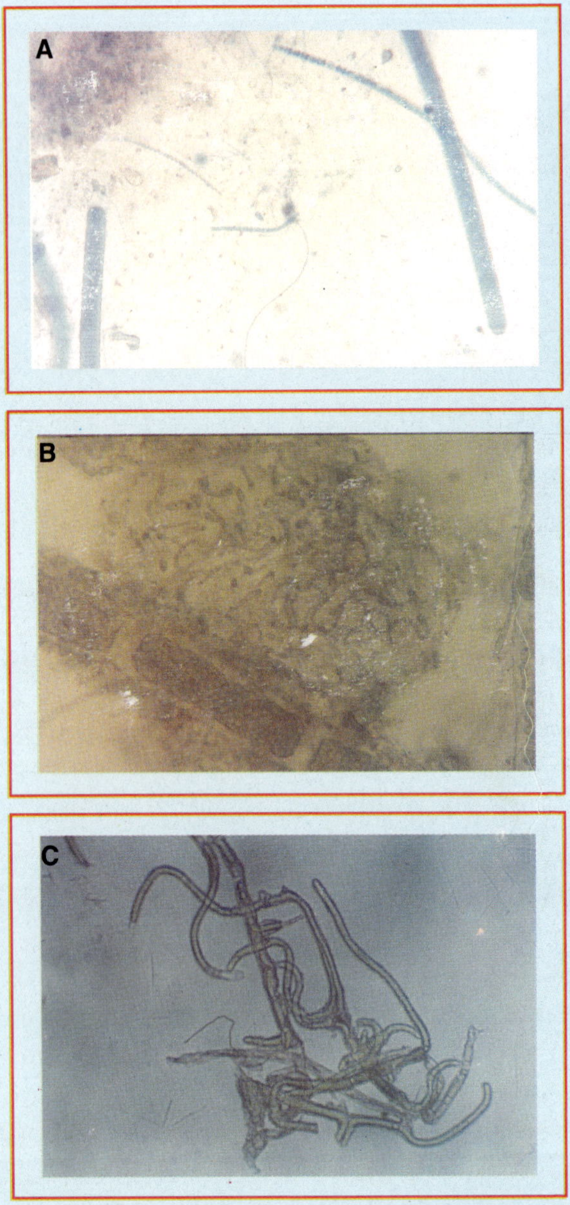

Plate - VI : A. (a) *Spirulina major* Kutz. Ex Gomont & (b) *Anabaena doliolum* Bharadwaja (X 100); B. *Nostoc punctiforme* (Kutz.) Hariot (X 200); C. *Mastigocoleus testarum* Lagerh. (X 200).

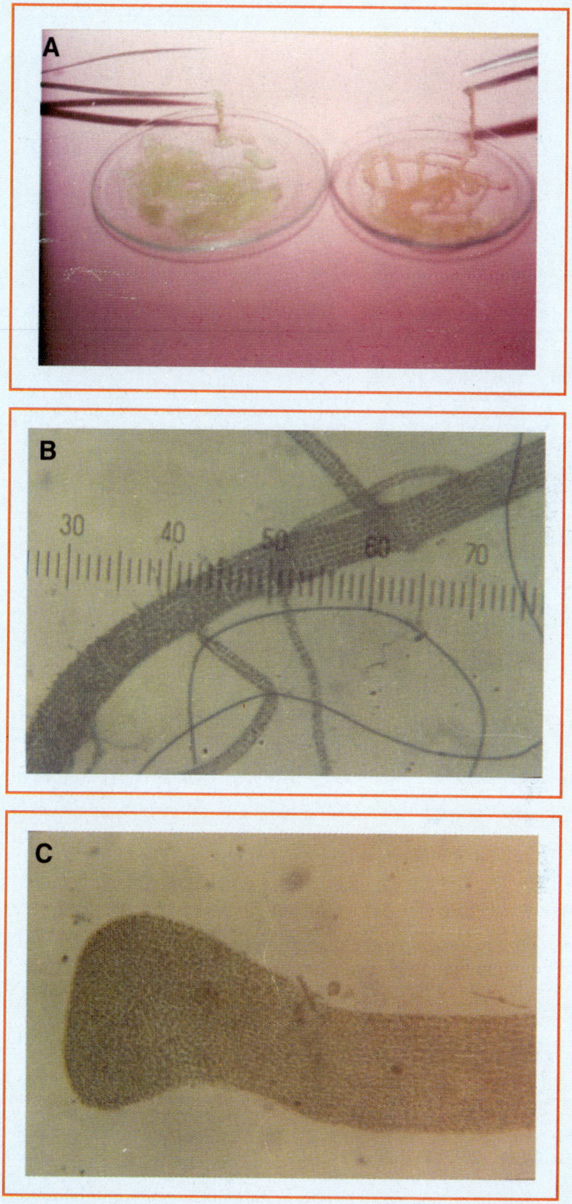

Plate - VII : A. (a) *Enteromorpha intestinalis* (Linnaeus) Link. & (b) *Enteromorpha clathrata* (Roth.) Greville; B. *Enteromorpha prolifera* (Muller) J. Agardh (X 20); C. *Enteromorpha compressa* (Linnaeus (Greville (X 20).

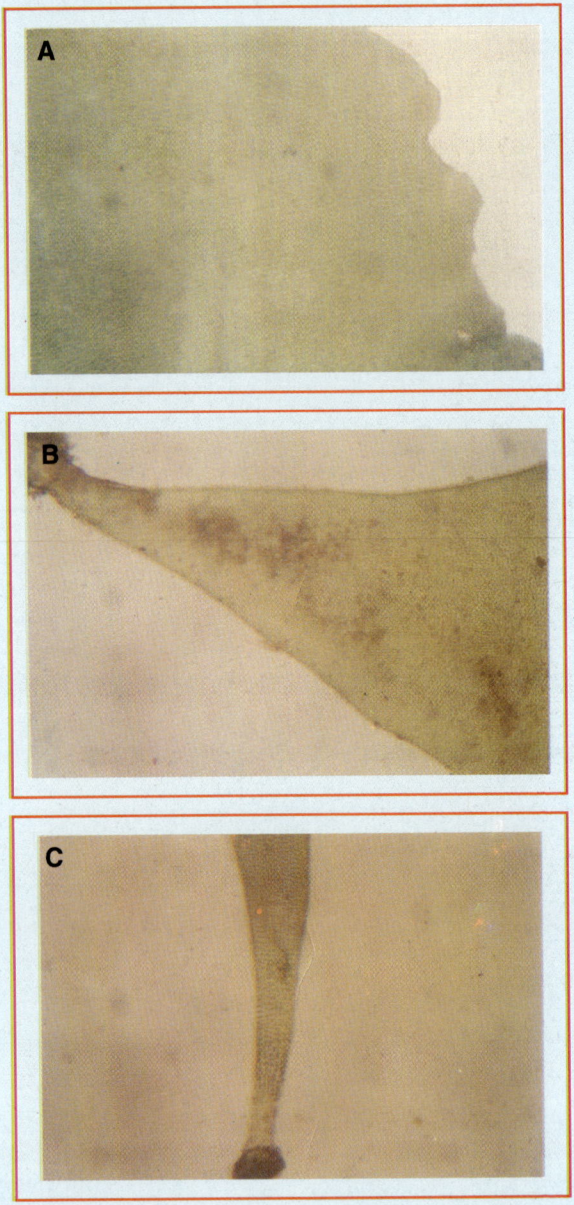

Plate - VIII : A. *Ulva lactuca* Linnaeus (x 20); B. & C. *Ulva patengensis* Salam & Khan (X 20).

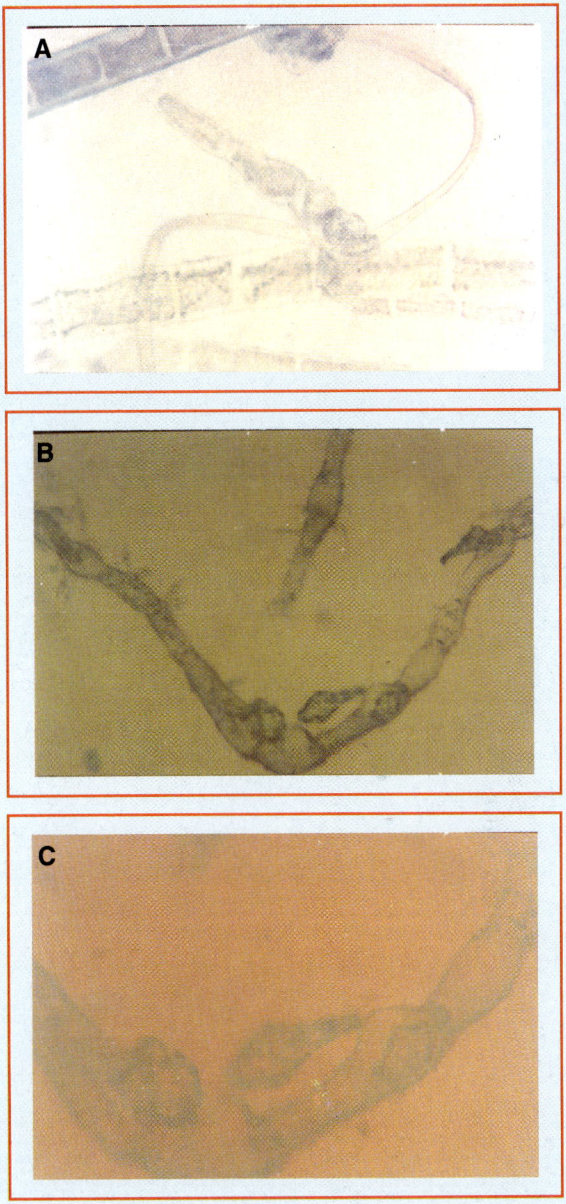

Plate - IX : A. & B. Dwarf antheridial filament of *Oedgonium undulatum* (Brebisson) Al. Braun. (X 20); C. Dwarf antheridial filament of *Oedgonium undulatum* (Brebisson) Al. Braun. (X 50).

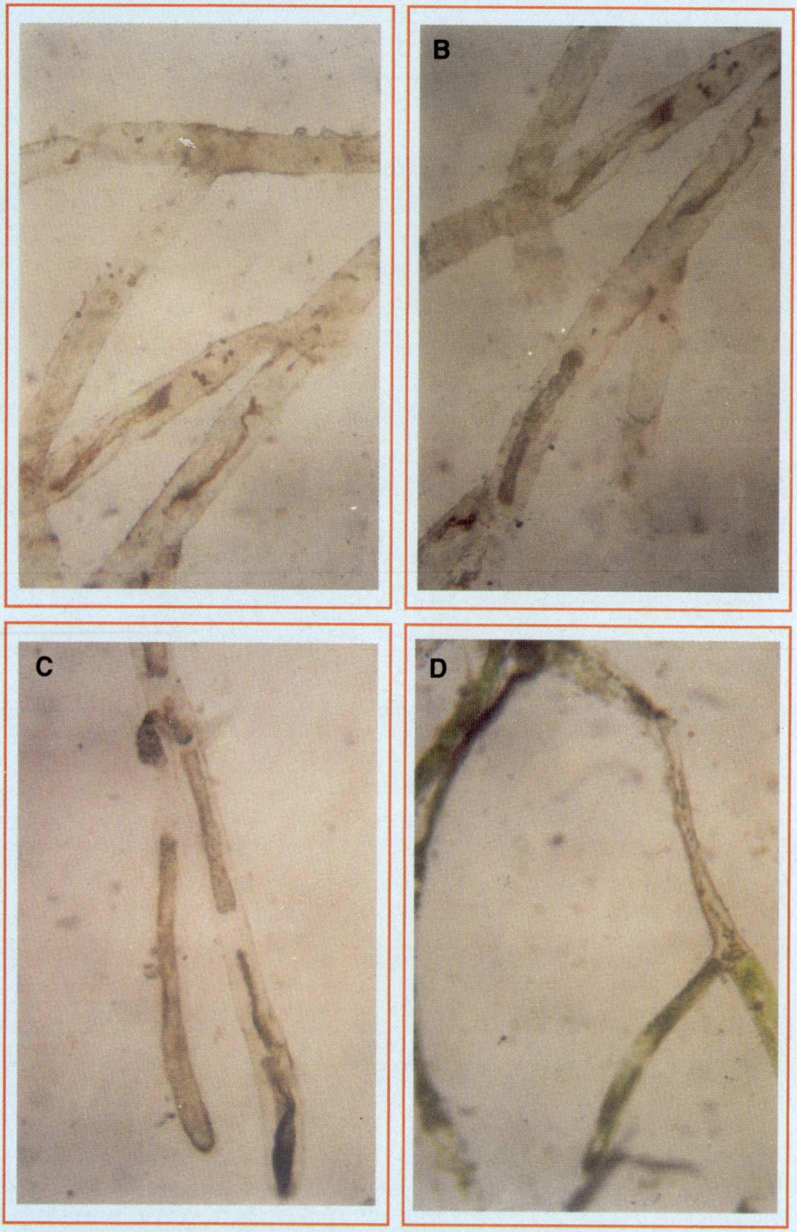

Plate - X : A., B. & C. *Cladophorella sundarbanesis* Islam. (X 20); D. *Cladophora echinus* (Bias.) Kutz. (X 20).

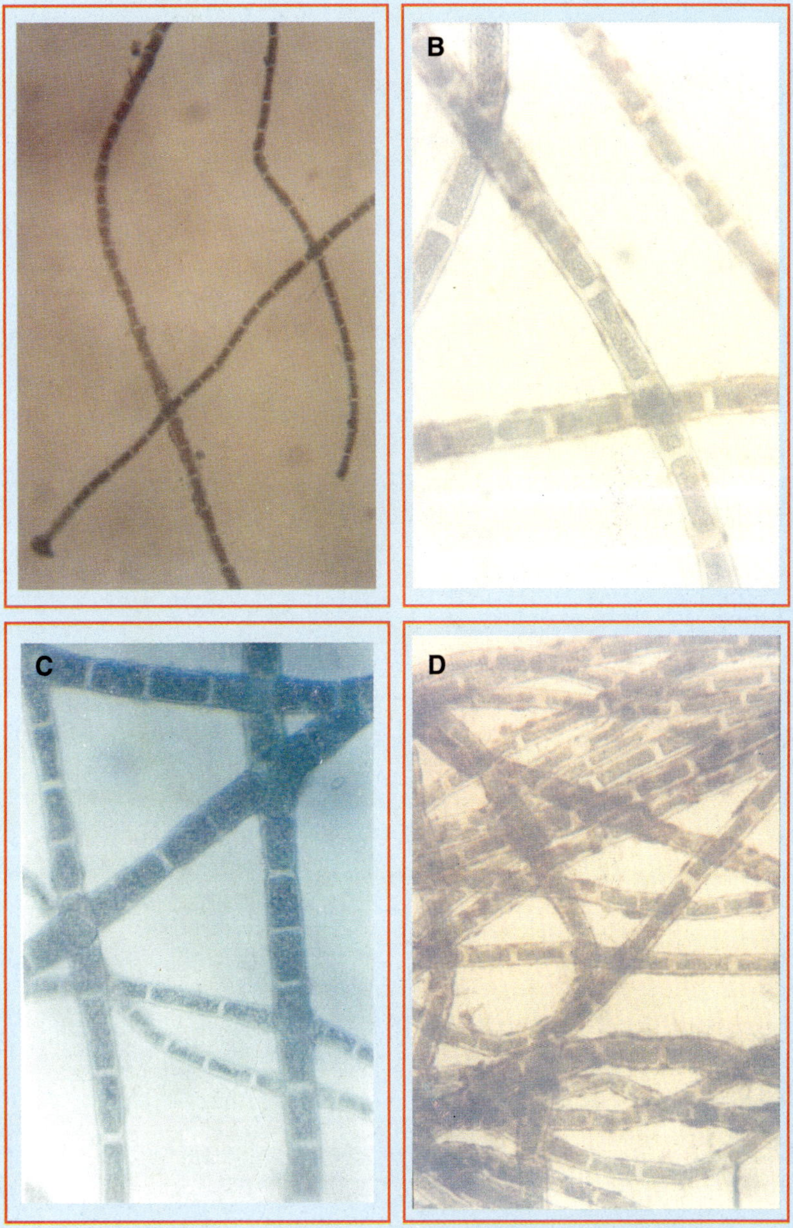

Plate - XI : A. *Lola capillaris* (Kutzing) (X 20); B. & C. *Lola implexa* (Harv.) Hamel Hamel (X 50); D. *Lola turtuosa* (Dillw.) Chapman (X 50).

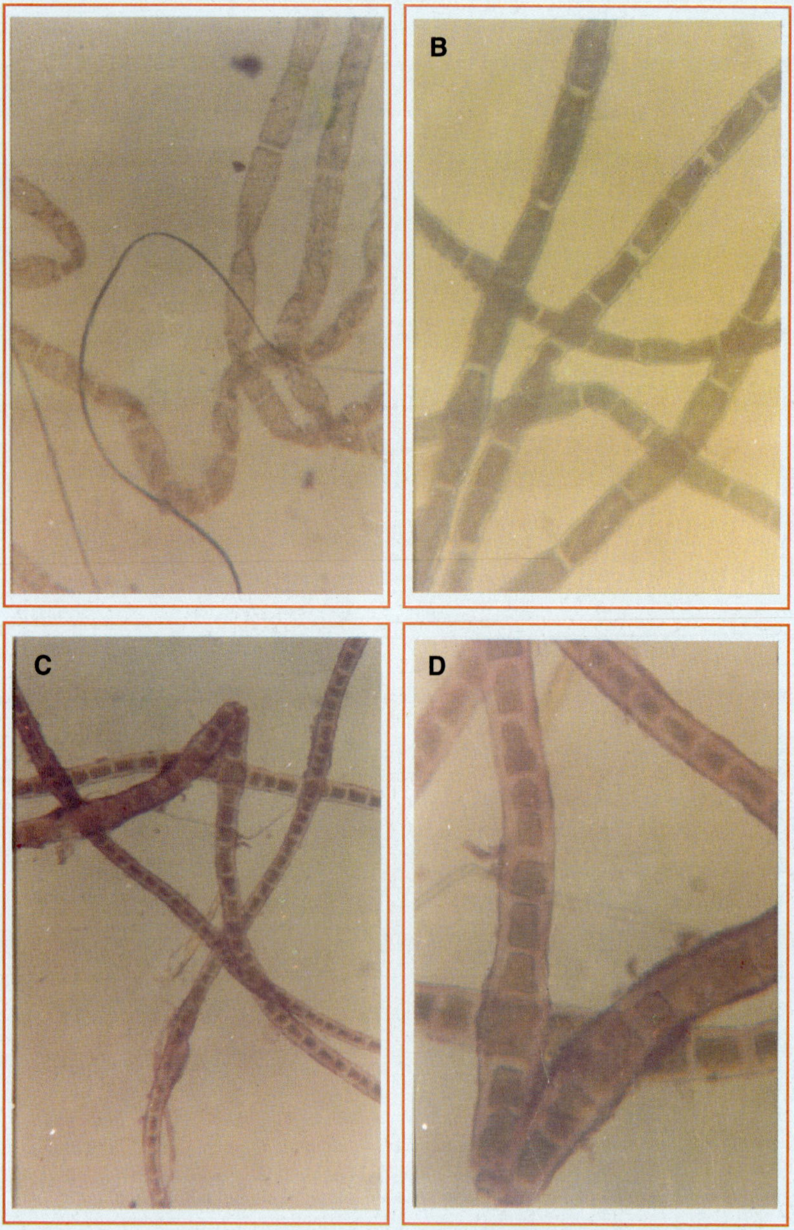

Plate - XII A. *Chaetomorpha gracilis* Kutzing (X 50); B. *Chaetomorpha brachygona* Harvey (X 50); C. & D. *Chaetomorpha aerea* (Dillwyn) Kutzing (X 50).

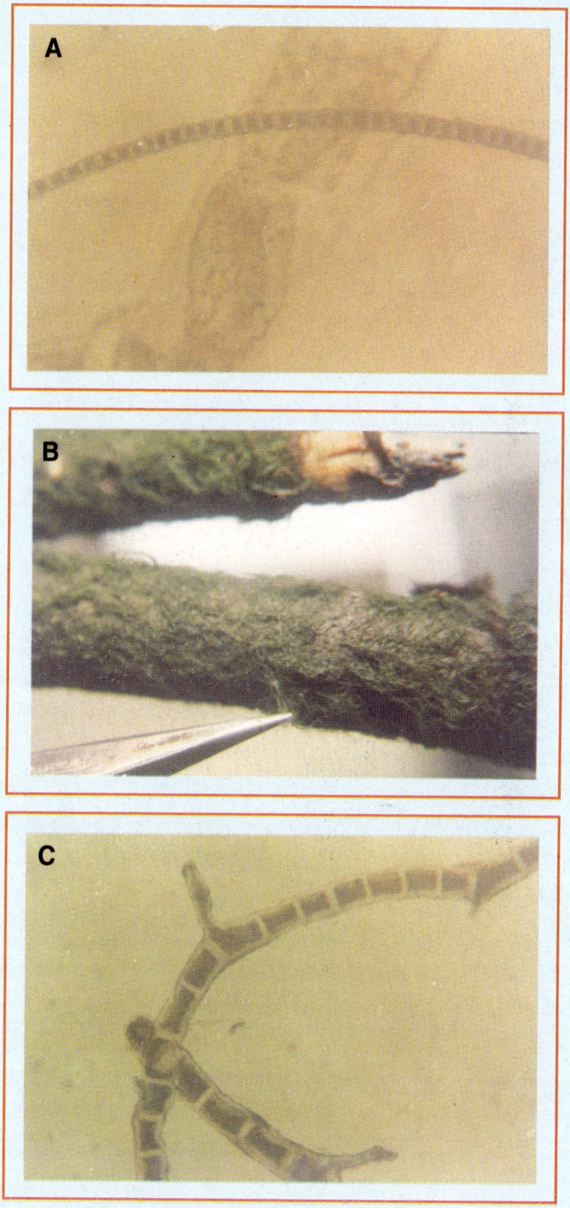

Plate - XIII : A. *Uronema confervicolum* Lagerheim (X 200); B. *Rhizoclonium* spp. as periphyte on pneumatophore; C. *Rhizoclonium riparium* (Roth.) Harvey (X 50).

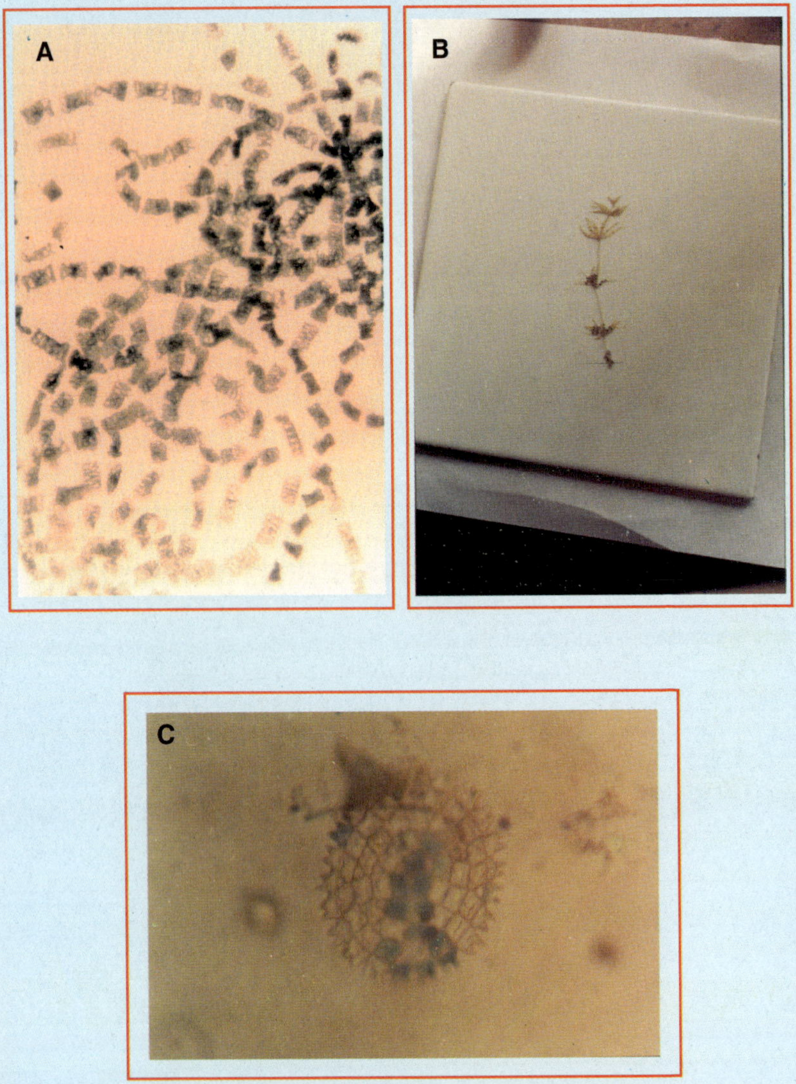

Plate - XIV : A. *Spirogyra dubia* Kg. (X 20); B. *Pediastrum duplex* (Meyan) (X 200); C. *Chara zeylanica* Willdenow.

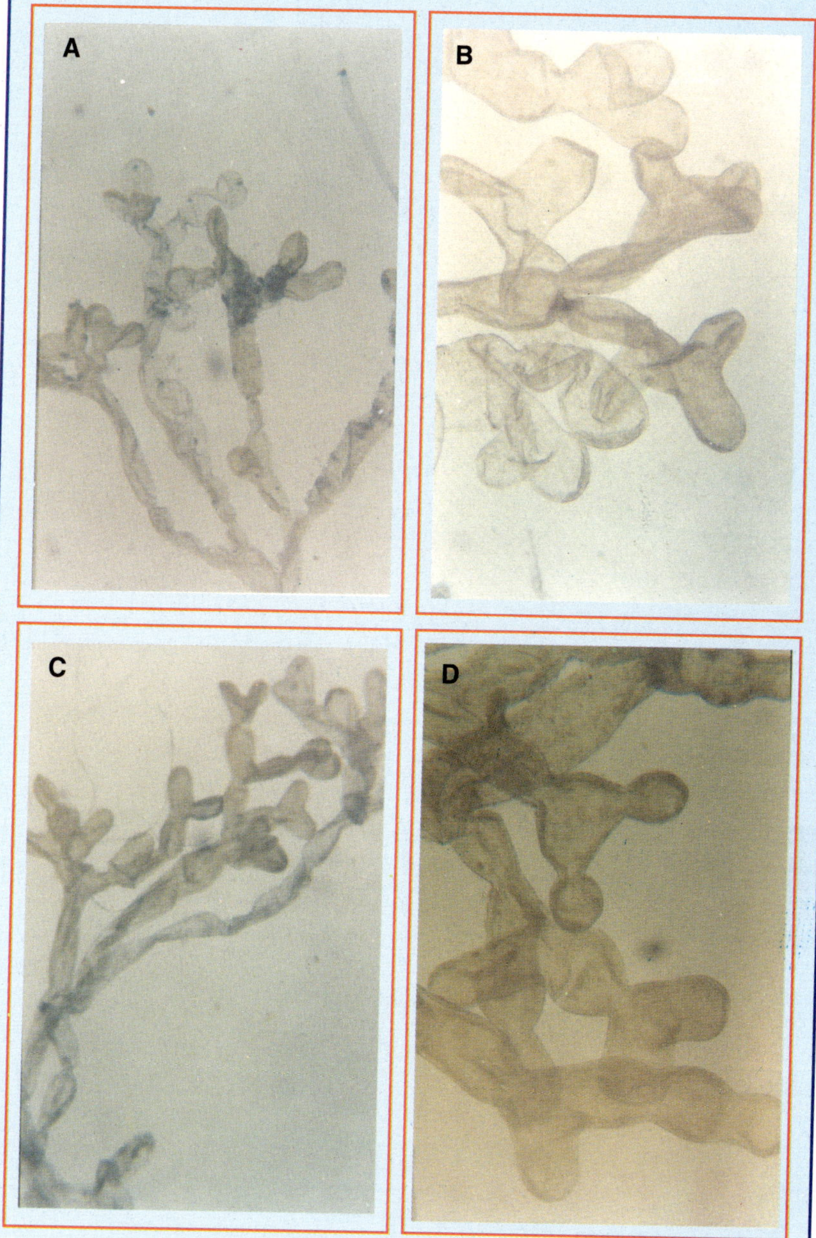

Plate - XV : A., B., C. & D. *Boodleopsis sundarbanensis* Islam (X 50).

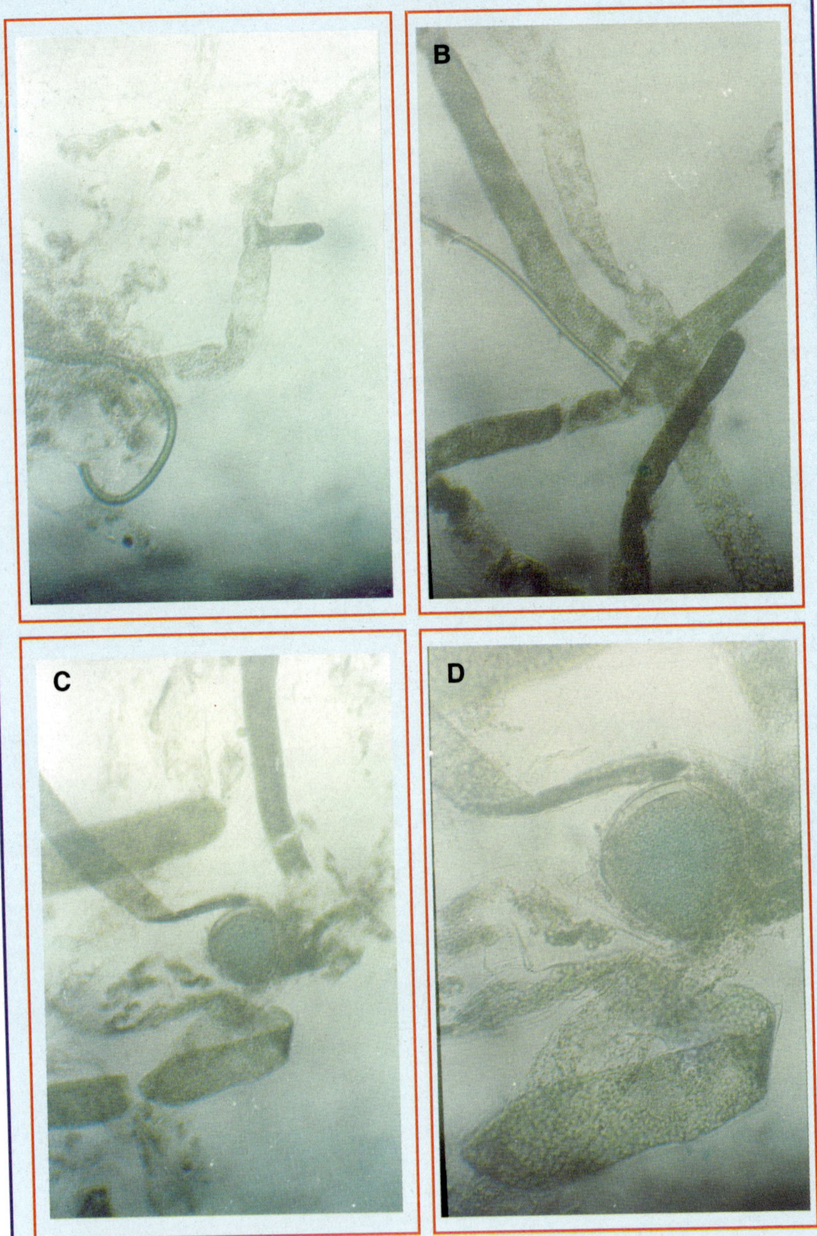

Plate - XVI : A. & B. *Vaucheria* sp. (X 50); C. *Vaucheria prescottii* Islam (X 50); D. *Vaucheria prescottii* Islam (X 100).

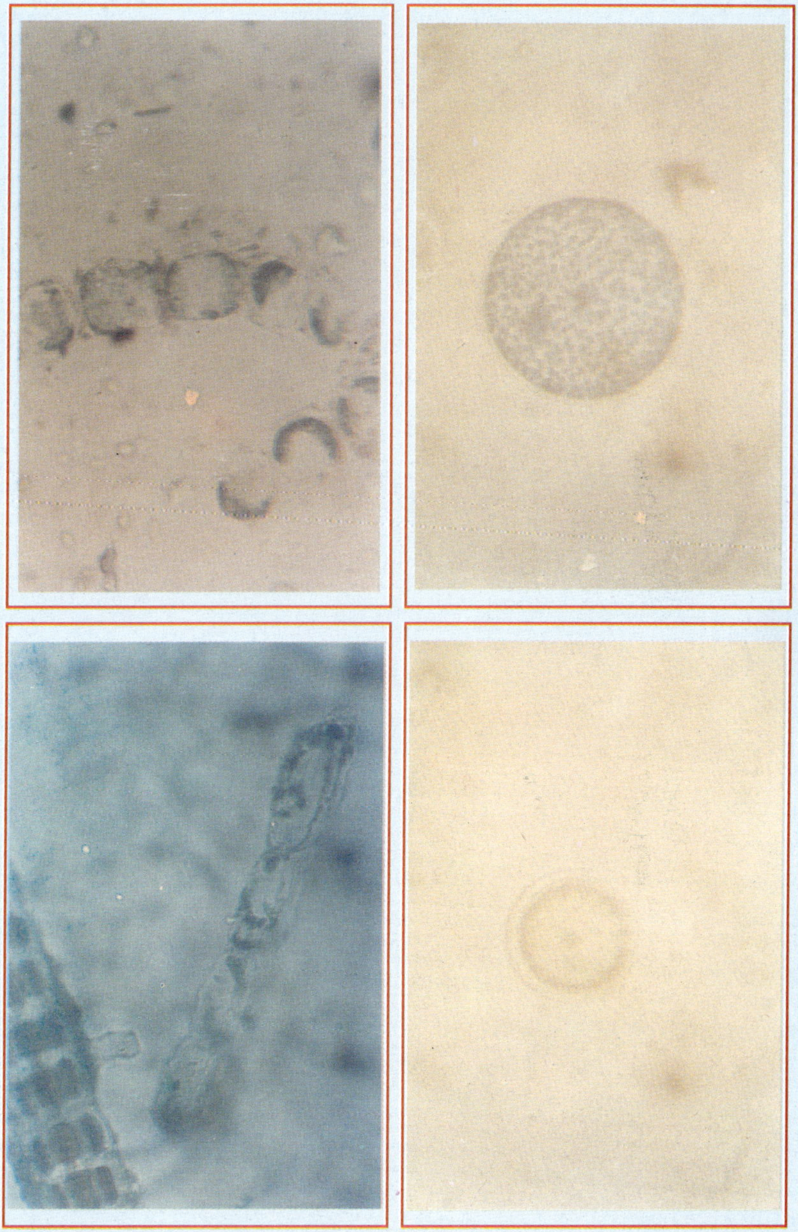

Plate - XVII : A. *Melosira moniliformis* (Muel.) Agardh (X 200); B. *Melosira sol* (Ehr.) Kuetz. (X 200); C. *Stephanopyxis palmeriana* (Grev.) Grun. (X 200); D. *Cyclotella glomerata* Bachmann (X 200).

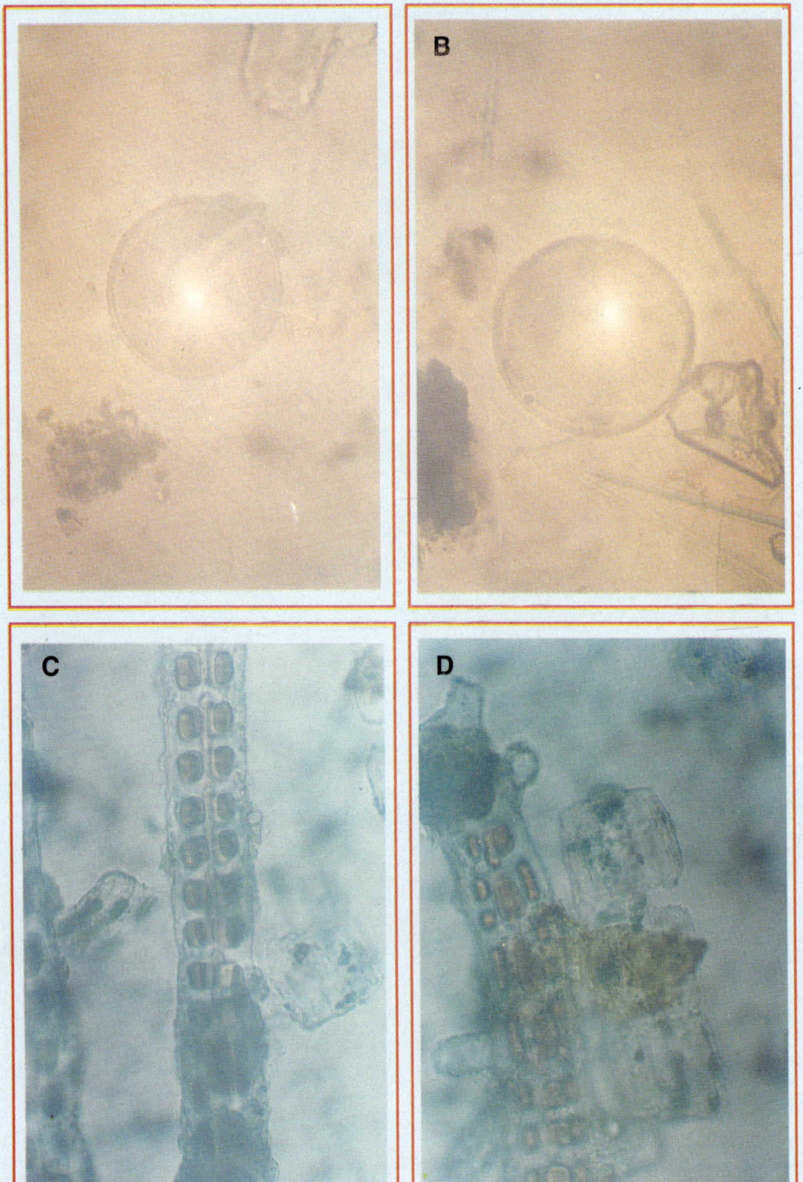

Plate - XVIII : A. *Coscinodiscus granii* Gough (X 200); B. *coscinodiscus gigas* Ehr. (X 200); C. & D. *Achnanthes microcephala* (Kutz.) Grunow (X 200).

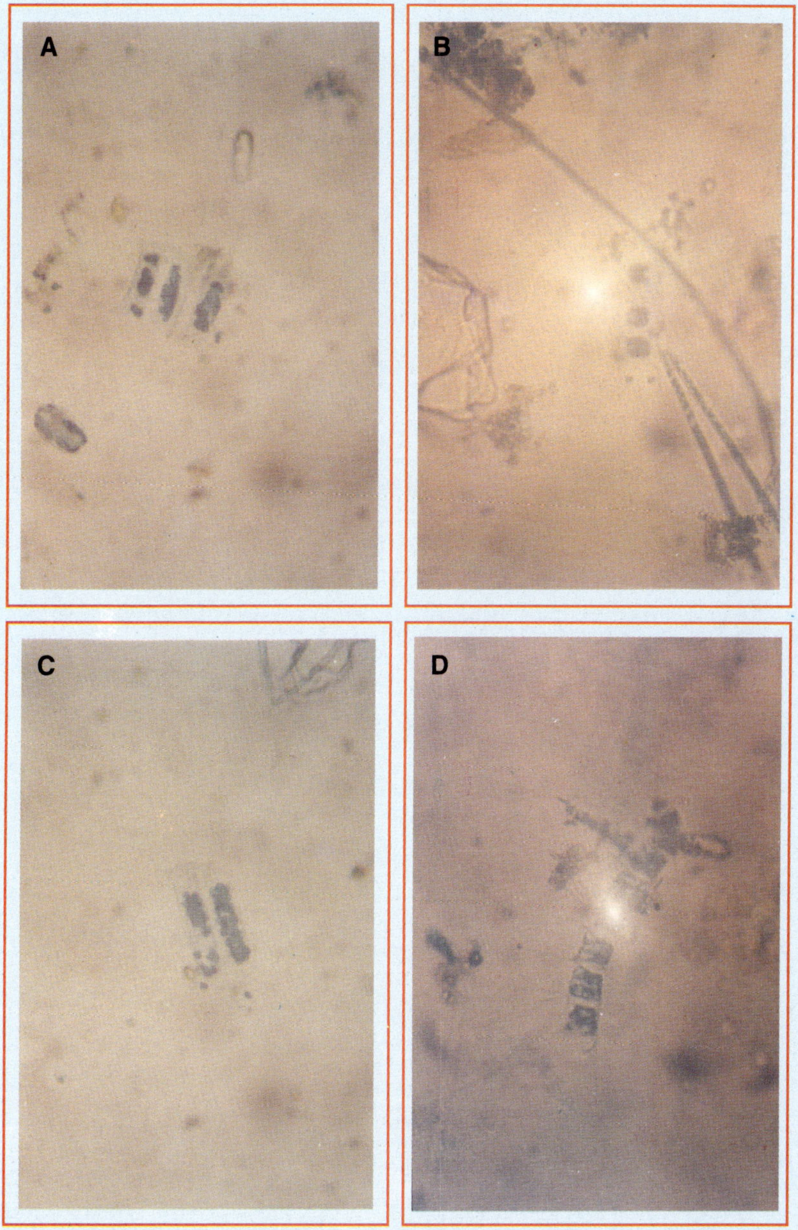

Plate - XIX : A. *Climacodium frauenfeldianum* Grunow (X 200); B. *Chaetoceros subsecundus* (Grunow) Hustedt (X 200); C. *Climacodium frauenfeldianum* Grunow (X 200); D. *Chaetoceros curvisetus* Cleve (X 200).

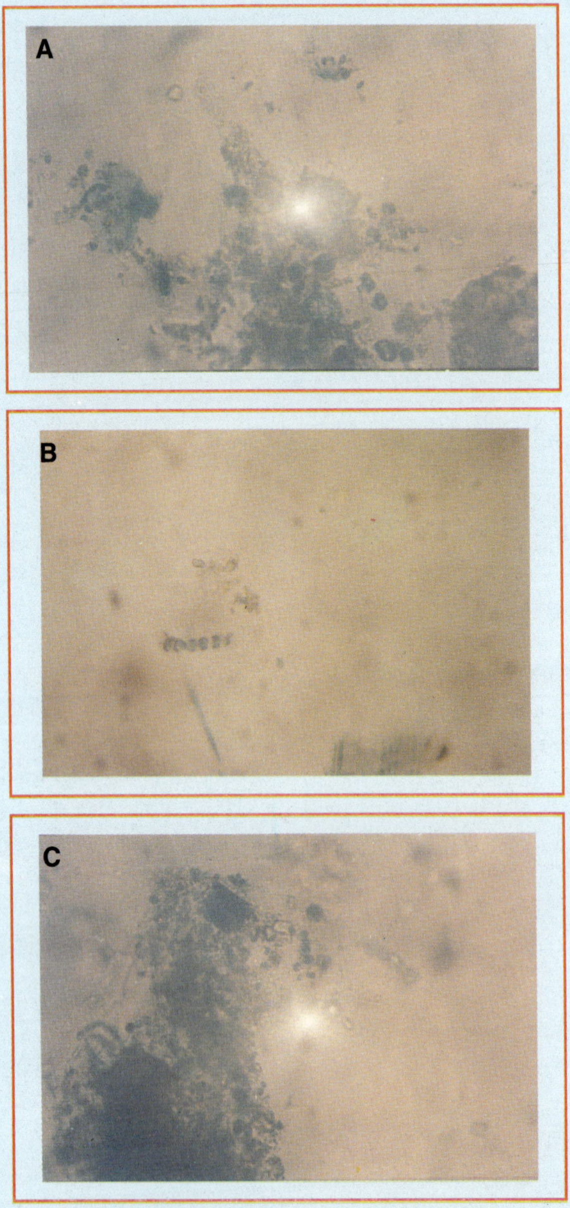

Plate - XX : A. *Bacteriastrum delicatulum* Cleve (X 200);
B. & C. *Bacteriastrum varians* Lauder (X 200).

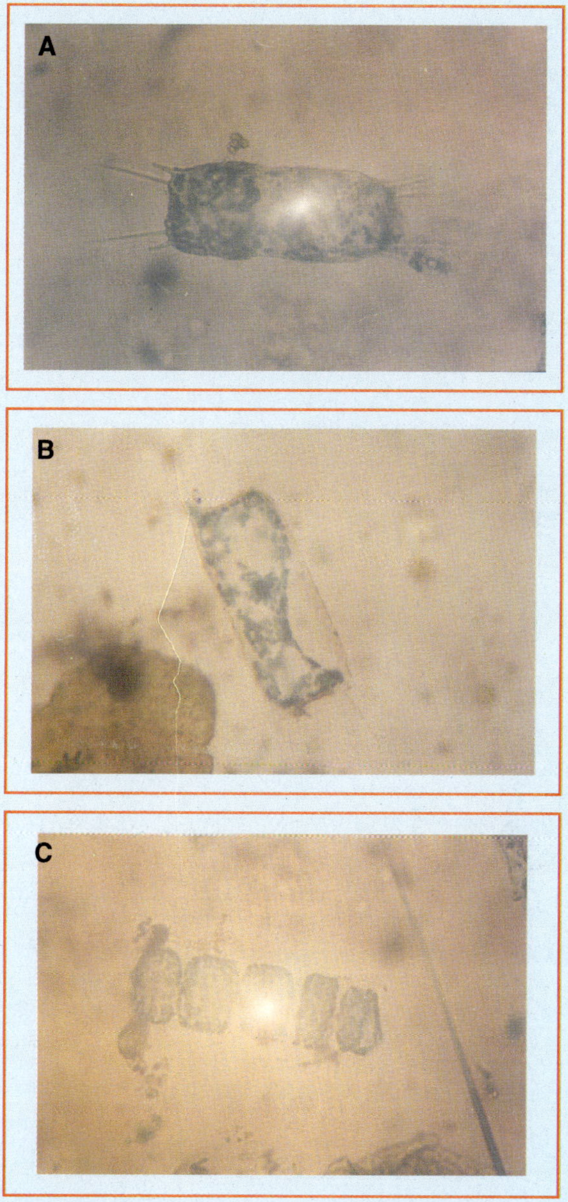

Plate - XXI : A. & B. *Biddulphia sinensis* Greville (X 200); C. *Biddulphia mobiliensis* Bailey (X 200).

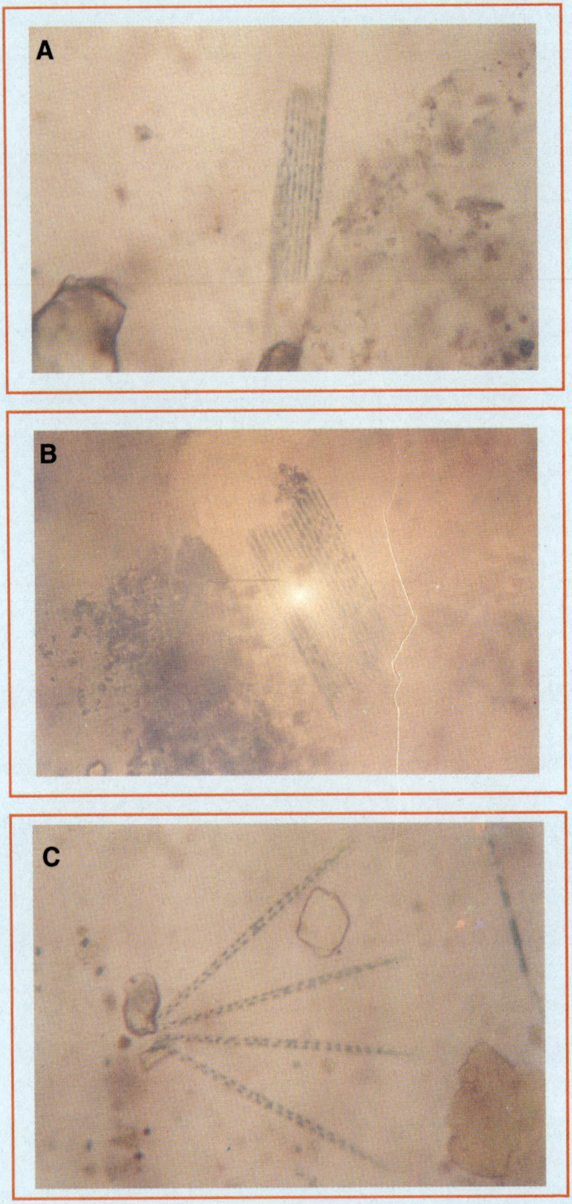

Plate - XXII : A & B. *Fragilaria vaucheriae* (Kutz.) Peterson, A. Cl. (X 200); C. *Asterionella japonica* Cleve (X 200).

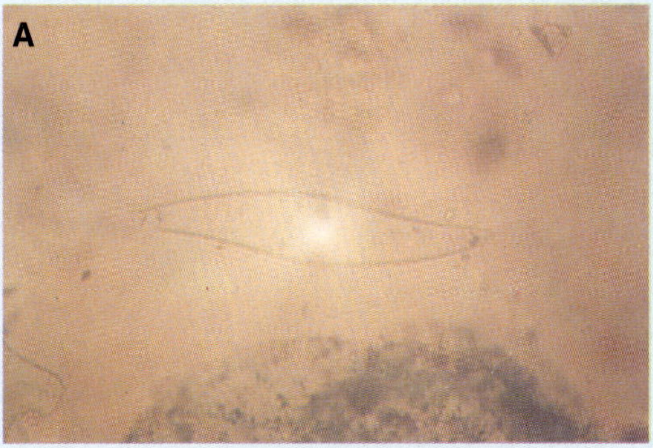

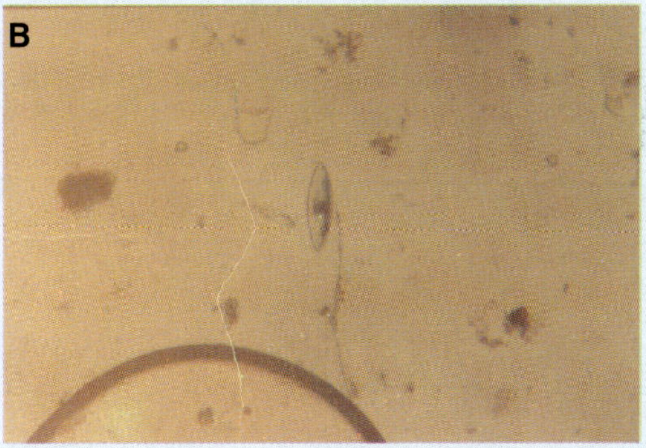

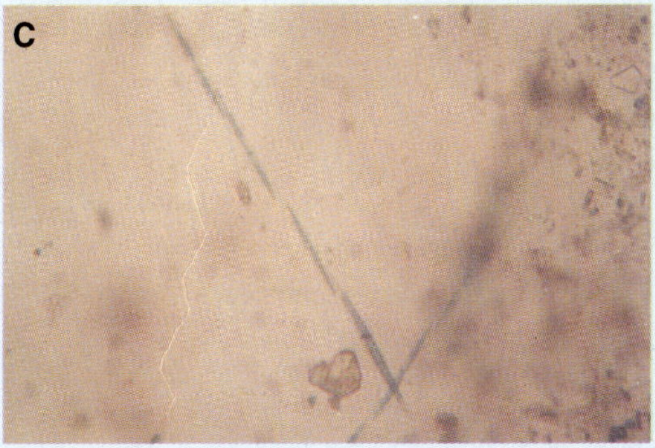

Plate - XXIII : A. *Gyrosigma acuminatum* (Kutz.) Rabenh. (X 200);
B. *Navicula cryptocephala* Kutz., C. *Nitzschia acicularis* W. Sm. (X 200).

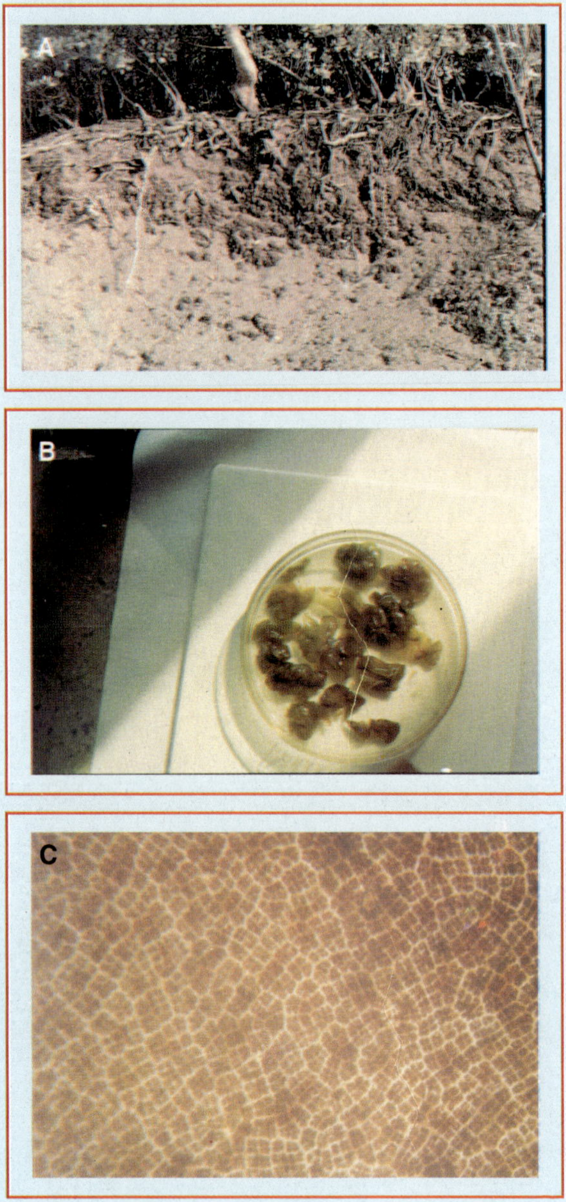

Plate - XXIV : A. *Colpomenia sinuosa* (Roth.) Derb. & Sol. On soil in the upper littoral zone of mangrove forests in Sundarbans, W. B.; B. *Colpomenia sinuosa* (Roth.) Derb. & Sol.; Superficial cells of *Colpomenia sinuosa* (Roth.) Derb. & Sol.

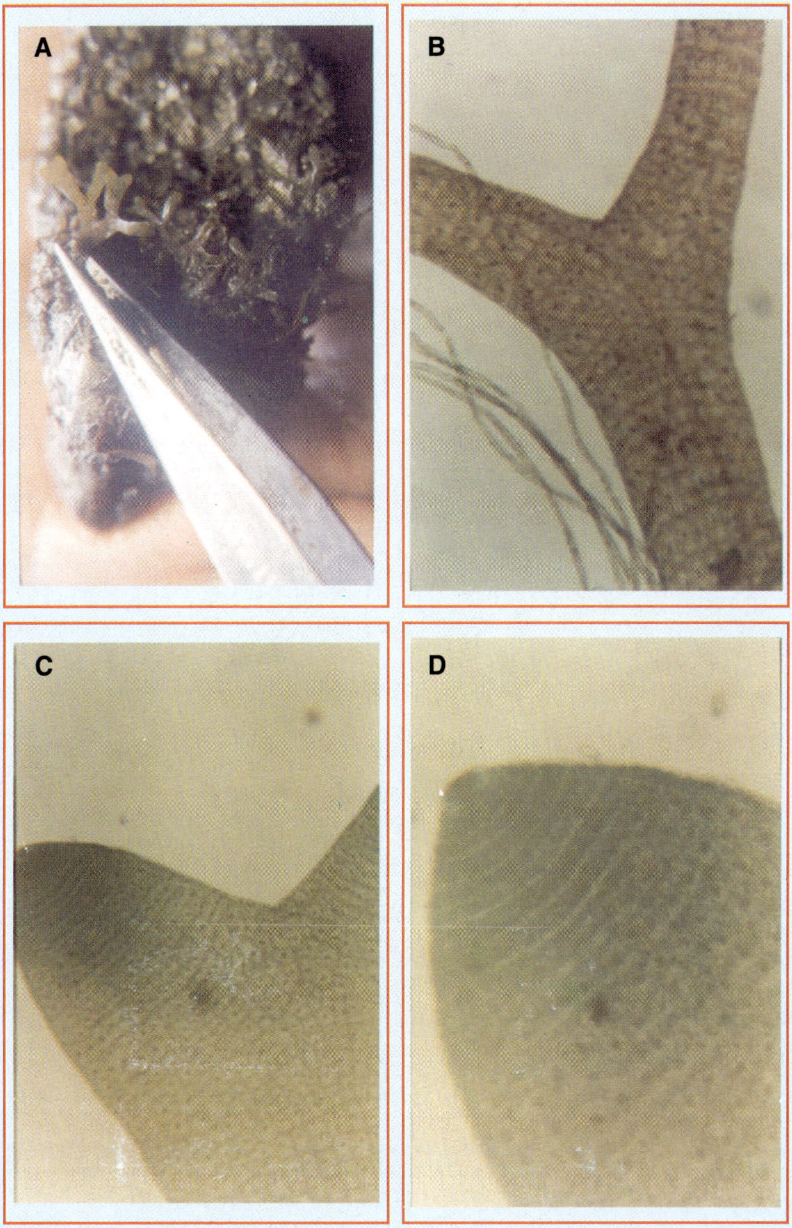

Plate - XXV : A. Thallus of *Dictyota ceylanica* Kutzing; B. Cellular detail of *Dictyota ceylanica* Kutzing showing bifurcation (X 20); C. & D. Tip of the thallus of *Dictyota ceylanica* Kutzing (X 20).

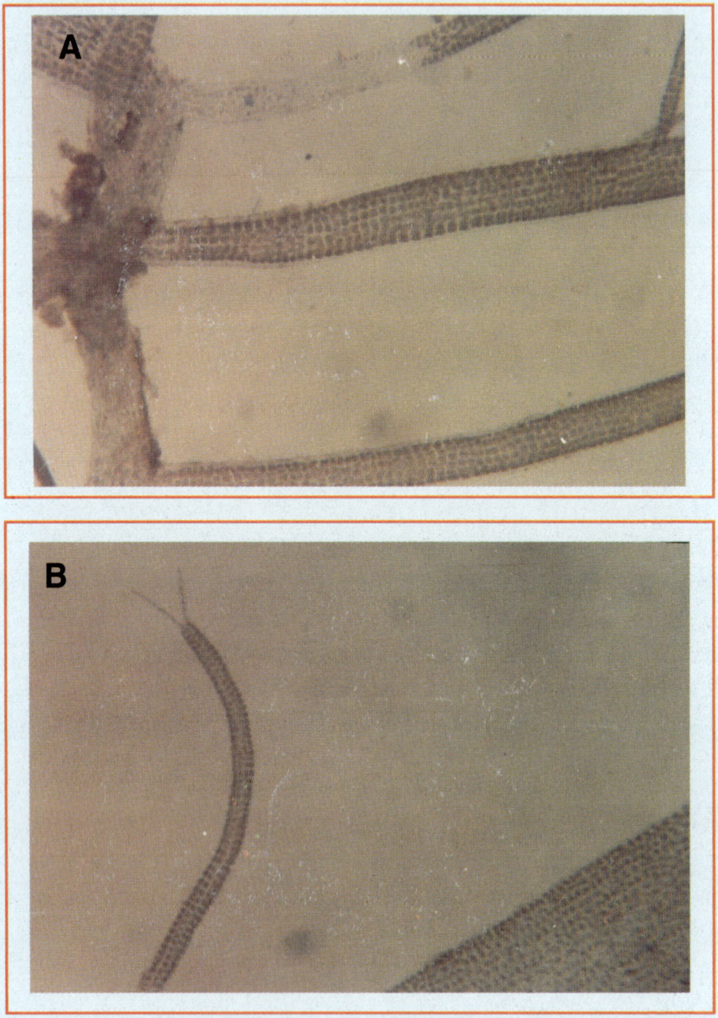

Plate - XXVI : A. *Compsopogon coeruleus* (Balbis) Montagne (X 20); B. Tip of a branchlet of *Compsopogon coeruleus* (Balbis) Montagne (X 20).

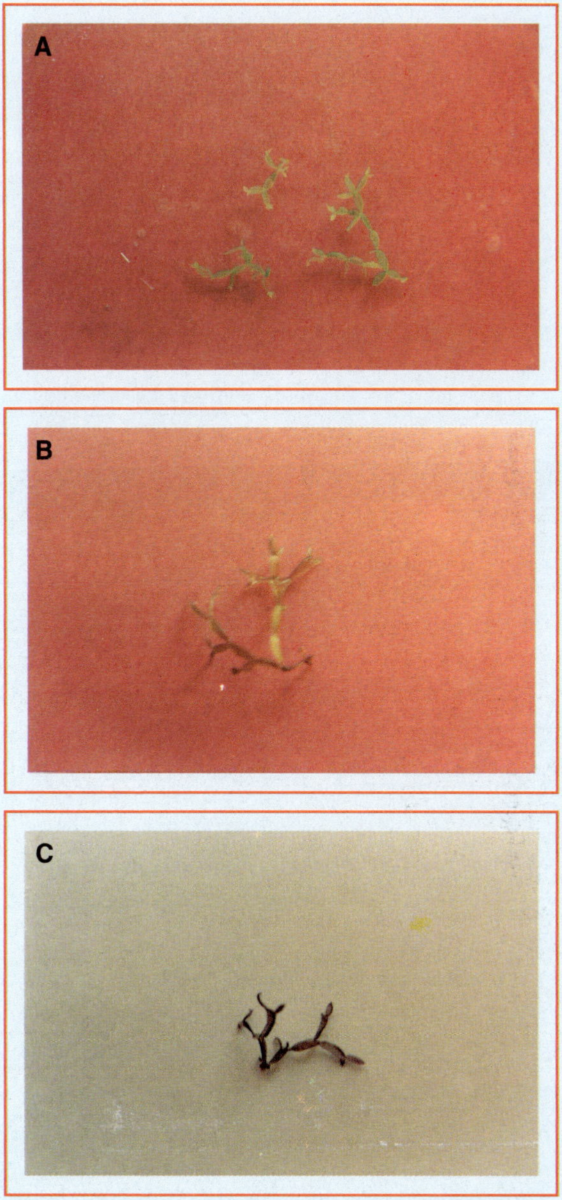

Plate - XXVII : A. *Catenella repens* (Lightfoot) Batters; B. *Catenella impudica* (Mont.) J. Ag.; C. *Catenella nipae* Zanardini.

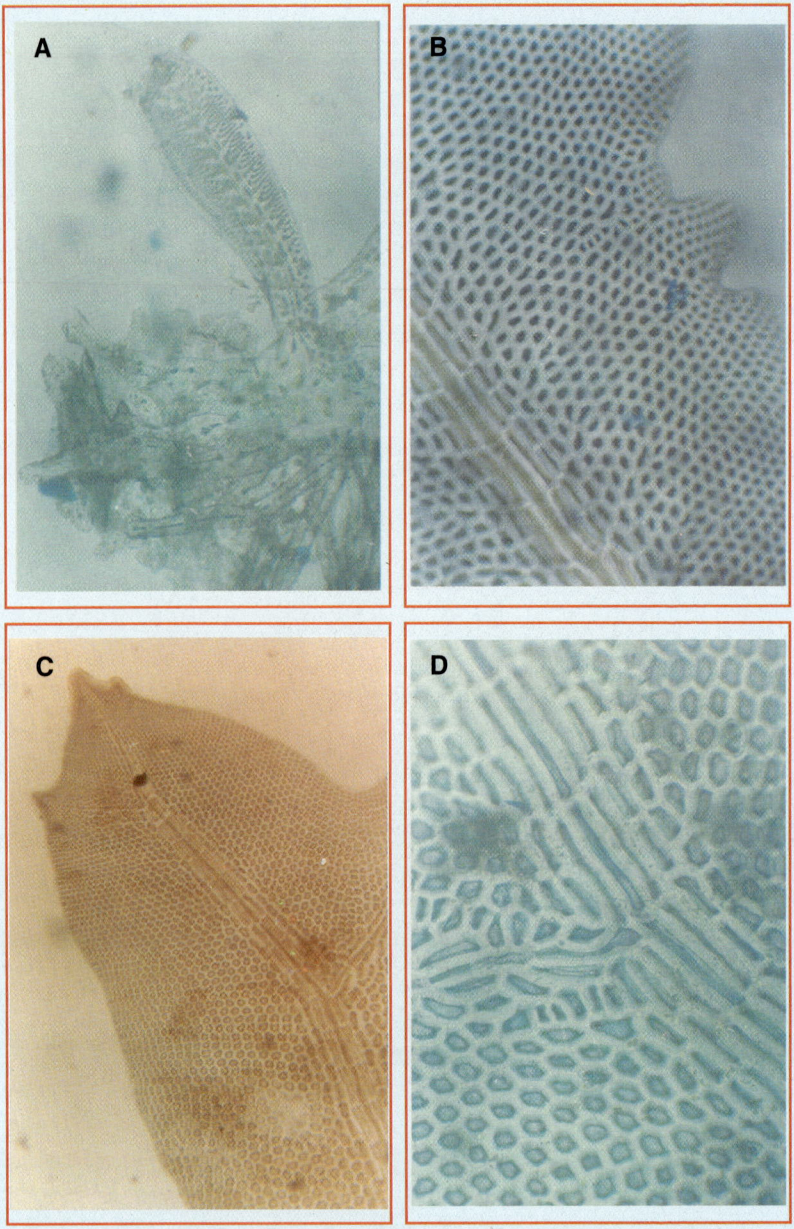

Plate - XXVIII : A. *Caloglossa leprieurii* (Montagne) J. Ag. (X 20); B. *Caloglossa leprieurii* (Montagne) J. Ag. (X 50); C. *Caloglossa adnata* (Zanardini) De Toni (X 20); D. *Caloglossa adnata* (Zanardini) De Toni (X 50).

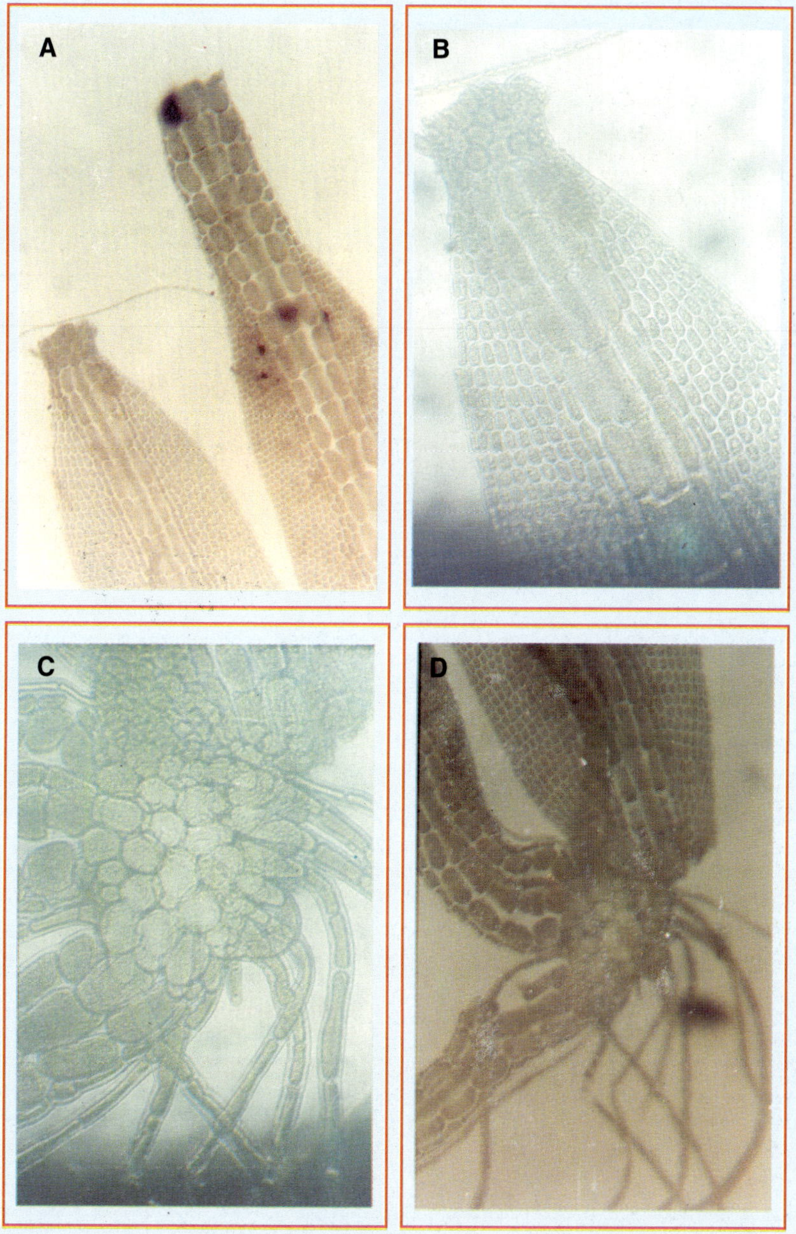

Plate - XXIX : A. *Bostrychia radicans* (Montagne) Mont. (X 20); B. *Bostrychia radicans* (Montagne) Mont. (X 50); C. Cortical cells at the base of *Bostrychia radicans* (Montagne) Mont. (X 50); D. Base of the thallus of *Bostrychia radicans* (Montagne) Mont. (X 20).

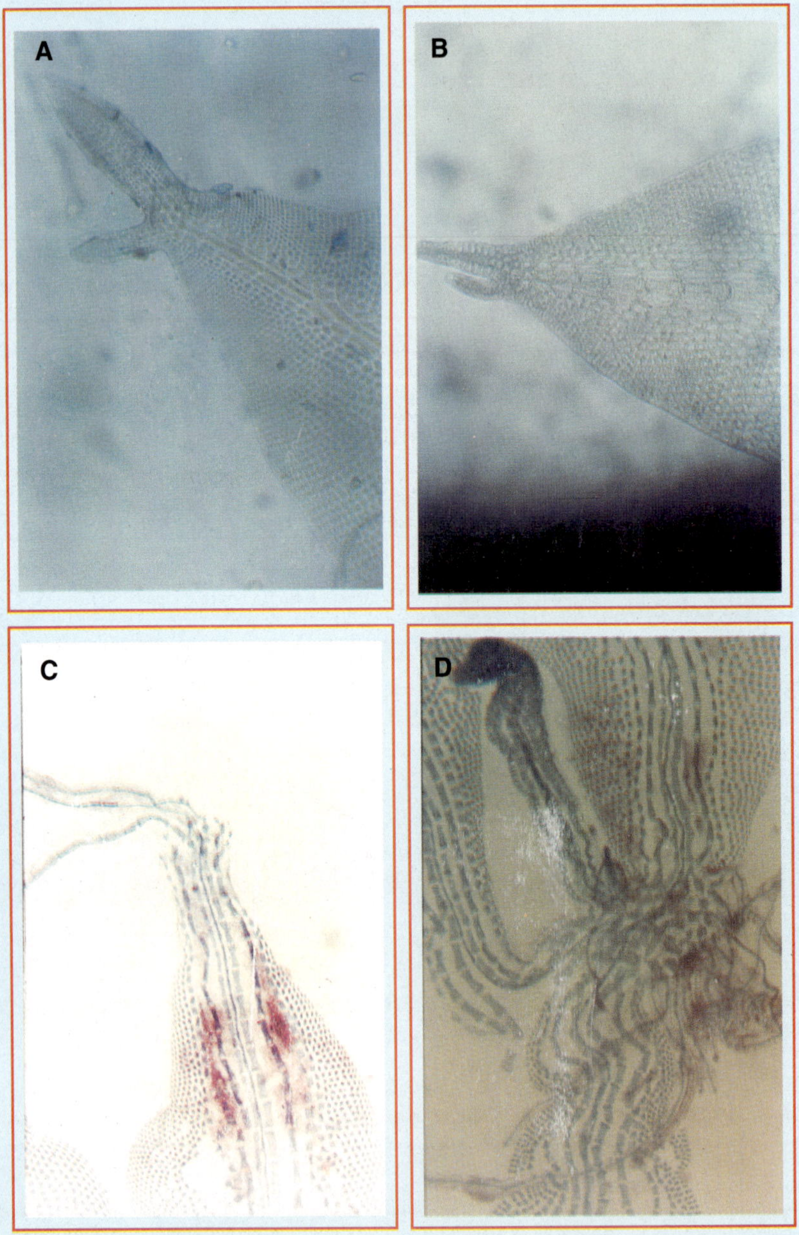

Plate - XXX : A. & B. *Bostrychia tenella* (Vahl.) (X 20); C. Tip of the thallus of *Bostrychia tenella* (Vahl.) (X 20); D. Base of the thallus of *Bostrychia tenella* (Vahl.) (X 20).

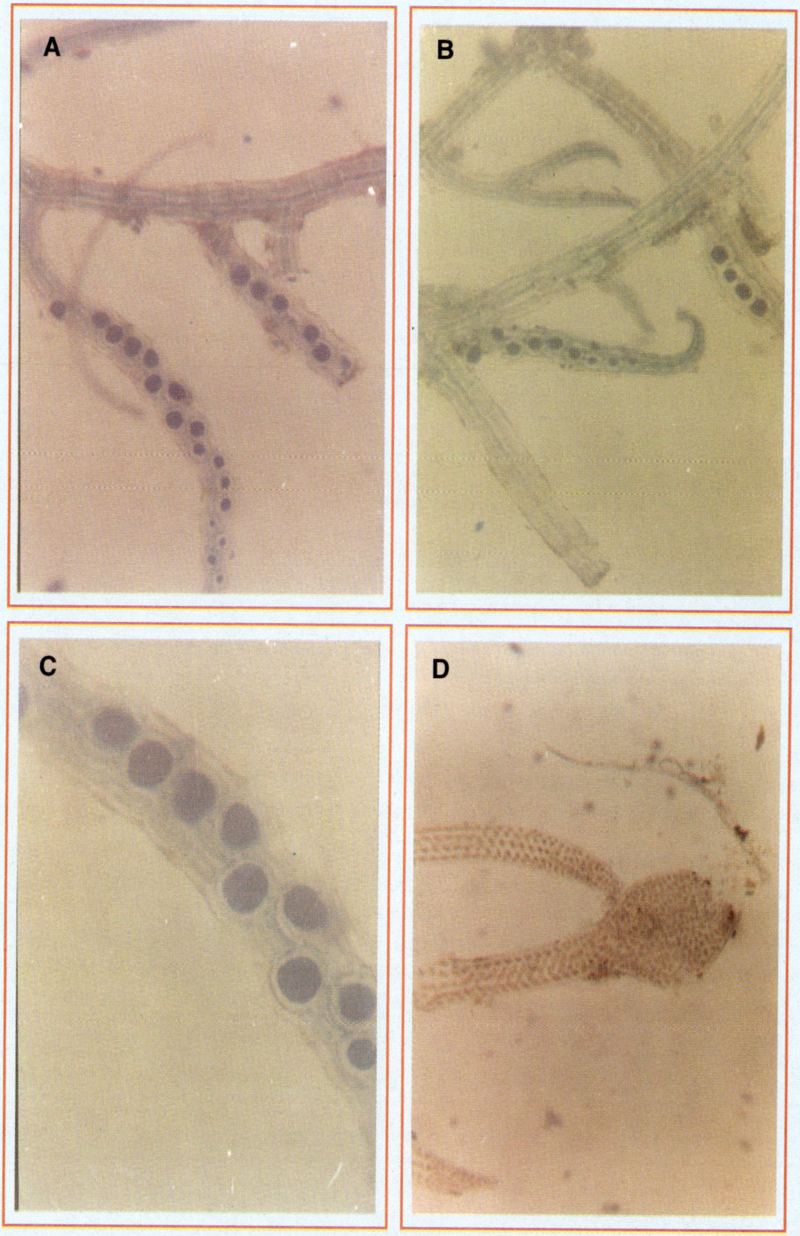

Plate - XXXI : A. Tetrasporophyte of *Polysiphonia mollis* Hooker & Harvey (X 20); B. *Polysiphonia mollis* Hooker & Harvey (X 20); C. Tetrasporophyte of *Polysiphonia mollis* Hooker & Harvey (X 20); D. Carposporophyte of *Polysiphonia mollis* Hooker & Harvey (X 20).

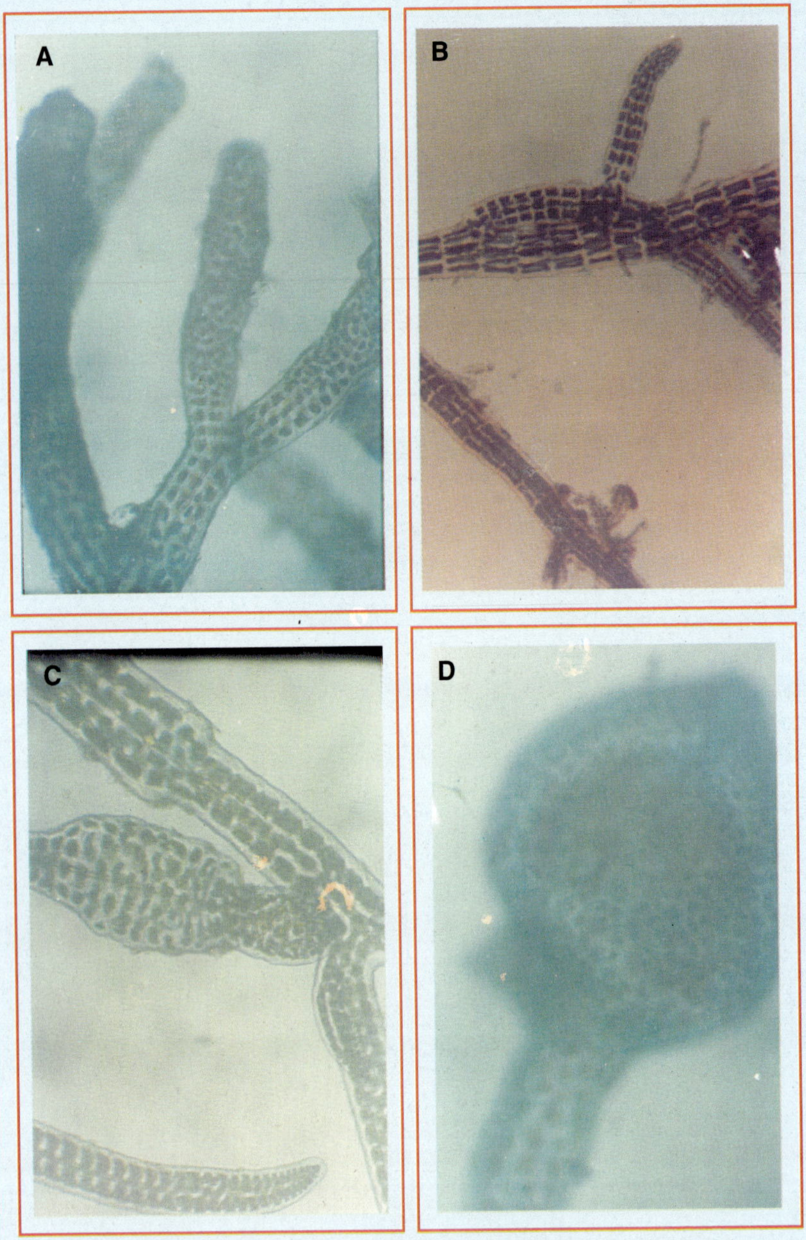

Plate - XXXII : A., B. & C. *Polysiphonia denudata* (Dillwyn.) Kutzing (X 20); D. Carposporophyte of *Polysiphonia denudata* (Dillwyn.) Kutzing (X 20).

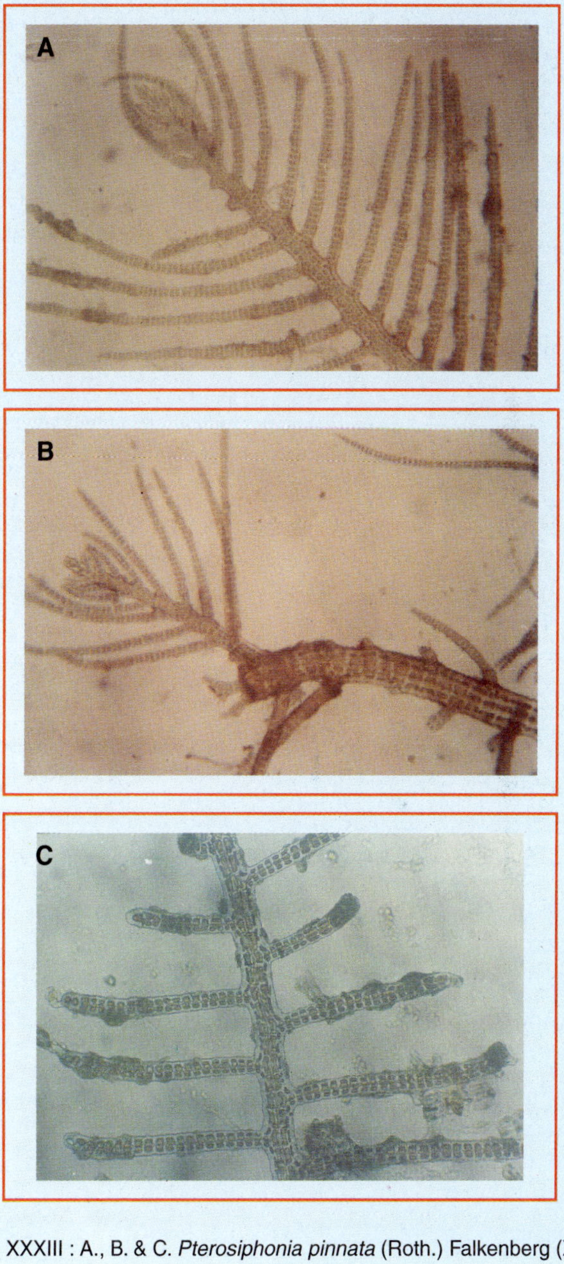

Plate - XXXIII : A., B. & C. *Pterosiphonia pinnata* (Roth.) Falkenberg (X 20).

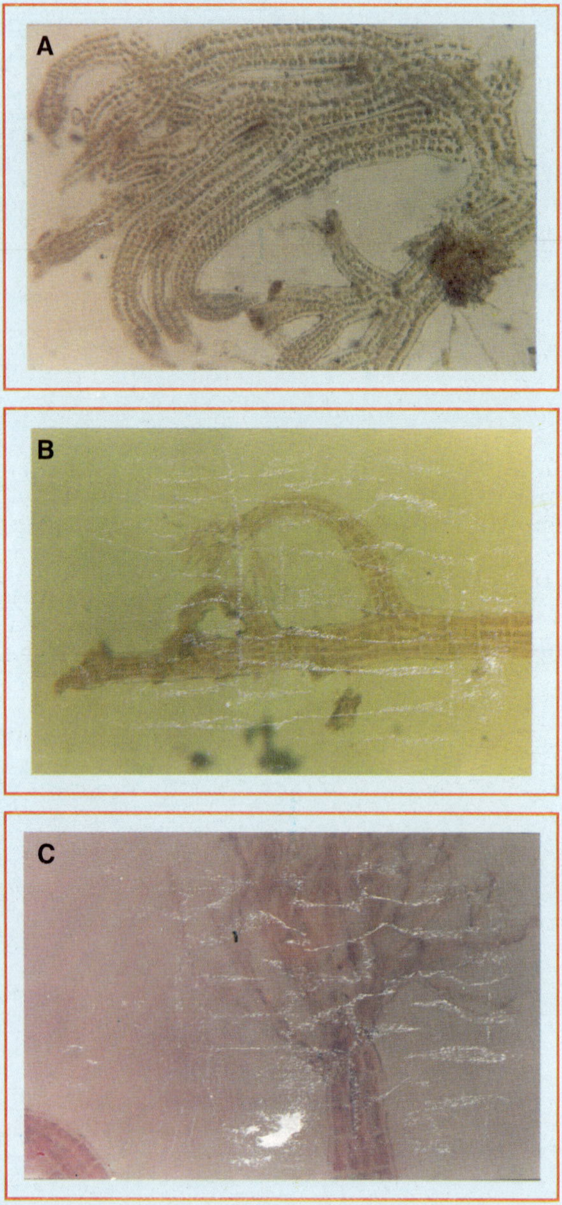

Plate - XXXIV : A. *Heterosiphonia* sp. (X 20); B. *Herposiphonia dendroidea* var. *Minor* Hollenberg (X 20); C. Trichoblast of *Herposiphonia dendroidea* var. *minor* Hollenberg (X 50).

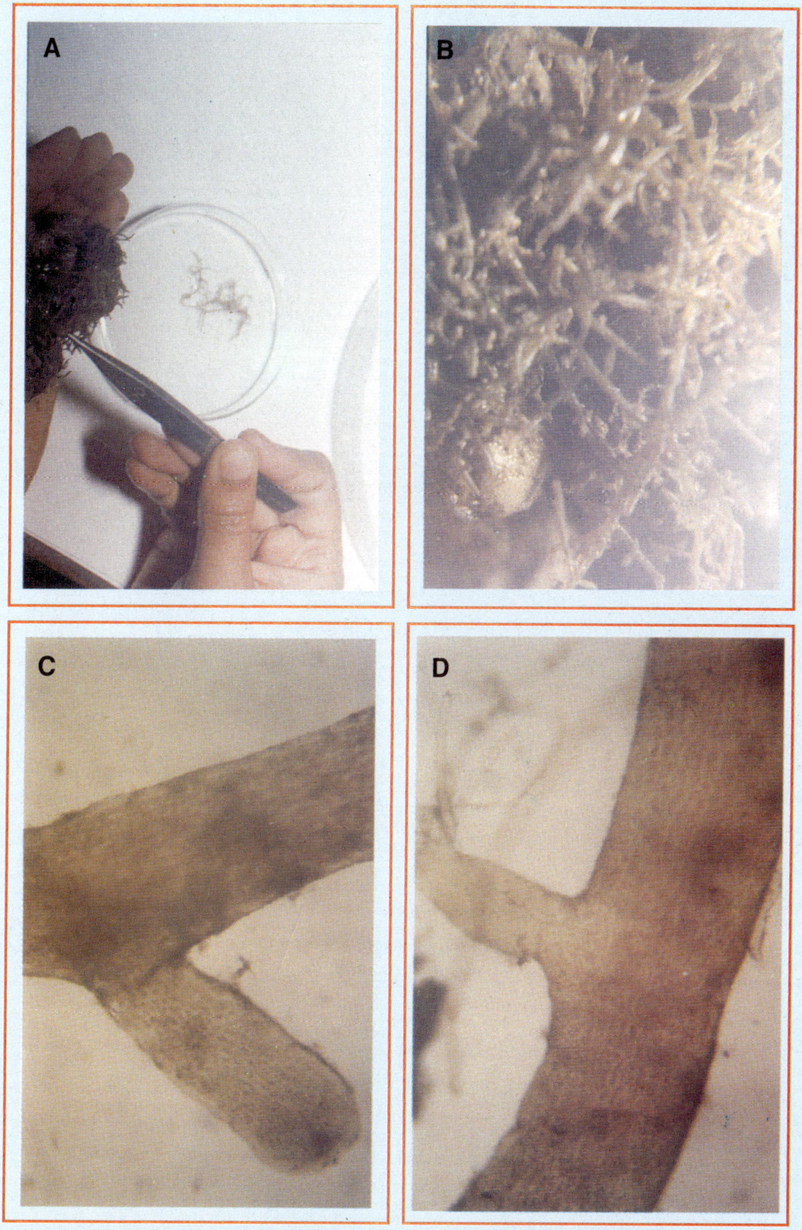

Plate - XXXV : A. & B. *Gelidiella acerosa* (Forsskal) Feldmann et Hamel;
C. Branchlet of *Gelidiella acerosa* (Forsskal) Feldmann et Hamel (X 20);
D. Haptera of *Gelidiella acerosa* (Forsskal) Feldmann et Hamel (X 20).

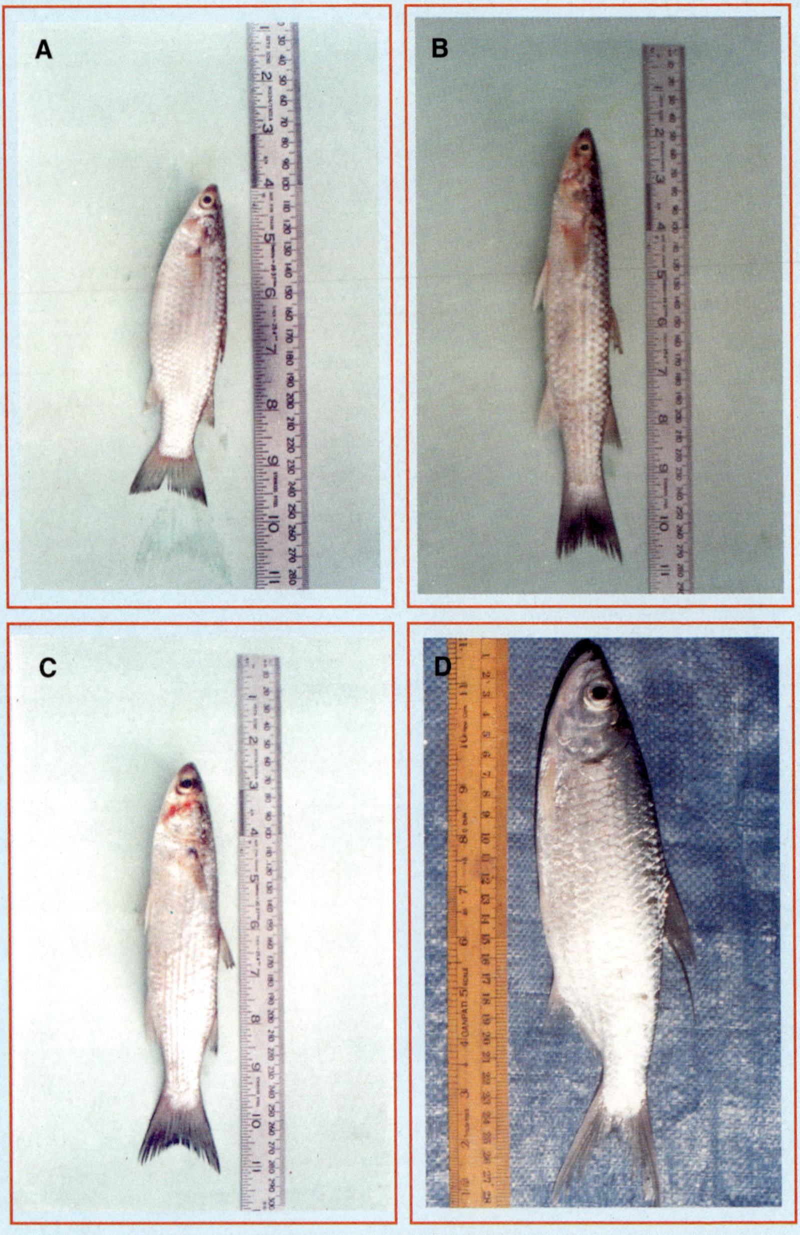

Plate - XXXVI : Fin-fish species collected from Indian Sundarbans and studied for Gut-Content Analysis:
A. *Liza parsia;* B. *Liza tade;* C. *Mugil cephalus;* D. *Valamugil cunnesius*

Plate - XXXVI : Fin-fish & Shell-fish species collected from Indian Sundarbans and studied for Gut-Content Analysis :
A. *Scatophagus argus;* B. *Penaeus monodon;* C. *Uca* sp.; *D. Telescopium* sp.

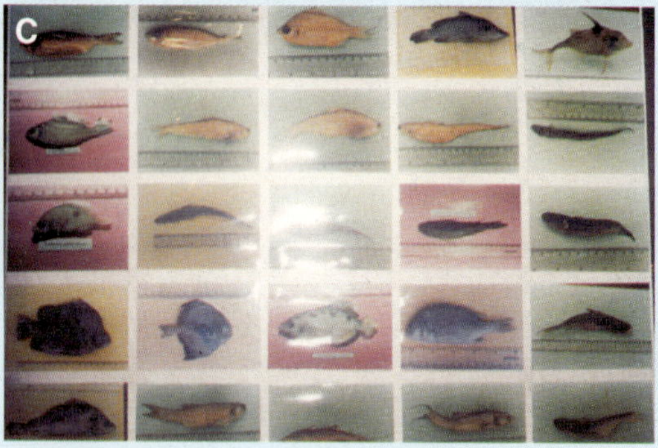

Plate - XXXVIII : A. Algal species collected from the Indian Sundarbans mangals. B. Ecological investigation on the algal habitats in the laboratory. C. Fish species of economic importance collected from the Indian Sundarbans.

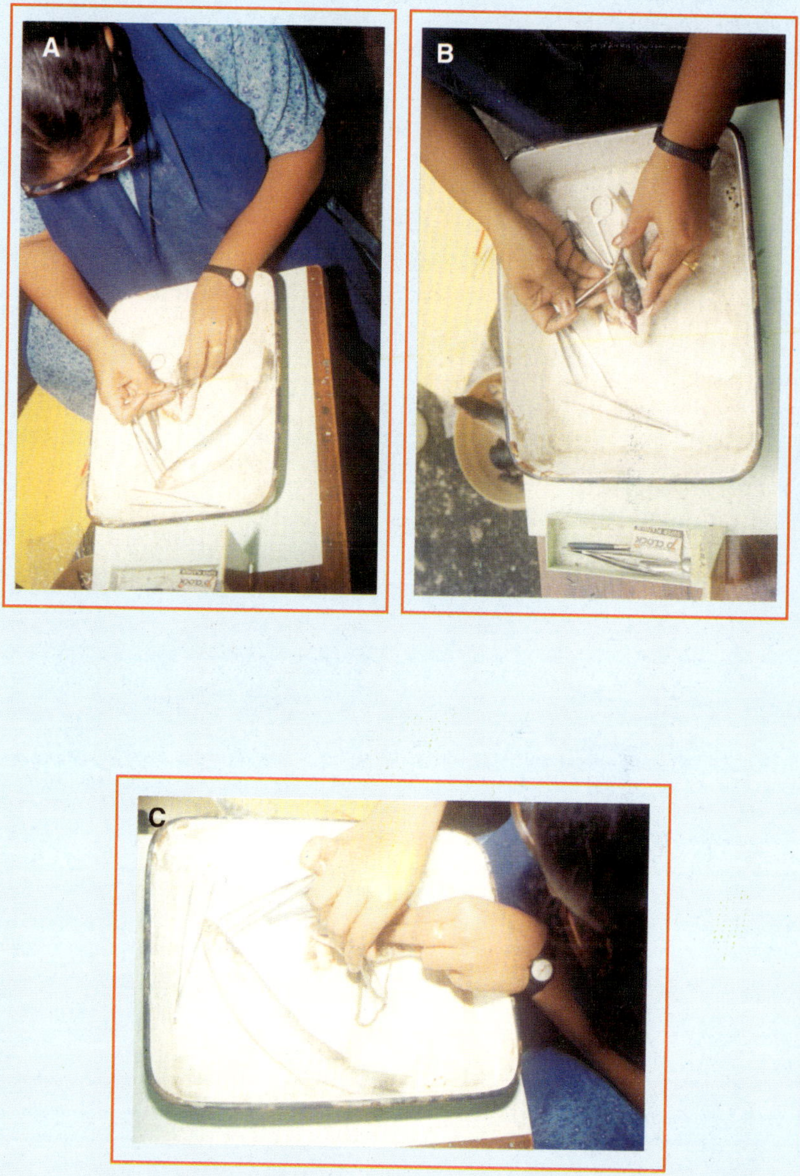

Plate - XXXIX : Gut Content Analysis of Fin-fish Species in the Laboratory.

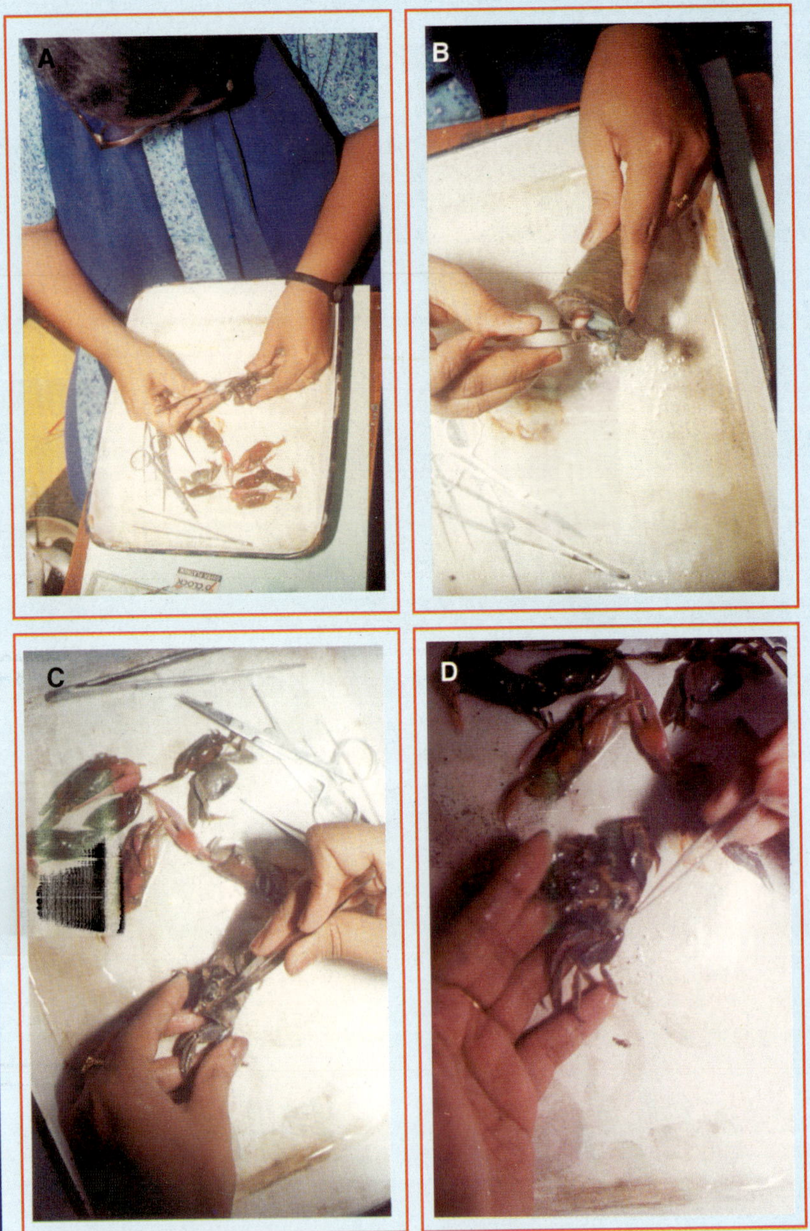

Plate - XL : Gut Content Analysis of Shell-fish Species in the Laboratory.

131. ***Nitzschia obtusa*** **W. Sm.**

Hustedt, Osterreichs und der Schweiz, p. 422, fig. 817d, 1930; Venkateswarlu, Biblio. Phyco., p. 37, pl. IV, fig. 70, 1983.

Valves 115 - 120μ long, 9μ - 10μ broad, keel punctate 7 - 8 in 10μ striae 29-30.

Habitat: Planktonic from brackish water fishery at Minakhan, Sundarbans, 24 Parganas (N), West Bengal.

Specimen No.: N.S. 50 B. (Plate XX, Fig. 131).

132. ***Nitzschia sublinearis*** **Hustedt**

Hustedt, Osterreichs und der Schweiz, Bacill., p. 411, fig. 786,1930; Venkateswarlu, Biblio. Phyco., pl. IV, fig. 73, 1983.

Valves of length 45μ - 48.5μ, 3μ -4.8μ broad, keel punctate 9-10μ in 10μ, striae 18-20 in 10μ.

Habitat : Plankton from Harwood Point, Sundarbans, 24 Parganas, West Bengal.

Specimen No. : N.S. 79. (Plate XX, Fig 132).

133. ***Nitzschia acicularis*** **W. Sm.**

Hustedt, Osterreichs und der Schweiz, Bacil., p. 423, fig, 821, 1930; Venkateswarlu, Biblio. Phyco., pl. IV, fig. 64, 1983; Suxena, Biblio. Phyco., pl. VII, fig. 143, 1983.

Valves 65μ -75μ long, 2μ -2.5μ broad, keel punctate 18-19 in 10μ.

Habitat : Plankton from Haroa, Minakhan and Golabari, Sundarbans, 24 Parganas, West Bengal.

Specimen No. : N. S. 50A, 50B, (Plate XX, Fig. 133). Photo Plate XXIII.

(b) PHAEOPHYTA

Key to the Orders

1. Plants filamentous, heterotrichous, crustose, cushionform, cylindrical or vesicular; thalli parenchymatous, or pseudoparenchymatous, growth diffuse **- Ectocarpales**
2. Plants thalloid, possess a basal attaching organ, internally little differentiated, upper portion richly branched, internally parenchymatous **- Dictyotales**

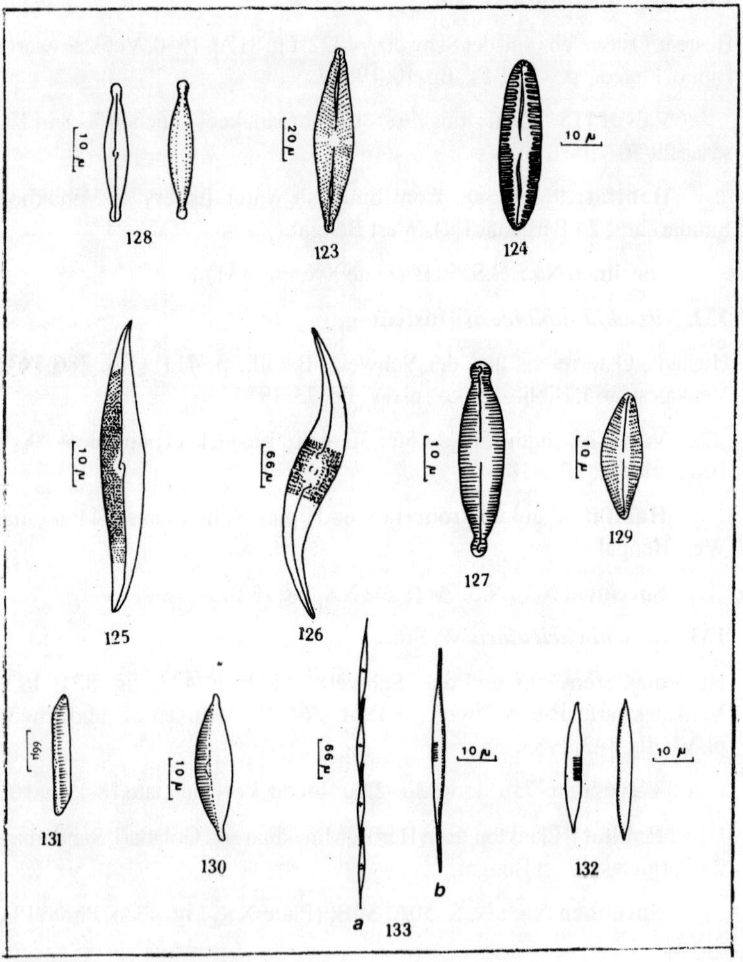

Plate XX

123. *Stauroneis phoenicenteron* (Nitzsch.) Ehr. ; 124. *Pinnularia viridis* (Nitzsch.) Ehr.; 125. *Pleurosigma angulatum* (Quekett.) Wm. Smith; 126. *Gyrosigma acuminatum* (Kütz.) Rabenh.; 127. *Gomphonema sphaerophorum* Ehr.; 128. *Anomoeneis exilis* (Grun.) Cleve; 129. *Cymbella ehrenbergii* Kütz.; 130. *Amphora veneta* Kütz.; 131. *Nitzschia obtusa* W. Sm.; 132. *Nitzschia sublinearis* Hustedt; 133. *Nitzschia acicularis* W. Sm.

Order-ECTOCARPALES

Family - ENCOELIACEAE

Plants vesicular, hollow or cylindrical when mature, dichotomously or irregularly branched, thallus parenchymatous, differentiated into medulla and cortex, sporangia associated with hair, cryptostomata present.

Colpomenia Derbès et Solier

Compt. Rend. Acad, Sci. 1: 11, 1856.

Plants sinuous, vericular, irregularly lobed, hollow, even surface, and internally composed of 5-6 layers of cells, cells of outermost layer polygonal and closely packed, inner cells become larger inwardly, thin walled, nearly round in shape-cryptostomata abundant, with numerous colourless hair, plurilocular sporangia scattered over the surface, sporangia cylindrical, mostly uniseriate associated with periphyses and hair.

134. *Colpomenia sinuosa* (Roth.) Derb & Sol.

Compt. Rend. Acad. Sci., 1:11, pl. 22, figs. 18 - 20, 1856; De Toni, Syll. Alg. 3 : 489, 1895; Barton, J. Linn. Soc. Bot. 35 (247), 1903: 479; Okamura, Icon. Japan, Alg. 1: 86, pl. 19 figs. 11- 12, pl. 20, figs. 10-12, 1907; Boergesen, Dansk. Bot. Arkiv., 2 : 176, Fig. 138, 1914a; Setchell & Gardner, Univ. alif. Pub. Bot. 8 : 539, pl. 45, figs. 82-86, 1925; Taylor, Carneg. Inst. Wash. 379 : 110, pl. 7, fig. 1, pl. 19, figs. 3-4, 1928; Marine Algae of the eastern trop and subtropical coasts of Americas, 260, pl. 36, fig. 1, 1960; Dixit, J. Univ. Bombay, 8(5) : 12, 1940; Fritsch, Struck. Reprod. Agr. 2 : 109, figs. 39 A-D, F, G, I, 1945; Srinivasan, Phyco. Indica, pl. XXVIII, 28 pl. 1, 1969; Islam, Biblio. Phyco., pl. 42, fig. 250, 1976.

Plants sessile, irregularly lobed, sinuous hollow vesicles, several plants growing together closely in small groups or globose masses, vesicles 2 - 10 cm diam., light yellow brown, thallus of large plants often irregularly torn, flattened and expanded, upper layer composed of single layered rectangular epidermal cells 3.8-5μ broad, cryptostomata present, below this 4 -6 layers of large thin walled cells 70 - 145μ diam., multicellular hair given out, on lower side rhizoids present.

Habitat: On soil, substratum mid littoral zone during low tide at Pirkhali deep forest zone, Sundarbans, 24 Parganas, West Bengal.

Specimen No.: N. S. 63. (Plate XXI, Fig. 14a, 134b, 134c, 134d). Photo Plate XXIV.

Order- DICTYOTALES

Family - DICTYOTACEAE

Plants flattened, ribbon shaped, fan - shaped or foliaceous, dichotomously or irregularly branched, thallus parenchymatous, little differentiated.

***Dictyota* Lamouroux**

Nouv. Bull. Sci. Soc. Philom. 1: 331 ; 1809b, J. bot. 2 : 38, 1809a.

Plants erect, flat, ribbon-like, more or less bushy, attached to the substratum by means of a small cuneate basal rhizomatous disc, upper portion of thallus richly brand dichotomously or pinnately, branches narrow or slightly broader, margins entire or dentate, apices more or less broad, possessing a broad lenticular apical cell. Thallus two layered, outer layer consisting of small thin - walled cells and middle layer consists of large thick cells.

135. *Dictyota ceylanica* Kützing

Tab. Phycol. 9 : pl. 25, fig. 1, 1859; Weber V. Bosse, Siboga Exped. 59a : 185, pl. 3, fig. 7, 1913; Boergesen, Ceylon, J. Sci. A. Bot. 12(2) : 78, fig. 7, 1936; Yamada, Sci. Pap. Inst. Alg. Res. Pac. Sci., Hokkaido Univ. 3(2) : 186, fig. 4, 1950.

Greenish brown membranous thalloid algae upto 4 cm tall, spreading or creeping on the substratum, upper portion branched dichotomously, branches uniformly broad, about 1 mm in breadth, margins entire, apices broad, rounded. Rhizoids growing frequently from the margins of the branches, thus binding the neighbouring branches. Thallus composed of superficial layers enclosing a middle layer in between, cells of the former small, thin walled, more or less squorish in shape, cells of the middle layer thick walled, squarish to rectangular.

Habitat : Creeping on the inundated forest floor in slightly upper littoral zones at Dhanchi Island, at Jharkhali on pneumatophores along with other algae in the Sundarbans, south 24 Parganas, West Bengal.

Specimen No.: N. S. 84. (Plate XXI, Fig. 135a, 1.35b). Photo Plate XXV.

The above algal species belonging to Chrysophyta and Phaeophyta below according to their systematic position along with their habit and habitat.

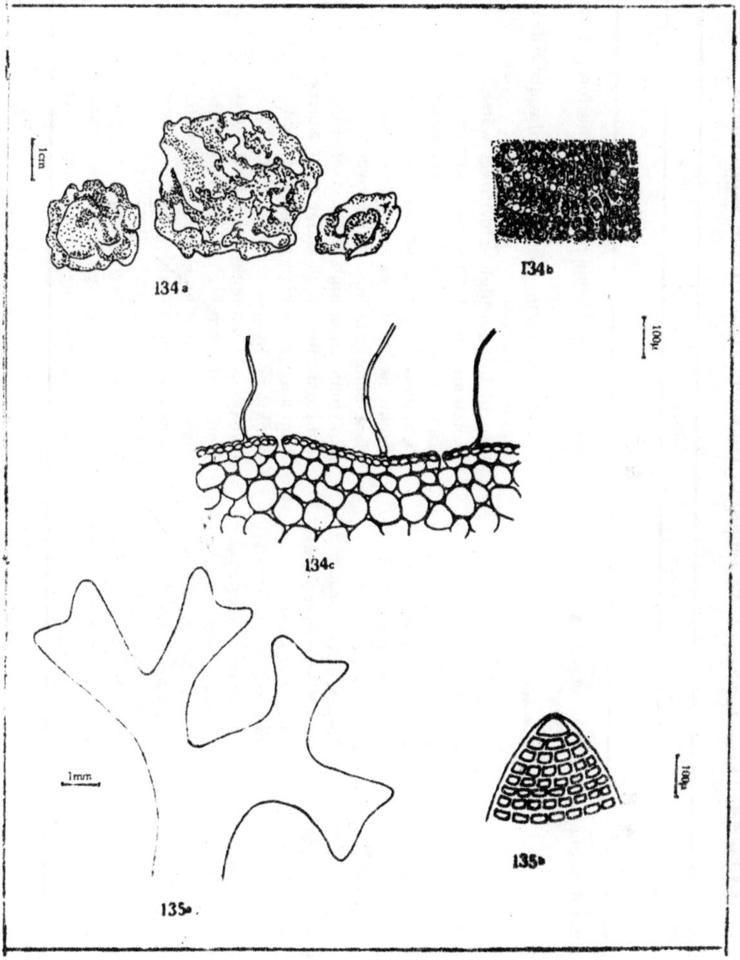

Plate XXI

134a. Habit sketch of *Colpomenia sinuosa* (Roth.) Dertb. & Sol. ; 134b. Superficial cells of *Colpomenia sinuosa* (Roth.) Derb. & Sol.; 134c. T. S. thallus of *Colpomenia sinuosa* (Roth.) Derb. & Sol.; 135a. Thallus of *Dictyota ceylanica* Kützing; 135b. Cellular detail of *Dictyota ceylanica* Kützing.

Table 18 : Habiat of Identified Chrysophyta & Phaeophyta Members From Indian Sundarbans.

Class, Order & Sub Order	*Family*	*Name of the Speies*	*Habitat & Habitat*
CHRYSOPHYTA			
Xanthophyceae	Heteroisphonales	Vaucheria pressotti	Dense mat on decomposed matter on forest floor in high saline area.
		Vaucheria sp.	Among the bottom biota of brackishwater fishery at mid saline area.
Bacillariophyceae		*Melosira moniliformis*	Planktonic from high saline river water.
Biddulphiales	Melosiraceae	*Melosira sol*	Planktonic from high saline river water.
Coscinodiscineae		*Stephanopyxis palmeriana*	Planktonic from high saline river water.
	Thalassiosiraceae	*Cylotella glomerata*	Planktonic from high saline river water.
		Coscinodiscus excentricus	Planktonic from high saline river water.
	Coscinodiscaceae	*Coscinodiscus gigas*	Planktonic from high saline river water.
		Coscinodiscus granii	Planktonic from high saline river water.
	Hemidiscaceae	*Hemidiscus cuneiformis*	Planktonic from high saline river water.
	Leptocylindraceae	*Corethron hystrix*	Planktonic from high saline river water.
Rhizosoleniineae		*Rhizosolenia imbricata*	Planktonic from high saline river water.
	Rhizosoleniaceae	*Rhizosolenia setigera*	Planktonic from high saline river water.
Biddulphineae	Hemiaulaceae	*Climacodium*	Planktonic from high saline river water.
		frauenfeldianum	Planktonic from high saline river water.
	Biddulphiaceae	*Biddulphia mobiliensis*	Planktonic from high saline river water.
		Biddulphia sinensis	Planktonic from high saline river water.
		Chaetoceros curvisetus	Planktonic from high saline river water.
		Chaetoceros flexuosus	Planktonic from high saline river water.
		Chaetoceros laciniosus	Planktonic from high saline river water.

Contd...

Class, Order & Sub Order	*Family*	*Name of the Speies*	*Habitat & Habitat*
	Chaetocerotaceae	*Chaetoceros tenuissimus*	Planktonic from high saline river water.
		Chaetoceros subsecundus	Planktonic from high saline river water.
		Bacteriastrum varians	Planktonic from high saline river water.
		Bacteriastrum cosmosum	Planktonic from high saline river water.
Fragilariineae		*Bacteriastrum delicatulum*	Planktonic from high saline river water.
		Fragilaria vaucheriae	Planktonic from high saline river water.
	Fragilariaceae	*Asterionella japonica*	Planktonic from high saline river water.
		Synedra ulna	Planktonic from low-high saline river water/ brackishwater fisheries.
	Diatomaceae	*Diatoma vulgare*	Planktonic from high saline river water.
		Achnanthes microcephala	Epiphytic on red algae in high saline areas.
		Achnanthes minutissima	Epiphytic on red algae in high saline areas.
Bacillariineae		*Coccineis placentula*	Epiphytic on red algae in high saline areas.
		Navicula cryptocephala	Planktonic from high saline river water.
		Navicula radiosa	Planktonic from low-high saline river water/ brackishwater fisheries.
		Anomoeoneis exilis	Planktonic from high saline river water.
		Stauroneis phoenicenteron	Planktonic from high saline river water.
	Naviculaceae	*Pinnularia virdis*	Planktonic from low-high saline river water/ brackishwater fisheries.
		Pleurosigma angulatum	Planktonic from high saline river water.
		Gyrosigma acuminatum	Planktonic from low-high saline river water/ brackishwater fisheries.
		Gomphonema sphaerophorum	Planktonic from high saline river water.

Contd...

Class,Order & Sub Order	Family	Name of the Speies	Habitat & Habitat
	Cymbellaceae	*Amphora veneta*	Planktonic from high saline river water.
		Cymbella ehrenbergii	Planktonic from high saline river water.
		Nitzschia acicularis	Planktonic from low-high saline river water/ brackishwater fisheries.
		Nitzschia obtusa	Planktonic from low-high saline river water/ brackishwater fisheries.
		Nitzschia sublinearis	Planktonic from low-high saline river water/ brackishwater fisheries.
PHAEOPHYTA			
Dictyotales	Dictyotaceae	*Dictyota ceylanica*	Periphytic on pneumatophores alongwith other red algae at high saline areas.
Ectocarpales	Encoceliaceae	*Colpomenia sinuosa*	On highly consolidated soil substratum in the littoral zone at high saline areas.

6.5 DIVISION-RHODOPHYTA

The red algae (Rhodophyta) owe their characteristic red colour to the pigment r-phycoerythrin in their chromatophores. The red algae prefer habitats that support a rhizoidal system in marine, intertidal and sublittoral regions. They prefer to grow on hard substrata in the forms of rocks and well consolidated soil strata. But the Indian part of Sundarbans are devoid of any rocky substratum and this is responsible for the presence of a relatively less number of red algae as compared to the Sundarbans of Bangladesh, where rocky substratum is common, *viz.,* the Chattagram coastal areas at Chakaria Sundarbans.

Table-19 gives an account of the identified Rhodophyta members from the Indian Sundarbans. A total of 15 red algae were collected and identified from the Indian Sundarbans, belonging to 2 sub-class, *viz.,* Bangioideae and Floridae, 3 orders, *viz.,* Compsopogonales, Gigartnales and Gelidiales, 6 families, *viz.,* Erythrotrichiaceae (1 species), Rhabdoniaceae (3 species), Delesseriaceae (2 species), Rhodomelaceae (4 genus, 6 species), Dasyaceae (1 species) and Gelidiaceae (2 genus) (Pie Charts–IIa, IIb, IIIc). All these forms are found as periphytic forms attached to or creeping on the pneumatophores, stilt roots, prop roots, knee roots root butresses, lower trunk regions of mangroves and on bricks or other hard substraum or as littoral forest flora in the sea ward fringes of high saline areas. Only 1 species, *viz., Compsopogon coeruleus* is found attached to submerged twigs and other plant parts in the brackish water areas of low to medium salinity in Indian Sundarbans.

Key to the Sub Classes

1a. Cells not connected by protoplasmic, strands, with axial or parietal chromatophores and generally with pyrenoids *-Bangioideae*

1b. Cells uninucleate to plurinucleate, often connected by a protoplasmic bridge, rarely with axial or stellate chromatophores, with or without pyrenoids *-Florideae*

Sub Class - BANGIOIDEAE

Order - COMPSOPOGONALES

Plants fairly large, branched, ultimately bushy appearance, dull violet to greyish green in colour.

Family - ERYTHROTRICHIACEAE

Plants bushy, often large, greyish to greenish, filaments slender, alternately tapering to uniseriate tips.

Compsopogon Montagne, 1846

Plants filamentous, large, bushy, alternately branched, cells with with numerous small peripheral chromatophores of irregular shapes and several nuclei, older parts of thallus corticated, outer segments divided to form a single superficial layer of small cells.

136. *Compsopogon coeruleus* (Balbis) Montague

Taylor, Papers Michigan Acad. Sci., Arts & lett. 39 : 85 - 109, 1954; Taylor, Marine Algae of the eastern trop. and subtropical coasts of Americas, 1960, p. 296; Santra & Pal, Indian pl. IV, fig. 38 - 39, 1988.

Thallus dark green, much branched, main axis erect, attached to substratum with the help of a basal disc, about 10 cm, long, cells of main axis irregularly arranged, polygonal and of varied sizes, and shapes. Cells of branches (multiseriate and uniseriate) regularly arranged, 5.5-12.5μ long and 7.8-23μ broad. The apical cells of uniseriate branches hemispherical, breadth of multiseriate branches 60-180μ .

Habitat : Attached to twigs submerged at Ramgopalpur, Sundarbans, 24 Parganas, W. Bengal.

Specimen No. : N.S. 3. (Plate XXII, Fig.136a, 136b). Photo Plate XXVI.

Sub Class - FLORIDAE

Key to the Orders

1a. Tetrasporangia at or below the thallus surface **-Gelidiales**

1b. Tetrasporangia scattered just below the thallus surface or in restricted areas on branchlets **-Gigartinales**

1c. Tetrasporangia external or more or less covered by the cortex of ordinary branches **-Ceramiales**

Order -GIGARTINALES

Family-RHABDONIACEAE

Plants cylindrical or flattened, radially or bilaterally branched, medulla, filamentous, cortex of close radial rows of cells, inner large, outer small and rich in chromatophores, tetrasporangia scattered over the thallus in the outer cortex.

Catenella Greville, 1830

Plants creeping, branched, articulate branches of terete or compressed, broadly oval segments, attached to the substrate branches by haptera at intervals, medulla a loose network of slender filaments, cortex with large inner cells and small outer cells.

Key to the species

1a. Haptera formed from normal though somewhat attenuated segments of the thallus -*C. impudica*
1b. Haptera formed as more or less flagellar outgrowths from points of branching, as regular branch segments -*C. repens*
1c. Dichotomously branched, deeply constricted with basal disc like haptera -*C. nipae*

137. *Catenella repens* (Lightfoot) Batters

Feldmann & Lami, Essai d'ecologie et geographie botanique, xiii + 284 pp., 1935; Taylor, 1942; Joly, 1951, 1957; Islam Bangladesh J. Bot., pl. 5, figs. 5-8, photomicrograph A(2), 1973; Islam, Biblio. Phyco., pl. 52, fig. 301-304, 1976; Santra & Pal, Indian Biol., pl. VII, figs. 44-47, 1988.

Thallus deep pink to dull purple, creeping on pneumatophores and barks, dorsiventrally compressed segments, with repent branches, branches ditrichotomous below, axis and branches divided into dorsiventrally compressed ellipsoid to ovate segments, haptera rhizoidal at points of forking and not in the plane of branching of the thallus, pinnate above, each segment about 6-8mm long and 1-2mm broad.

Habitat: Creeping on the bark of *Excoecaria agallocha* at Bakkhali and on pneumatophores of *Avicennia* spp. at Jharkhali, both in the upper littoral regions.

Specimen No.: N. S. 33. (Plate XXII, Fig. 137). Photo Plate XXVII.

138. *Catenella impudica* (Mont.) J. Ag.

Montagne, Ann. Sci. Nat., Bot., iii, 14 : 283 - 309, 1850 ; Möbius, Hedwigia, 28 : 309-347, pls. 10, 1, 1889; Murray, Jour. Bot., Brit. For. 27 : 237-242, 257-262, 298-305, 1889; Post, Rev. Algol, 9:1 - 84, 4 figs, 1936; Taylor Marine

Algae of the eastern trop. and subtropical coasts of Americas, p. 462, 1960; Islam, pl. 52, fig. 305-306, 1976; Santra & Pal, Indian Biol., pl. VIII, fig. 48-51, 1988.

Thallus dull violet coloured, creeping on substratum, 2-4cm tall, ditrichotomously branched, expanded, younger segments slender, terminal ones relatively more acute, older segments broader and flattened, thallus attached by the projections at the terminal points of segments, segments 5- 6 mm long and about 2 mm broad, elliptical to oblong.

Habitat: Creeping on pneumatophores of *Avicennia* spp. and other barks of halophytic plants at Jharkhali, and on bark of *Excoecaria agallocha* at Bakkhali, both in the upper littoral zones.

Specimen : No.: N. S. 35, N. S. 48. (Plate XXII, Fig. 138). Photo Plate XXVII.

139. *Catenella nipae* Zanardini

Islam, Bangladesh J. Botany, pl. 5, fig. 1-4, 1973; Islam, Biblio. Phyco., pl. 52, fig. 307-308, 1976.

Plants stout, robust, 2-3 cm high, each segment 5-8 mm long and about 1.5-2mm broad and swollen, dark purplish pink colour dichotomously branched, deeply constricted with basal disc like haptera, each segment at terminal or subterminal position, thallus creeping on substratum.

Habitat : Attached to bark of *Exoecaria agallocha* in the Upper littoral zones at Bakkhali, Sundarbans, 24 Parganas (South), West Bengal.

Specimen No. : N. S. 33, N. S. 48. (Plate XXII, Fig. 139). Photo Plate XXVII.

Order - CERAMIALES

Key to the Families

1a. Not foliaceaous and rarely with flat branches -2
1b. Foliaceous or, if bushy with flat branches - **Delesseriaceae**
2a. Axes covered with uniseriate, branched chromatophorous filaments - **Dasyaceae**
2b. Axes naked, corticated or covered with branchlets, axes polysiphonous -**Rhodomelaceae**

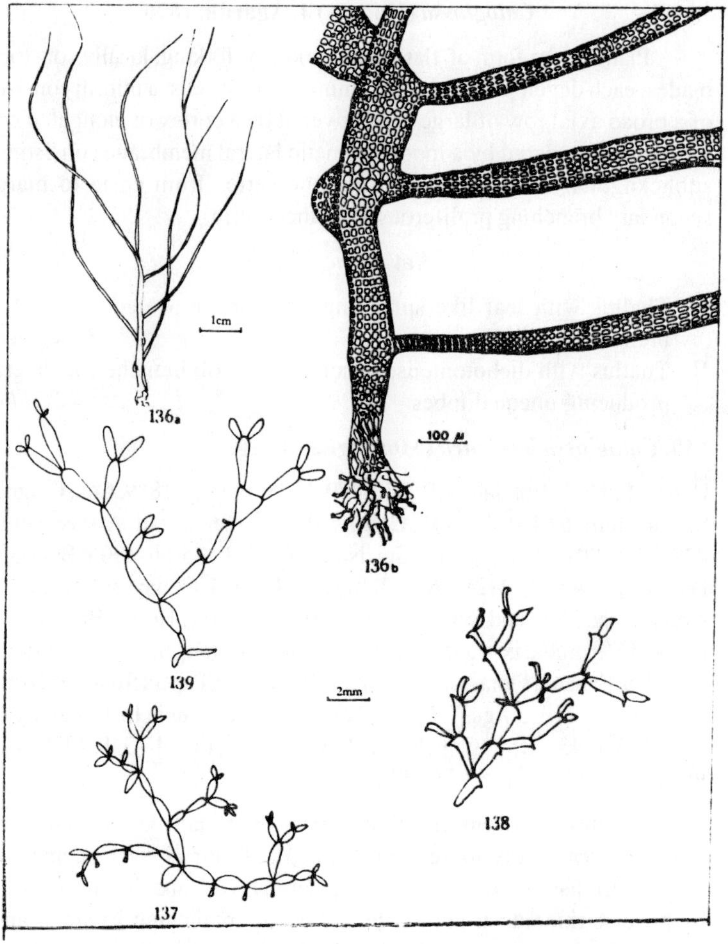

Plate XXII

136a. Habit sketch of *Compsopogon coeruleus* (Balbis) Montagne; 136b. Cells of *Compsopogon coeruleus* (Balbis) Montagne; 137. *Catenella repens* (Lightfoot) Batters ; 138. *Catenella impudica* (Mont.) J. Ag.; 139. *Catenella nipae* Zanardini.

Family - DELESSERIACEAE

Caloglossa (Harvey) J. Agardh, 1876

Plants in the form of flat, dichotomously forking, locally constricted blades, each developing from a prominent apical cells, a midrib consisting of a broad axial row of large cells covered by a cortex of elongated cells, this midrib bordered by a monostromatic lateral membrane consisting of subhexagonal cells running in oblique series from costa to margin, secondary branching proliferous from the midrib.

Key to the species

1a. Thallus with leaf like spreading branches, repeatedly, irregularly branched in clusters - *C. leprieurii*

1b. Thallus with dichotomous branching, mid rib branches off laterally producing unequal lobes - *C. adnata*

140. *Caloglossa leprieurii* (Montagne) J. Ag.

Hauck, Engler's Bot. Jahrb., 9 : 457-470, 1888; Murray, 1889, Cox, Contri. U. S. Nat. Herb. 6 : 1-921, 1901; Collins, Proc. Amer. Acad. Arts & Sci. 37 : 229-270, 1901, Vickers, Ann. Sci. Nat. Bot., ix, 1 : 45-66, 1905; Boergesen, Dansk. Bot. Arkiv., 2 (2) 1-66 + 2, figs. 1-44, 1914; Collins & Hervey, Proc. Amer. Acad. Arts and Sci., 53: 1-195, pls. 1-6, 1917; Howe, Britton, N. L., Flora of Bermuda, ix + 585 pp., 1918; Howe, Britton, N.L. and Millspaugh, C.F., The Bahama Flora, vii + 695 pp., 1920; Taylor, Papers from, the Tortugas lab., 25 : v + 219, 3 figs., 37 pls. 1928; Taylor, Carnegie Inst. Wash. Publ., 461 : 115-124, 1935 ; Islam, Biblio. Phyco., pl. 57, fig. 343-348, 1976. Santra, Indian Biol. pl. X, fig. 56-58, 1988.

Thallus dorsiventrally flattened, 4 -5 cm tall, leaf like spreading branches, repeatedly forked, constricted at forkings, each segment lanceolate, reddish -violet coloured, upto 2mm broad, attached to substratum by profuse rhizoids at underside of forking, branching irregular and in clusters, midrib consists of 3-5 longitudinal rows of cylindrical cells.

Habitat: Attached to pneumatophores and bark of mangroves at Bakkhali.

Specimen No.: N. S. 48. (Plate XXIII, Fig. 140a, 140b). Photo Plate-XXVIII.

141. ***Caloglossa adnata*** **(Zanardini) De Toni**

Islam, Biblio. Phyco., pl. 56, fig. 338, 1976.

Thallus dorsiventrally flattened, branched or dichotomously forking, about 2 cm long, light pinkish brown to violet coloured, attached to pneumatophores by profuse rhizoids from the lower surface at the points of forking. Prominent midrib, consisting of broad axial row of large cells, bordered on similar sub-hexagonal cells. Midrib branches off laterally producing unequal lobes.

Habitat: On pneumatophores and barks at Ramgopalpur, mixed with *Catenella* sp. and *Bostrychia* spp.

Specimen No.: N.S. 12. (Plate XXIII, Fig. 141a, 141b). Photo Plate - XXVIII.

Family - RHODOMELACEAE

Key to the Genera

1a. Ultimate branchlets (ramelli), polysiphonous -2

1b. Ultimate branches at least the tips monosiphonous but chromatophorous, thus not trochoblasts, ramelli persistent and clothes upper branches, dorsiventral organization - *Bostrychia*

2a. Thallus radially organized, axes filamentous, apical cells not in pits, branchlets not distinct from lesser branches - *Polysiphonia*

2b. Thallus erect and branchlets distichous in arrangement - *Pterosiphonia*

2c. Thallus dorsiventral, creeping, erect axes sharply determinate unbranched, trichoblasts spirally arranged - *Herposiphonia*

Bostrychia **Montagne, 1838**

Plants filiform, black or dull purple, stoloniferous, rhizoidal and erect branches often distinguishable, even rhizoidal branches polysiphonous, ordinarily regularly, bilaterally branched, branches near apex incurved, most part polysiphonous, several cells of equal length being disposed about the central axis or these pericentral cells regularly transversely divided, ramellar branchlets often monosiphonous the tips.

Key to the Species

1a. Primary haptera consisting of transformed branches formed in ordinary branch positions, ecorticate, branchlets od last order chiefly polysiphonous only the lat 1-3 segments monosiphonous, plants compact moss like - *B. radicans*

1b. Primary haptera on special flagellar outgrowths from the axes generally on the lower side and at a fork but not in ordinary position and succession of branching, corticate parenchymatous, branch systems distinctly divided into long axial branches and short 1-3 pinnate lateral branches of rather limited growth, the branchlets less regularly pinnate, ramelli altogether monosiphonous - *B. tenella*

142. *Bostrychia radicans* (Montagne) Mont.

Montagne, Ann. Sci. Nat., Bot., iii, 14 : 283 - 309, 1850; Mobius, Hedwigia, 28 : 309-347, pls. 10, 11, 1890, 1895; Post, Rev. Algol, 9 : 1-84, 4 figs., 1936, Taylor; Papers, Mich Sci., 39 : 85 - 109, 1954a; Marine Algae of the eastern trop. and subtropical coasts of Americas, p. 595, 1960; Islam, Bangladesh J. Botany, 20, 1973; Islam, Biblio. Phyco., 66, 1976.

Plants pink to pale purple, creeping, attached by haptera, part of the thallus erect, repeatedly branched, cortical cells at the base and at the tip of the branches, but polysiphonous throughout the frond, branches end in cortical cells near the tips.

Habitat: Attached to pneumatophores alongwith *Catenella* spp. and *Caloglossa* spp. at Bakkhali and Ramgopalpur in the littoral and above littoral zones.

Specimen No. N. S. 12. (Plate - XXIII, Fig. 142a, 142b). Photo Plate XXIX.

143. *Bostrychia tenella* (Vahl.) Ag. J.

Maze & Schramm, Essai de Classification des algues de la Ghuadeloups, xix + 283 + iii pp., 1870-77; Murray, Journ. Bot., Brit. & For. 26 : 193-196, 1889; Mobius, Hedwigia, 28, 309-347, pls. 10, 11, 1890, 1895; Collins, Proc. Amer. Acad. Arts and Sci, 37: 229-270, 1901, Boergesen, Dansk. Bot. Arkiv. 1 (4) : 1-158, 1913-1920, Collins & Hervey, Proc. Amer. Acad. Arts and Sci., 53 : 1-195, 1917; Howe, Algae, 489-540, 1918; Feldmann & Lami, Comptes Rendus Acad. Sci., 203 (17) : 883-885, 1937; Islam, Bangladesh J. Bot., 20, 1973; Islam, Biblio. Phyco., 66, 1976.

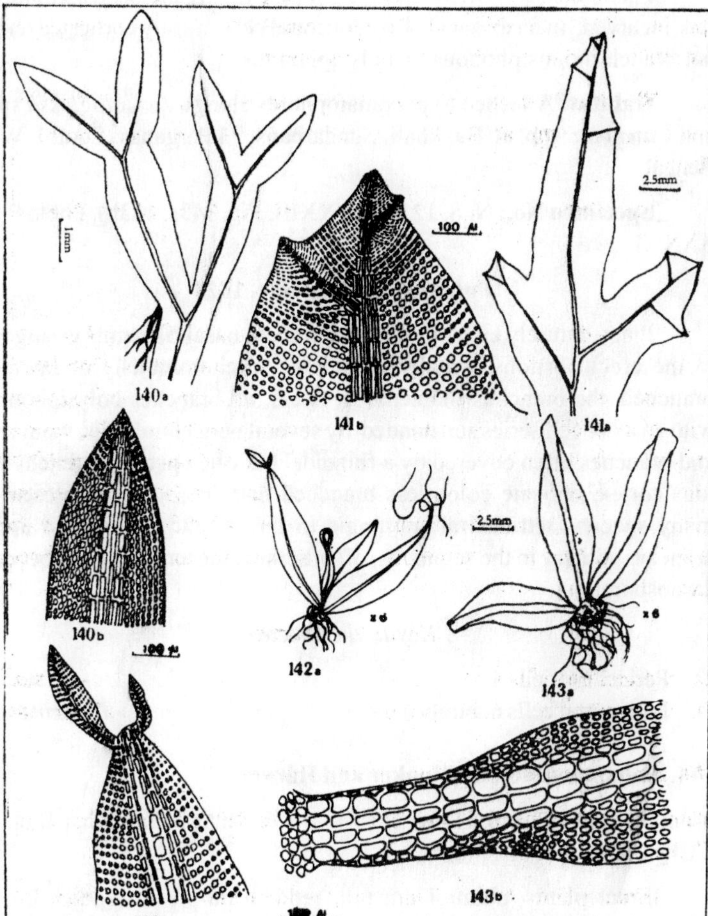

Plate XXIII

140a. Thallus of *Caloglossa leprieurii* (Montagne) J. Ag.; 140b. Cells of *Caloglossa leprieurii* (Montagne) J. Ag.; 141a. Thallus of *Caloglossa adnata* (Zanardini) De Toni; 141b. Cells of *Caloglossa adnata* (Zanardini) De Toni ; 142a. Thallus of *Bostrychia tenella* (Vahl.); 142b. Cells of *Bostrychia tenella* (Vahl.); 143a. Thallus of *Bostrychia radicans* (Montagne) Mont.; 143b. Cells of *Bostrychia radicans* (Montagne) Mont.

Plants densely tufted, erect, light purple tip pale brown coloured, tips incurved, main axis with 4-6 pericentral cells, mostly corticated terminal branch monosiphonous or polysiphonous.

Habitat: Attached to pneumatophores alongwith *Caloglossa* spp., and *Catenella* spp. at Bakkhali, Sundarbans, 24 Parganas (South), West Bengal.

Specimen No.: N. S. 12. (Plate XXIII, Fig. 143a, 143b). Photo Plate XXX.

Polysiphonia Greville, 1824

Plants entirely erect, or with decumbent basal filaments giving rise to the erect portions, usually abundantly dichotomously or laterally branched, the branches filamentous, terete, all branches polysiphonous with an axial cell series surrounded by several pericentral cells, main axes and branches often covered by a rhizoidal but often pseudoparenchymatous cortex, delicate colourless branched hair (trichoblasts) present in many species, tetrahedral sporangia in the slightly thickened upper branches solitary in the segments, often seriate, the terminal cells become caroosporangia.

Key to the species

1a. Pericentral cells 4 - *P. mollis*
1b. Pericentral cells 6, rarely 5 to 8 - *P. denudata*

144. *Polysiphonia mollis* Hooker and Harvey

Islam, Biblio. Phyco. pl. 53, fig. 311-313, 1976; Santra & Pal, Indian Biol., pl. VI, figs. 40-43, 1988.

Erect plants, about 1 cm tall, reddish purple to reddish brown coloured, with discoid holdfasts, cells about 16-17μ broad and 40-45μ long, main axis 74-146μ diameter, tetraspores 30-45μ diameter, tetraspores tetrahedrally arranged.

Habitat: Thallus attached on pneumatophores in dense tufts forming a velvety mat at Bakkhali and Jharkhali, Sundarbans, 24 Parganas, West Bengal.

Specimen No.: N. S. 48. (Plate XXIV, Fig. 144a, 144b). Photo Plate XXI.

145. *Polysiphonia denudata* (Dillwyn) Kützing

Hoyt, Bull. Bur. Fisheries [v.s.] 36 - 367 - 356, 1920; Murray, Journ. Bot., Brit. & For. 26 : 193 - 196, 1889; Boergesen, Dansk. Bot. Arkiv. 1(4) : 1-158, 1913-1920; Taylor, Marine Algae of the eastern trop. and subtropical coasts of Americas, p. 580, 1960.

Plants upto 25 cm tall, base disciform, lower branches strongly divergent when young, the upper branches more irregular and erect, colour dark reddish-purple, in the upper branches and branchlets, soft , below slender but stiff and colourless, sometimes brownish, lower portion without branchlets, 340-750μ diam., segments 250-500μ long nearly ecorticate, sec. branches 100-170μ diam., segments 400-900μ long, branches 33-45μ diam. segments 45- 80μ long throughout with 6 pericentral cells, trichoblasts absent, rudimentary moderately developed tetrasporangia oval to spherical, 60-100μ diam.

Specimen No. : N. S. (Plate XXIV, Fig. 145). Photo Plate XXXII.

***Pterosiphonia* Falkenberg, 1889**

Plants rhizomatous, erect branches sparingly divided, alternately pinnately branched, short branchlets simple or once forked, pericentral cells 5-10, naked or with a parenchymatous cortex, vegetative trichoblasts absent, tetrasporangia single in each fertile segment, lying along the face of the branch axis.

146. *Pterosiphonia pinnata* (Roth) Falkenberg

Williams, Amer. Journ. Bot., 35 (10) : 683 - 695, 20 figs., 1949; Taylor, Marine Algae of the eastern trop. and subtropical coasts of Americas, p. 593, 1960; Islam, Bangladesh J. Botany, 20, photomicrograph B (l - 4), 1973b; Islam, Biblio. Phyco., 67, 1976, Santra & Pal, Indian Biol., pl. IX, figs. 52-53, pl. IX, 1988.

Plants tufted, soft, and fragile, 2-4cm tall, dark purple colour, distichously pinnate, at times bipinnate, main axis consists of 8 - 9 pericentral cells, branchlets 1-1.5mm long, corticated below, incurved when young, obliquely erect when mature.

Habitat: Attached to pneumatophores in the above littoral zone at Jharkhali, Sundarbans, West Bengal.

Specimen No.: N. S. 35. (Plate XXIV, Fig. 146). Photo Plate XXXIII.

Herposiphonia Nägeli, 1846

Plants creeping, the wide ranging and freely branched rhizomatous filaments attached by unicellular rhizoids, through some branches or even large portions of the plants often free from the substratum, axes terete or compressed, with several pericentral dorsiventral, bearing erect branchlets on the dorsal side, lateral branches or their rudiment alternating with branchlets in a defintie sequence from the nodes, erect axis of limited growth, simple, polysiphonous like the rhizomatous axes in the young state with the tips inrolled, on the dorsal side the erect axes bearing a tuft of deciduous trichoblasts.

147. ***Herposiphonia dendroidea*** **var.** ***minor*** **Hollenberg**

Islam, Biblio. Phyco., pl. 55, figs. 330 -332, 1976.

Plant pinkish red, delicate, creeping, main axis prostrate, prostrate branches 45-65μ diam., segments 75 - 98μ long, attached to substratum by short uniseriate rhizoidal structures, branches erect, trichoblast present on mature branch tips.

Habitat: On the forest floor in the littoral and above littoral zones at Jharkhali, Sundarbans, 24 Parganas (South), West Bengal.

Specimen No.: N.S. 1. (Plate XXIV, Fig. 147). Photo Plate XXXIV.

Family - DASYACEAE

Hetersiphonia

148. ***Heterosiphonia*** **sp.**

Srinivasan, Phyco. Indica, pl. VII, 1969; Islam, Biblio. Phyco., 61, 1976.

Thallus erect, densely tufted, rose red to dark brown coloured, about 2-2.5 cm long, much branched with shaggy, feathery stems, ramuli in two rows, one on each side of branches, containing one central row of 4 - 8 pericentral cells. Cells in main axis 15-30μ broad and upto 60μ long.

Description matches well with species *H. muelleri* (Sond.) De Toni, described by Srinivasan, except for the dimensions of cells and the plant, which is much smaller in the species collected from Jharkhali.

Habitat : On the bark of *Aegialitis rotundifolia* at Jharkhali, Sundarbans, 24 Parganas (S), West Bengal.

Specimen No. : N. S. 56. (Plate XXIV, Fig.148). Photo Plate XXXIV.

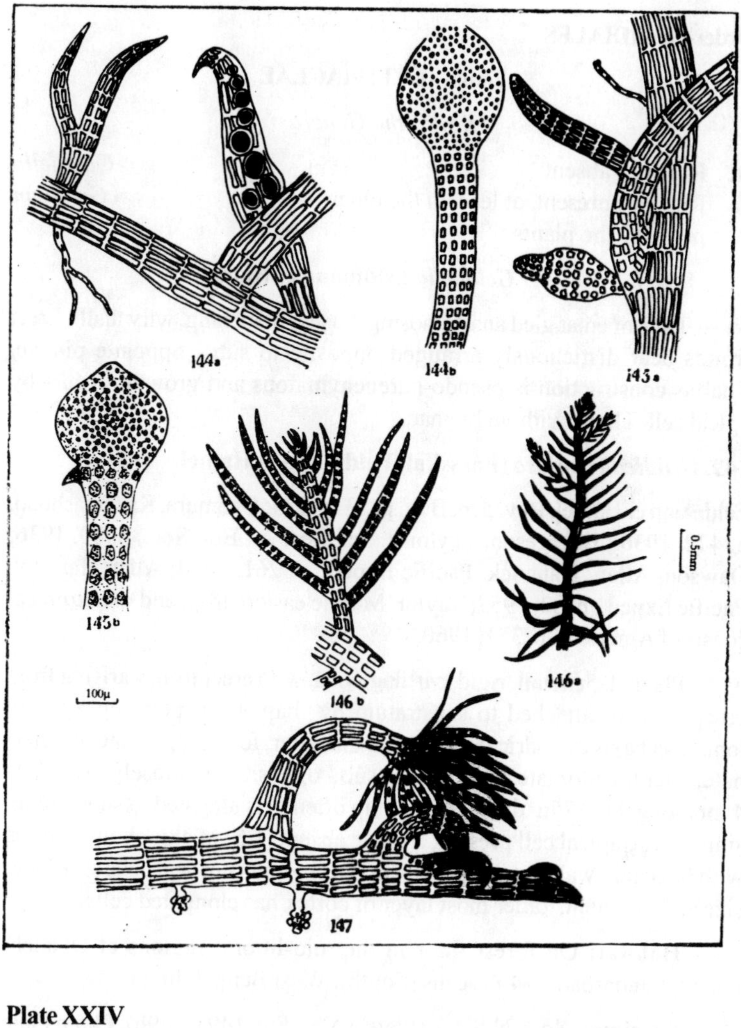

Plate XXIV

144a. Tetrasporophyte of *Polysiphonia mollis* Hooker & Harvey; 144b. Carposporophyte of *Polysiphonia mollis* Hooker & Harvey; 145a. *Polysiphonia denudata* (Dillwyn.) Kützing ; 145b. Carposporophyte of *Polysiphonia denudata* (Dillwyn.) Kützing; 146a. *Pterosiphonia pinnata* (Roth.); 146b. Cells of *Pterosiphonia pinnata* (Roth:); 147. *Herposiphonia dendroidea* var. *minor* Hollenberg.

Order-GELIDIALES

Family - GELIDIACEAE

Key to the Genera

1a. Rhizines absent - *Gelidiella*
1b. Rhizines present, at least in the older parts of the plants - *Gelidium*

Gelidiella Feldmann et Hamel

Tufts of entangled anastomosing, bushy, branching, wiry thalli. Erect fronds bear distichously arranged opposite to sub - opposite pinnae. Thallus construction is pseudo-parenchymatous and growth occurs by apical cell. Thalli without hyphae.

149. *Gelidiella acerosa* (Forsskal) Feldmann et Hamel

Feldmann et Hamel, Rev. gen. Bot., p. 533, 1934; Okamura, Kazamashobo, p. 472, 1936 ; Boergesen, Ceylon J. Sci. Sec. A., Bot. Soc., p. 80, 1936; Dawson, Allan Hancock Pacific Exped., p. 261, 1944; Allan Hancock Pacific Exped., p. 82, 1952; Taylor, Marine eastern trop. and subtropical coasts of Americas, p. 351, 1960.

Plants 1-5cm tall, rigid, cartilaginous, with erect fronds arising from creeping axes attached to substratum by haptera, haptera upto 5 mm long; erect axis cylindrical upto 1 mm diameter, few lower branches alternate, later the ultimate branches are sub - opposite and closely placed, 1-4 mm long, 80-275μ diam., branch tip often gets attached to substratum, single large apical cell present, hyphae absent, T.S. of axis show medulla with large thin walled cells 10-28μ diameter, outer cortical cells smaller in size and pigment, outer most layer of cortex has elongated cells.

Habitat: On forest floor in the mid-littoral regions at Dhanchi Island, Sundarbans, 24 Pargans (South), West Bengal, India.

Specimen No. : N.S. 84 (Plate XXV, Fig. 149a, 149b). Photo Plate XXXV.

Gelidium Lamouroux, 1813

Plants moderately small sized with main axis erect and laterally branched, branches firm, cylindrical or flattened, branches developed from an apical cell, axial row of cells giving rise to pericentral cells, cortex with hyphae (rhizines) in the axis between the larger cells of the cortex.

150. *Gelidium pusillum* (Stackh.) Le Jolis

De Toni, Sylloge Algarum, p. 134, 1897; Okamura, Kazamashobo, p.11, 1910; Boergesen, Dansk Bot. Arkiv., p.280, 1920; Kew. Bull. Misc. Infn., p. 42, 1935; Feldmann et Hamel, Rev. algol. p. 114, 1936; Rao, Phykos, p. 73, fig. 7 C-D, 1976.

Plants upto 1.4 cm tall, saxicolous, disc like holdfast, branches creeping, thallus with narrow erect axes, lateral ramuli slightly flattened, expanded with a rounded end.

Habitat : On forest floor at Dhanchi Island, Sundarbans, 24 Parganas (S), West Bengal.

Specimen No. : N. S. 85. (Plate XXV, Fig. 150a, 150b, 150c).

The above algal species belonging to Rhodophyta are tabulated below according to their systematic position along with their habit and habitat.

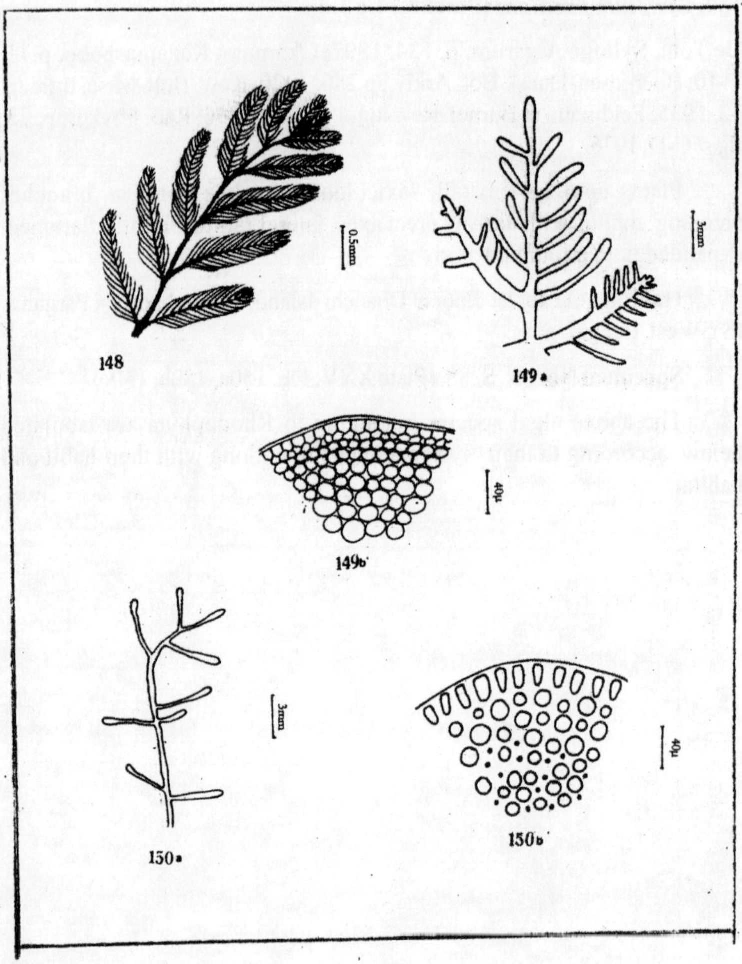

Plates XXV

148. *Heterosiphonia* sp., 149a. Thallus of *Gelidiella acerosa* (Forsskal) Feldmann et Hamel ; 149b. T. S. of *Gelidiella acerosa* (Forsskal) Feldmann et Hamel; 150a. Thallus of *Gelidium pusillum* (Stackh.)Le Jjolis; 150b. T. S. of *Gelidium pusillum* (Stackh.) Le Jolis showing hyphae.

Table 19: Habitat of Identified Rhodophyta Members From Indian Sundarbans.

Sub Class & Order	*Family*	*Name of Species*	*Habit & Habitat*
Bangioideae			
Compsopogonales	Erythrotrichiaceae	*Compsopogan coeruleus*	Attached to twigs submerged in brackishwater fisheries low-mid saline areas.
Floridae			
Gigartinales	Rhabdoniaceae	*Catenella nipae*	On the barks of *E. agallocha* in the upper littoral zone at Bakkhali.
		Catenella repens	Creeping on pneumatophores of mangroves at high saline areas.
		Catenella impudica	Creeping on pneumatophores of mangroves & bark of *E.agallocha* at high saline areas.
Ceramiales	Delesseriaceae	*Caloglossa adnata*	On pneumatophores and bark of mangroves at high saline areas.
		Caloglossa leprieurii	On pneumatophores and bark of mangroves at high saline areas.
		Polysiphonia mollis	On pneumatophores in dense tufts at high saline areas.
		Polysiphonia denudata	Creeping on pneumatophores & bark of mangroves at high saline.

Contd...

Sub Class & Order	*Family*	*Name of Species*	*Habit & Habitat*
	Rhodomelaceae	*Pterosiphonia pinnata*	On pneumatophores in the above littoral zone at high saline areas.
		Bostrychia radicans	On pneumatophores in littoral and above littoral zones.
		Bostrychia tenella	On pneumatophores at high saline areas.
		Herposiphonia dendroidea var. minor	On the forest floor in the littoral and above littoral zone at high saline areas.
	Dasyaceae	*Heterosiphonia sp.*	On the bark of *Aegialitis rotundifolia* & other mangroves at high saline areas.
Gelidiales	Gelidiaceae	*Gelidiella acerosa*	As littoral flora & also colonising the bark of mangroves in high saline areas.
		Gelidium sp.	As littoral flora on hard substratum in high saline areas alongwith other red algae.

Chapter 7

Ecology

7.1 Ecology of the Algal Habitat

The Sundarbans is a unique ecosystem that is in fact a transition zone with distinct ecological characteristics of its own. Conditions are so suited that this region is packed with life, so diverse and unique in forms, shapes, sizes and niches that go on to making the most naturally fertile regions of the world. Three major life forms of autotrophs are often intermixed in these transitional zones, i.e., estuaries and in shore marine waters, an play varying roles in maintaining a high gross production rate; these are (1) phytoplankton; (2) benthic microflora-algae living in and on mud, sand, rocks or other hard surfaces, *viz.,* bodies or shells of animals; and (3) macroflora-large attached plants-the seaweeds, emergent marsh grasses and the tropical mangrove trees (Odum, 1975). Estuaries are the only naturally subsidized solar-powered ecosystems, i.e., they are naturally productive systems of nature that not only have high life support capacity but also produce excess organic matter that may be exported to other systems or stored (Odum, 1975). These estuaries are efficient nutrient traps which enhances the estuary's capacity to absorb nutrient in wastes thus providing excellent "nursery grounds" for numerous coastal shell fish and fin fish species which are not only harvested in the estuaries but offshore also.

7.2 Algal Habitat

The distribution, abundance and habitat ecology of the 150 algal species mentioned earlier have been highlighted in Table 20. The physicochemical parameters and the algal species encountered in various zones of Indian Sundarbans during different seasons have been highlighted in

Tables 21, 22 & 23. The habitat ecology has been classified according to the Spectral Ranges as described by Lowe (1974) and Fjerdingstad (1964). The Spectral Ranges described by Lowe (1974) is as follows :

1. **pH spectrum**

Acidobiontic :	Organisms occuring below 5.5 pH.
Acidophilous :	Organisms occuring between 5.5 and 6.5 pH.
Indifferent	Organisms occuring between 6.5 and 7.5 pH.
Alkaliphilous :	Organisms occuring between 7.5 and 9.0 pH.
Alkalibiontic :	Organisms occuring in alkaline water with pH above 9.0.

2. **Temperature spectrum**

Euthermal :	Organisms occuring at temperatures above 30° C.
Mesothermal:	Organisms occuring in temperatures between 15 - 30° C.
Oligothermal :	Organisms occuring below 15° C.

3. **Salinity spectrum**

Euryhaline:	Organisms occurring inhigh saline zones (18ppt or above)
Mesohaline:	Organisms occurring in mid saline zones 95ppt-18ppt)
Oligohaline:	Organisms occurring in low saline zones (below 5ppt)

Spectral Ranges as described by Fjerdingstad (1964)

1. **Saprobien spectrum**

Saprobiontic :	Organisms occuring in heavily polluted water.
Saprophilic :	Organisms usually in polluted water but also in clean waters.
Saproxenous :	Organisms usually in clean water but also in polluted water.
Saprophobic :	Organisms occuring in clean water habitats only.

2. **Current spectrum**

Limnobiontic :	Organisms of lentic environment only.
Limnophilous :	Found in lentic habitats but also present in lotic habitats.
Indifferent :	Common in both lotic and lentic water.
Rheophilous:	Found in lotic habitats but also present in lentic habitats.
Rheobiontic:	Found in lotic environment only.

Table 20 : Distribution Abundance and Habitat Ecology of the Studied Algal Species from Indian Sundarbans.

Species	Zone I	Zone II	Zone III	Habit & Habitat
Volvox sp.	+	—	—	alkaliphilous, mesothermal, oligohaline, saprophobic, limnophilous
Pandorina morum	+	—	—	alkaliphilous, mesothermal, oligohaline, saprophobic, limnophilous
Uronema confervicola	+	—	—	alkaliphilous, mesothermal, oligohaline, saprophobic, limnophilous
Enteromorpha clathrata	++	—	—	alkaliphilous, mesothermal, oligohaline, saprophobic, limnophilous
Enteromorpha compressa	—	++	—	alkaliphilous, mesothermal, oligohaline, saprophobic, limnobiontic
Enteromorpha intestinalis	+++	+++	—	alkaliphilous, mesothermal, oligohaline, mesohaline, saprophilic, limnobiontic
Enteromorpha prolifera	+++	+++	—	alkaliphilous, mesothermal, oligohaline, mesohaline, saprophilic, limnobiontic
Ulva lactuca	—	+++	—	alkaliphilous, mesothermal, oligohaline, mesohaline, saprophilic, limnobiontic
Ulva patengensis	—	+++	—	alkaliphilous, mesothermal, oligohaline, mesohaline, saprophilic, limnobiontic
Ulva fasciata	—	+	—	alkaliphilous, mesothermal, oligohaline, mesohaline, saprophilic, limnobiontic
Chaetomorpha aerea	++	++	++	alkaliphilous, mesothermal, mesoeuryhaline, saproxenous, limnobiontic
Chaetomorpha gracilis	—	+++	++	alkaliphilous, mesothermal, mesoeuryhaline, saproxenous, limnophilous

Contd...

Species	*Zone I*	*Zone II*	*Zone III*	*Habit & Habitat*
Chaetomorpha brachygona	++	+++	++	alkaliphilous, mesothermal, mesoeuryhaline, saproxenous, limnophilous
Lola capillaris	+	+++	+	alkaliphilous, mesothermal, mesoeuryhaline, saprophobic, limnophilous
Lola implexa	+	+	+	alkaliphilous, mesothermal, mesoeuryhaline, saproxenous, limnophilous
Lola tortuosa	+	++	+++	alkaliphilous, mesothermal, mesoeuryhaline, saprophobic, rheobiontic
Rhizoclonium grande	—	+	+++	alkaliphilous, mesothermal, euryhaline, saprophobic, rheobiontic
Rhizoclonium hookeri	—	+	+++	alkaliphilous, mesothermal, euryhaline, saprophobic, rheobiontic
Rhizoclonium riparium	—	+	+++	alkaliphilous, mesothermal, euryhaline, saprophobic, rheobiontic
Cladophora echinus	+	—	—	alkaliphilous, mesothermal, oligohaline, saprophobic, limnobiontic
Cladophorella sundarbanensis	—	+	++	alkaliphilous, mesothermal, euryhaline, saprophobic, rheobiontic
Oedogonium undulatum	+	++	—	alkaliphilous, mesothermal, oligohaline, saprophobic, rheobiontic
Spirogyra dubia	++	++	—	alkaliphilous, mesothermal, oligomesohaline, saproxenous, limnobiontic

Contd...

Species	Zone I	Zone II	Zone III	Habit & Habitat
Spirogyra setiformis	++	—	—	alkaliphilous, mesothermal, oligohaline, saproxenous, limnobiontic
Spirogyra ternata	—	++	—	alkaliphilous, mesothermal, mesohaline, saproxenous, limnobiontic
Triplastrum simplex	+	—	—	alkaliphilous, mesothermal, oligohaline, saproxenous, limnobiontic
Triplastrum abbreviatum	+	—	—	alkaliphilous, mesothermal, oligohaline, saproxenous, limnobiontic
Cosmarium striolatum	+	+	—	alkaliphilous, mesothermal, oligomesohaline, saproxenous, limnobiontic
Cosmarium depressum	+	+	—	alkaliphilous, mesothermal, oligomesohaline, saproxenous, limnobiontic
Closterium acutum	+	—	—	alkaliphilous, mesothermal, oligomesohaline, saproxenous, limnobiontic
Chlorella vulgaris	+	—	—	alkaliphilous, mesothermal, oligohaline, saproxenous, limnobiontic
Radiococcus sp.	+	—	—	alkaliphilous, mesothermal, oligohaline, saproxenous, limnobiontic
Pediastrum boryanum	+	+	+	alkaliphilous, mesothermal, oligo-mesoeuryhaline, saprophobic, limnobiontic
Pediastrum duplex	+	+	+	alkaliphilous, mesothermal, oligo-mesoeuryhaline, saprophobic, limnobiontic
Pediastrum tetras	+	+	—	alkaliphilous, mesothermal, oligohaline, saprophobic, limnobiontic

Contd...

Species	Zone I	Zone II	Zone III	Habit & Habitat
Scenedesmus bijuga	+	--	--	alkaliphilous, mesothermal, oligohaline, saprophobic, limnobiontic
Scenedesmus quadricauda	+	--	--	alkaliphilous, mesothermal, oligohaline, saprophobic, limnobiontic
Boodleopsis sundarbanensis	--	+	--	alkaliphilous, mesothermal, meso-euryhaline, saprophilic, benthic
Chara zeylanica	--	+	--	alkaliphilous, mesothermal, oligohaline, saprophobic, limnobiontic
Vaucheria prescotti	--	++	+	alkaliphilous, mesothermal, meso-euryhaline, saprophilic, benthic
Vaucheria sp.	--	+	--	alkaliphilous, mesothermal, meso-euryhaline, saprophilic, benthic
Melosira moniliformis	--	+	--	alkaliphilous, mesothermal, mesohaline, saprophobic, rheophilous
Melosira sol.	--	+	--	alkaliphilous, mesothermal, mesohaline, saprophobic, rheophilous
Cyclotella glomerata	--	--	+	alkaliphilous, mesothermal, mesohaline, saprophobic, rheophilous
Coscinodiscus excentricus	++	++	++	alkaliphilous, mesothermal, oligo-mesoeuryhaline, saproxenous, limnophilous
Coscinodiscus gigas	++	++	++	alkaliphilous, mesothermal, oligo-mesoeuryhaline, saproxenous, limnophilous

Contd...

Species	Zone I	Zone II	Zone III	Habit & Habitat
Coscinodiscus granii	++	++	++	alkaliphilous, mesothermal, oligo-mesoeuryhaline, saproxenous, limnophilous
Stephanopyxis palmeriana	--	--	+	alkaliphilous, mesothermal, euryhaline, saprophobic, rheobiontic
Hemidiscus cuneiformis	--	--	+	alkaliphilous, mesothermal, euryhaline, saprophobic, rheobiontic
Biddulphia mobiliensis	--	--	+	alkaliphilous, mesothermal, euryhaline, saprophobic, rheobiontic
Biddulphia sinensis	--	--	+	alkaliphilous, mesothermal, euryhaline, saprophobic, rheobiontic
Chaetoceros curvisetus	--	--	+	alkaliphilous, mesothermal, euryhaline, saprophobic, rheobiontic
Chaetoceros flexuosus	--	--	+	alkaliphilous, mesothermal, euryhaline, saprophobic, rheobiontic
Chaetoceros laciniosus	--	--	+	alkaliphilous, mesothermal, euryhaline, saprophobic, rheobiontic
Chaetoceros tenuissimus	--	--	+	alkaliphilous, mesothermal, euryhaline, saprophobic, rheobiontic
Chaetoceros subsecundus	--		+	alkaliphilous, mesothermal, euryhaline, saprophobic, rheobiontic
Climacodium frauenfeldianum	--	--	+	alkaliphilous, mesothermal, euryhaline, saprophobic, rheobiontic

Contd...

Species	Zone I	Zone II	Zone III	Habit & Habitat
Bacteriastrum varians	—	—	+	alkaliphilous, mesothermal, euryhaline, saprophobic, rheobiontic
Bacteriastrum cosmosum	—	—	+	alkaliphilous, mesothermal, euryhaline, saprophobic, rheobiontic
Bacteriastrum delicatulum	—	—	+	alkaliphilous, mesothermal, euryhaline, saprophobic, rheobiontic
Corethron hystrix	—	—	+	alkaliphilous, mesothermal, euryhaline, saprophobic, rheobiontic
Rhizosolenia imbricata	—	—	+	alkaliphilous, mesothermal, euryhaline, saprophobic, rheobiontic
Rhizosolenia setigera	—	—	+	alkaliphilous, mesothermal, euryhaline, saprophobic, rheobiontic
Fragilaria vaucheriae	—	+	+	alkaliphilous, mesothermal, meso-euryhaline, saprophobic, rheobiontic
Asterinella japonica	—	—	+	alkaliphilous, mesothermal, euryhaline, saprophobic, rheobiontic
Synedra ulna	+	+	+	alkaliphilous, mesothermal, oligo-meso-euryhaline, saprophobic, rheobiontic
Diatoma vulgare	—	+	+	alkaliphilous, mesothermal, meso-euryhaline, saprophobic, rheobiontic
Achnanthes microcephala	—	+	+	alkaliphilous, mesothermal, meso-euryhaline, saprophobic, rheobiontic

Contd...

Species	Zone I	Zone II	Zone III	Habit & Habitat
Achnanthes minutissima	--	+	+	alkaliphilous, mesothermal, meso-euryhaline, saprophobic, rheobiontic
Cocconeis placentula	--	+	+	alkaliphilous, mesothermal, meso-euryhaline, saprophobic, rheobiontic
Navicula cryptocephala	--	+	+	alkaliphilous, mesothermal, meso-euryhaline, saprophobic, rheobiontic
Navicula radiosa	+	+	--	alkaliphilous, mesothermal, oligo-mesoaline,saprophobic, rheobiontic
Anomoeoneis exilis	+	+	--	alkaliphilous, mesothermal, oligo-mesoaline,saprophobic, rheobiontic
Stauroneis phoenicenteron	--	+	--	alkaliphilous, mesothermal, mesoaline, saprophobic, rheobiontic
Pinnularia viridis	+	--	--	alkaliphilous, mesothermal, oligohaline,saprophobic, rheobiontic
Pleurosigma angulatum	--	+	+	alkaliphilous, mesothermal, meso-euryhaline, saprophobic, rheobiontic
Gyrosigma acuminatum	+	+	+	alkaliphilous, mesothermal, oligo-meso-euryhaline,saprophobic, rheobiontic
Gomphonema sphaerophorum	--	--	+	alkaliphilous, mesothermal, euryhaline, saprophobic, rheobiontic
Amphora veneta	--	--	+	alkaliphilous, mesothermal, euryhaline, saprophobic, rheobiontic

Contd...

Species	Zone I	Zone II	Zone III	Habit & Habitat
Cymbella ehrenbergii	--	--	+	alkaliphilous, mesothermal, euryhaline, saprophobic, rheobiontic
Nitzschia acicularis	+	+	++	alkaliphilous, mesothermal, oligo-mesoaline,saprophobic, rheobiontic
Nitzschia obtusa	+	+	+	alkaliphilous, mesothermal, oligo-meso-euryhaline, saprophobic, rheobiontic
Nitzschia sublinearis	+	+	+	alkaliphilous, mesothermal, oligo-meso-euryhaline, saprophobic, rheobiontic
Colpomenia sinuosa	--	--	++	alkaliphilous, mesothermal, euryhaline, saprophobic, benthic (as littoral flora)
Dictyota ceylanica	--	+	+	alkaliphilous, mesothermal, euryhaline, saprophobic, periphytic
Gloeocapsa decoroticans	+	--	--	alkaliphilous, mesothermal, oligohaline, saproxenous, limnobiontic
Gloeocapsa punctata	--	+	--	alkaliphilous, mesothermal, oligohaline, saproxenous, limnobiontic
Gloeocapsa aeruginosa	--	+	--	alkaliphilous, mesothermal, euryhaline, saprobiontic, rheophilous
Gloeocapsa rupestris	+	--	--	alkaliphilous, mesothermal, mesohaline, saprobiontic, rhephilous
Gloeocapsa kuetzingiana	+	--	--	alkaliphilous, mesothermal, mesohaline, saprobiontic, rhephilous

Contd...

Species	*Zone I*	*Zone II*	*Zone III*	*Habit & Habitat*
Microcystis bengalensis	+	+	--	alkaliphilous, mesothermal, oligo-meso-euryhaline, saprobiontic, limnobiontic
Aphanocapsa littoralis	--	--	+	alkaliphilous, mesothermal, euryhaline, saprobiontic, limnobiontic
Aphanocapsa pulchra	--	+	+	alkaliphilous, mesothermal, meso-euryhaline, saprobiontic, limnobiontic
Merismopedia tenuissima	--	+	--	alkaliphilous, mesothermal, mesohaline, saproxenous, rhephilous
Johannesbaptistia pellucida	+	--	--	alkaliphilous, mesothermal, oligohaline, saproxenous, limnophilous
Chamaesiphon curvatus	--	+	+	alkaliphilous, mesothermal, meso-euryhaline, saproxenous, epiphytic
Dermocarpa leibleiniae	+	+	+	alkaliphilous, mesothermal, oligo-meso-euryhaline, saproxenous, epiphytic
Dermocarpa hemispherica	--	+	+	alkaliphilous, mesothermal, oligo-meso-euryhaline, saproxenous, epiphytic
Stichosiphon sansibaricus	+	+	+	alkaliphilous, mesothermal, oligo-meso-euryhaline, saproxenous, epiphytic
Xenococcus chaetomorphae	+	+	+	alkaliphilous, mesothermal, oligo-meso-euryhaline, saproxenous, epiphytic
Xenococcus cladophorae	+	+	+	alkaliphilous, mesothermal, oligo-meso-euryhaline, saproxenous, epiphytic

Contd...

Species	Zone I	Zone II	Zone III	Habit & Habitat
Spirulina princeps	+	--	--	alkaliphilous, mesothermal, oligohaline, saproxenous, limnophilous
Spirulina major	+	--	--	alkaliphilous, mesothermal, oligohaline, saproxenous, limnophilous
Arthrospira platensis	+	--	--	alkaliphilous, mesothermal, oligohaline, saproxenous, limnophilous
Oscillatoria nigrovirdis	+	+	+	alkaliphilous, mesothermal, oligo-meso-euryhaline, saprophilic, limnophilous
Oscillatoria limosa	++	++	+	alkaliphilous, mesothermal, oligo-meso-euryhaline, saprophilic, limnophilous
Oscillatoria subbrevis	+	+++	--	alkaliphilous, mesothermal, oligo-mesohaline, saprophilic, limnophilous
Oscillatoria curviceps	+	++	+	alkaliphilous, mesothermal, oligo-meso-euryhaline, saprophilic, limnophilous
Oscillatoria princeps	+	+++	+	alkaliphilous, mesothermal, oligo-meso-euryhaline, saprophilic, limnophilous
Oscillatoria chlorina	--	+	+	alkaliphilous, mesothermal, meso-euryhaline, saprophobic, rheophilous
Oscillatoria tenuis	++	+++	+	alkaliphilous, mesothermal, oligo-meso-euryhaline, saproxenous, limnophilous
Phormidium stagnina	--	+	--	alkaliphilous, mesothermal, mesohaline, saprobiontic, benthic

Contd..

Species	*Zone I*	*Zone II*	*Zone III*	*Habit & Habitat*
Phormidium fragile	--	+	--	alkaliphilous, mesothermal, euryhaline, saprobiontic, limnobiontic benthic
Lyngbya birgei	+	+	--	alkaliphilous, mesothermal, oligo-mesohaline, saprophilic limnobiotic
Lyngbya hieronymusii	+	+	--	alkaliphilous, mesothermal, oligo-mesohaline, saprophilic limnobiotic
Lyngbya lutea	+	++	++	alkaliphilous, mesothermal, oligo-meso-euryhaline, saproxenous limnobiotic
Lyngbya majuscula	--	++	++	alkaliphilous, mesothermal, meso-euryhaline, saproxenous limnobiotic
Lyngbya confervoides	+	+	--	alkaliphilous, mesothermal, oligo-mesohaline, saproxenous limnobiotic
Lyngbya semiplena	+	+	+	alkaliphilous, mesothermal, oligo-meso-euryhaline, saproxenous limnobiotic
Schizothrix lamyii	--	++	--	alkaliphilous, mesothermal, oligo-meso-euryhaline, saprophilic, limnobiotic/ benthic
Microcoleus chthonoplastes	--	+	+	alkaliphilous, mesothermal, meso-euryhaline, saprophilic, limnobiotic/ benthic
Hydrocoleum lyngbyaceum	--	+	+	alkaliphilous, mesothermal, meso-euryhaline, saprophilic, limnobiotic/ benthic

Contd...

Species	*Zone I*	*Zone II*	*Zone III*	*Habit & Habitat*
Nostoc punctiforme	+	+	+	alkaliphilous, mesothermal, oligo-meso-euryhaline, saproxenous, limnobiontic
Nostoc linckia	—	+	—	alkaliphilous, mesothermal, mesohaline, saprophobic, limnobiontic
Scytonema hofmanni	+	—	—	alkaliphilous, mesothermal, oligohaline, saproxenous, limnobiontic
Anabaena anomala	+	+	+	alkaliphilous, mesothermal, oligo-meso-euryhaline, saproxenous, limnobiontic
Anabaena doliolum	+	+	—	alkaliphilous, mesothermal, oligo-mesohaline, saproxenous, limnobiontic
Anabaena iyengarii	—	—	+	alkaliphilous, mesothermal, euryhaline, saproxenous, limnobiontic
Anabaena gelatinicla	—	+	—	alkaliphilous, mesothermal, euryhaline, saprobiontic, limnobiontic
Anabaena arnoldii	+	—	+	alkaliphilous, mesothermal, oligo-mesohaline, saprophobic, limnobiontic
Raphidiopsis curvata	+	—	—	alkaliphilous, mesothermal, oligo-mesohaline, saprophobic, limnobiontic
Raphidiopsis indica	+	—	—	alkaliphilous, mesothermal, oligohaline, saproxenous, limnobiontic

Contd...

Species	*Zone I*	*Zone II*	*Zone III*	*Habit & Habitat*
Calothrix contarenii	--	+	--	alkaliphilous, mesothermal, mesohaline, saproxenous, limnobiontic
Mastigocoleus testarum	--	--	+	alkaliphilous, mesothermal, euryhaline, saprophobic, indifferent
Compsopogon coeruleus	+	--	--	alkaliphilous, mesothermal, mesohaline, saproxenous, limnophilous
Catenella nipae	--	+	+	alkaliphilous, mesothermal, meso-euryhaline, saprophobic, rheophilous/ periphytic
Catenella impudica	--	+	+	alkaliphilous, mesothermal, meso-euryhaline, saprophobic, rheophilous/ periphytic
Catenella repens	--	++	++	alkaliphilous, mesothermal, meso-euryhaline, saprophobic, rheophilous/ periphytic
Caloglossa adnata	--	+	+	alkaliphilous, mesothermal, meso-euryhaline, saprophobic, rheophilous/ periphytic
Caloglossa leprieurii	--	+	+	alkaliphilous, mesothermal, meso-euryhaline, saprophobic, rheophilous/ periphytic
Polysiphonia mollis	--	+++	+++	alkaliphilous, mesothermal, meso-euryhaline, saprophobic, rheophilous/ periphytic

Contd...

Species	*Zone I*	*Zone II*	*Zone III*	*Habit & Habitat*
Polysiphonia denudata	--	++	++	alkaliphilous, mesothermal, meso-euryhaline, saprophobic, rheophilous/ periphytic
Pterosiphonia pinnata	--	++	++	alkaliphilous, mesothermal, meso-euryhaline, saprophobic, rheophilous/ periphytic
Bostrychia radicans	--	++	++	alkaliphilous, mesothermal, meso-euryhaline, saprophobic, rheophilous/ periphytic
Bostrychia tenella	--	++	++	alkaliphilous, mesothermal, meso-euryhaline, saprophobic, rheophilous/ periphytic
Herposiphonia dendroidea	--	+	+	alkaliphilous, mesothermal, meso-euryhaline, saprophobic, rheophilous/ periphytic
Heterosiphonia sp.	--	+	+	alkaliphilous, mesothermal, meso-euryhaline, saprophobic, rheophilous/ periphytic
Gelidiella acerosa	--	--	+	alkaliphilous, mesothermal, meso-euryhaline, saprophobic, rheophilous/ periphytic
Gelidium pusillum	--	+	--	alkaliphilous, mesothermal, meso-euryhaline, saprophobic, rheophilous/ periphytic

Table 21 : Range of Physico-Chemical Parameters & Algal Species Encountered in Zone-I, During Different Seasons.

Seasons	*pH*	*Water Temp.*	*D.O. (ppm)*	*Salinity (ppt) (ppm)*	*Alka-linity*	*Phosphate (ppm)*	*Nitrate (ppm)*	*Algal Species Encountered*
SUMMER								
March 97-June 97	7.7	28°C	8	5	135	0.63	0.28	*Enteromorpha* spp., *Chaetomorpha area, Chaetomorpha brachygoma, Lola* Spp. *Uronema* sp., *Cladophora* sp., *Closterium* sp., *Scenedesmums* spp., *Pediastrum* spp., *Melosira* spp., *Coscinodiscus* spp., *Synedra* sp., *Navicula radiosa, Pinnularia* sp. *Nitzschia* spp., *Microcystis* sp., *Dermocarpa* sp., *Arthrospira* sp., *Anabaenopsis* sp., *Scytonema* sp., *Stichosiphon* sp., *Xenococcus* spp., *Spirulina* spp., *Oscillatoria* spp., *Lyngbya* spp., *Nostoc* spp., *Raphidiopsis* sp., *Catenella* spp.
March 98-June 98	–	–	–	–	–	–	–	
March 99-June 99	7.9	29°C	9.8	20	172	0.76	0.32	

Contd...

Seasons	pH	Water Temp.	D.O. (ppm)	Salinity (ppt) (ppm)	Alka-linity	Phosphate (ppm)	Nitrate (ppm)	Algal Species Encountered
MONSOON								
June 97-Sept. 97	7.5	22.5°C	8.5	0.3	56	0.19	0.03	*Volvox* sp., *Pandorina* sp., *Enteromorpha* spp., *Oedogonium* sp., *Cosmarium* spp., *Chaetomorpha brachygona, Chaetomorpha aerea. Closterium* spp., *Chlorella* sp., *Radiococcus* sp., *Pediastrum,* sp., *Scenedesmus* spp., *Coscinodiscus* spp.,*Gyrosigma* sp., *Nitzs-chia* spp., *Microcystis* sp., *Ana-baenopsis* sp., *Johannes-baptistia* sp., *Xenococcus* sp., *Oscillatoria* spp., *Lyngbya* spp., *Comsopogon* sp., *Catenella* spp.
June 98-Sept. 98	–	–	–	–	–	–	–	
June 99-Sept. 99	7.6	23.5°C	10	10	98	0.31	0.14	
POST-MONSOON								
Oct 97-Nov. 97	7.5	23°C	9.5	1.8	61	0.06	0.02	*Enteromorpha* spp., *Spirogyra* spp., *Chaetomorpha aerea. Chaetomorpha brachygona, Triplastrum* spp., *Radiococcus* sp., *Pediastrum* spp., *Scenedesmus* spp., *Coscinodiscus* spp., *Anomoeneis* sp., *Nitzschia* spp., *Microcystis* sp., *Johannesbaptistia* sp., *Xenococcus* spp., *Oscillatoria* spp., *Lyngbya* spp., *Comsopogon* sp., *Catenella* spp.
Oct.98-Nov. 98	–	–	–	–	–	–	–	
Oct.99-Nov. 99	7.6	24°C	10.4	11	84	0.31	0.11	

Contd...

Seasons	*pH*	*Water Temp.*	*D.O. (ppm)*	*Salinity (ppt) (ppm)*	*Alka-linity*	*Phosphate (ppm)*	*Nitrate (ppm)*	*Algal Species Encountered*
WINTER								
Nov. 96- Feb. 97	7.5	18°C	8.8	6.2	82	0.35	0.03	*Enteromorpha* spp., *Chaetomorpha area, Chaetomorpha brachygona, Triplastrum* spp.,*Cosmarium* spp., *Pediastrum* spp., *Scenedesmus* spp., *Coscinodiscus* spp., *Synedra* spp., *Navicula radiosa, Pinnularia* sp., *Gyrosigma* sp. *Nitzschia* spp., *Microcystis* sp., *Dermocarpa* sp., *Stichosiphon* sp., *Xenococcus* sp., *Spirulina* spp., *Oscillatoria* spp., *Lyngbya* spp., *Nostoc* sp., *Compsopogon* Sp., *Catenella* spp.
Nov..97- Feb. 98	–	–	–	–	–	–	–	
Nov. 98- Feb. 99	7.6	19°C	9.6	17	110	0:63	0.16	

Contd...

Table 22 : Range of Physico-Chemical Parameters & Algal Species Encountered in Zone-II, During Different Seasons.

Seasons	*pH*	*Water Temp.*	*D.O. (ppm)*	*Salinity (ppt) (ppm)*	*Alka-linity*	*Phosphate (ppm)*	*Nitrate (ppm)*	*Algal Species Encountered*
SUMMER								
March 97-June 97	7.5	27°C	8.2	18	130	0.5	0.01	*Ulva* spp., *Uronema* sp., *Chaetomorpha* spp., *Lola* spp., *Rhizoclonium* spp., *Cladophora* sp., *Boodleopsis* sp., *Vaucheria* sp., *Melosira* spp., *Coscinodiscus* spp., *Fragilaria* sp., *Asterionella* sp., *Synedra* sp., *Achnanthes* spp., *Cocconeis* sp., *Aphanocapsa* sp., *Merismopedia* sp., *Chamaesiphon* sp., *Dermocarpa* sp., *Xenococcus* spp., *Oscillatoria* spp., *Phormidium* sp., *Lyngbya* spp., *Schizothrix* sp., *Microcoleus* sp., *Hydrocoleum* sp., *Nostoc* spp., *Anabaena* spp., *Calothrix* sp., *Gloeocapsa* sp., *Aphanothece* sp., *Catenella* spp., *Polysiphonia* spp., *Pterosiphonia* sp., *Bostrychia* spp., *Herposiphonia* sp., *Heterosiphonia* sp.
March 98-June 98	–	–	–	–	–	–	–	
March 99-June 99	8.0	28°C	8.8	30	186	1.17	0.14	

Contd...

Seasons	pH	Water Temp.	D.O. (ppm)	Salinity (ppt) (ppm)	Alka-linity	Phosphate (ppm)	Nitrate (ppm)	Algal Species Encountered
MONSOON								
June 97- Sept. 97	7.5	20°C	8.4	12	83.2	0.21	0.04	*Enteromorpha compressa*, *Ulva* spp., *Chaetomorpha* spp., *Lola* spp., *Rhizoclonium* spp., *Boodleopsis* sp., *Chara* sp., *Vaucheria* spp., *Melosira* spp., *Coscinodiscus* spp., *Achnanthes* spp., *Nitzschia* spp., *Merismopedia* sp. *Gloeocapsa* sp., *Dermocarpa* sp., *Xenococcus* spp., *Oscillatoria* spp., *Lyngbya* spp., *Catenella* spp. *Polysiphonia* spp., *Pterosiphonia* sp., *Bostrychia* spp.
June 98- Sept. 98	–	–	–	–	–	–	–	
June 99- Sept. 99	8.2	21°C	8.8	21	100.2	0.54	0.09	
POST-MONSOON								
Oct. 97- Nov. 97	7.8	18°C	9.6	16	91	0.6	0.06	*Enteromorpha compressa*, *Ulva* spp., *Chaetomorpha* spp., *Lola* spp., *Rhizoclonium* spp., *Boodleopsis* sp., *Chara* sp., *Vaucheria* spp., *Melosira* spp., *Coscinodiscus* spp., *Achnanthes* spp., *Nitzschia* spp., *Merismopedia* sp. *Gloeocapsa* sp., *Dermocarpa* sp., *Xenococcus* spp., *Oscillatoria* spp., *Lyngbya* spp., *Catenella* spp. *Polysiphonia* spp., *Pterosiphonia* sp., *Bostrychia* spp.
Oct. 98- Nov. 98	–	–	–	–	–	–	–	
Oct. 99- Nov. 99	8.4	20°C	10.8	26	108	1.14	0.09	

Contd...

Seasons	pH	Water Temp.	D.O. (ppm)	Salinity (ppt) (ppm)	Alka-linity	Phosphate (ppm)	Nitrate (ppm)	Algal Species Encountered
WINTER								
Nov. 96- Feb. 97	7.5	17.5°C	8.8	17	96	0.5	0.01	*Enteromorpha compressa*, *Ulva* spp., *Chaetomorpha* spp., *Lola* spp., *Rhizoclonium* spp., *Boodleopsis* sp., *Chara* sp., *Vaucheria* spp., *Melosira* spp., *Coscinodiscus* spp., *Fragilaria* sp., *Synedra* spp., *Diatoma* sp., *Achnanthes* spp., *Cocconeis* sp., *Navicula cryptocephala*, *Stauroneis* sp., *Gyrosigma* sp., *Nitzschia* spp., *Colpomenia* sp., *Microcystis* sp., *Aphanocapsa* sp., *Merismopedia* sp., *Chamaesiphon* sp., *Oscillatoria* spp., *Lyngbya* spp., *Schizothrix* sp., *Nostoc* spp., *Anabaena* spp., *Catenella* spp., *Polysiphonia* spp., *Pterosiphonia* sp., *Bostrychia* spp., *Herposiphonia* sp., *Heterosiphonia* sp., *Dictyota* sp.
Nov. 97- Feb. 98	–	–	–	–	–	–	–	
Nov. 98- Feb. 99	8.0	19°C	10.2	28	131	1.2	0.03	

Table 23 : Range of Physico-Chemical Parameters & Algal Species Encountered in Zone-III, During Different Seasons.

Seasons	*pH*	*Water Temp.*	*D.O. (ppm)*	*Salinity (ppt) (ppm)*	*Alka-linity*	*Phosphate (ppm)*	*Nitrate (ppm)*	*Algal Species Encountered*
SUMMER								
March 97-June 97	8.0	30°C	7.8	30	115	0.2	0.06	*Chaetomorpha* spp., *Lola* spp., *Rhizoclonium* spp., *Vaucheria* spp., *Coscinodiscus* spp., *Asterionella* sp., *Achnanthes* sp., *Pinnularia* sp., *Pleurosigma* sp., *Oscillatoria* spp., *Lyngbya* spp., *Microcoleus* sp., *Hydrocoleum* sp., *Nostoc* spp., *Anabaena* spp., *Catenella* spp., *Polysiphonia* spp., *Pterosiphonia* sp., *Bostrychia* spp.
March 98-June 98	–	–	–	–	–	–	–	
March 99-June 99	8.1	31°C	8.4	37	160	0.6	1.15	

Contd...

Seasons	*pH*	*Water Temp.*	*D.O. (ppm)*	*Salinity (ppt) (ppm)*	*Alka-linity*	*Phosphate (ppm)*	*Nitrate (ppm)*	*Algal Species Encountered*
MONSOON								
June 97- Sept. 97	7.7	26°C	8.8	20	51	0.2	0.01	*Chaetomorpha* spp., *Lola* spp. *Rhizoclonium* spp., *Spirogyra* spp., *Pediastrum* spp., *Scenedesmums* spp., *Coscinodiscus* spp., *Synedra* sp., *Nitzchia* spp., *Chamaesiphon* sp., *Xenococcus* spp., *Oscillatoria* spp., *Lyngbya* spp., *Catenella* spp. *Polysiphonia* spp., *Pterosiphonia* sp., *Bostrychia* spp.
June 98- Sept. 98	–	–	–	–	–	–	–	
June 99- Sept. 99	8.0	27°C	9.8	26	75	0.5	0.06	
POST-MONSOON								
Oct. 97- Nov. 97	7.3	22°C	9.8	22	53	0.2	0.01	*Chaetomorpha* spp. *Lola* spp., *Rhizoclonium* spp., *Spirogyra* spp., *Pediastrum* spp., *Scenedesmus* spp., *Coscinodiscus* spp., *Biddulphia* spp., *Synedra* sp., *Nitzschia* spp., *Chamaesiphon* sp., *Dermocarpa* sp., *Xenococcus* spp., *Oscillatoria* spp., *Lyngbya* spp., *Mastigocoleus* spp., *Catenella* spp., *Polysiphonia* spp., *Pterosiphonia* sp., *Bostrychia* spp.
Oct. 98- Nov. 98	–	–	–	–	–	–	–	
Oct. 99- Nov. 99	8.2	27°C	10.6	29	75	0.5	0.03	

Contd...

Seasons	pH	Water Temp.	D.O. (ppm)	Salinity (ppt) (ppm)	Alka-linity	Phosphate (ppm)	Nitrate (ppm)	Algal Species Encountered
WINTER								
Nov. 96- Feb. 97	8.0	19°C	9.6	29	99	0.3	0.01	*Chaetomorpha* spp., *Lola* sp., *Rhizoclonium* spp., *Pediastrum* spp., *Scenedesmus* sp., *Vaucheria* spp., *Melosira* spp., *Coscinodiscus* spp., *Stephanopyxis* sp., *Hemidiscus* sp., *Biddulphia* spp., *Chaetoceros* spp., *Climacodium* sp., *Bacteriastrum* spp., *Corethron* sp., *Rhizoslenia* spp., *Fragilaria* sp., *Asterionella* sp., *Synedra* spp., *Diatoma* sp., *Gomphonema* sp., *Amphora* spp.,*Cymbella* sp., *Nitzschia* spp., *Chamaesiphon* sp., *Dermocarpa* sp., *Xenococcus* spp., *Phormidium* sp., *Oscillatoria* spp., *Lyngbya* spp., *Nostoc* spp., *Anabaena* spp., *Catenella* spp., *Polysiphonia* spp., *Pterosiphonia* sp., *Bostrychia* spp., *Herposiphonia* sp., *Heterosiphonia* sp., *Gelidiella* sp., *Dictyota* sp.
Nov. 97- Feb. 98	–	–	–	–	–	–	–	
Nov. 98- Feb. 99	8.4	20°C	10.2	33	120	0.7	0.05	

All 150 species identified are alkaliphilous and mesothermal; 26 species are oligohaline, 17 species are mesohaline, 32 species are euryhaline, 13 species are oligo-mesohaline, 32 species are meso-euryhaline and the rest, *i.e.*, 30 are oligo-meso-euryhaline; 85 species are saprophobic, 40 species are saproxenous, 16 species are saprophilic, 9 species are saprobiontic; 20 species are limnophilous, 49 species are limnobiontic, 24 species are rheophilous and the rest are rheobiontic.

The observations made on distribution of phytoplankton reveals the fact that diatoms constitute the most important and dominant algae in both freshwater and marine habitats, whereas the desmids and chlorococcalean green algae are equally important in many freshwater habitats; planktonic desmids normally occur only in calcium-poor waters; diatoms are often abundant in calcareous water with mono: divalent ratio of less than 1.5 and addition of potassium phosphate changes the dominance of algal populations from blue-green to desmids and the blue-green or cyanophyceae blooms dominate the eutropic lakes but at times chlorococcales may also dominate, though inspite of all these observations it can be said that distribution is governed not by one but a combination of factors (Kumar, 1999).

The range of physico-chemical parameters and the algal forms encountered in different zones during different seasons is tabulated in Tables 21, 22 and 23. The studies were conducted on bimonthly basis covering all the four dominant seasons, *viz.*, summer (March-June), monsoon (June- September), post-monsoon (October-November) and winter (November-February), since November, 1996 to November, 1999.

In Table 21, the above mentioned details are tabulated for Zone - I, *i.e.*, sampling sites in the nothern parts of Indian Sundarbans. The average pH was found to vary between 7.5 and 7.8, with the highest being in summer and lowest during winter, the annual average water temperature varied between 19 - 28°C, the average D.O. varied between 8.5 to 10 ppm, with maximum noted during post monsoon and minimum during summer, the salinity ranges between 0.8 to 18 ppt, with highest during summer and lowest during post monsoon, the average alkalinity ranges between 72 to 150 ppm, with lowest during monsoon and highest during summer, the average phosphate varies between 0.26 to 0.76 ppm. with highest during summer and lowest during post monsoon and the nitrate content varies between 0.05 to 0.28 ppm, with low values observed during monsoon, post - monsoon, as well as, winter and high value during summer. The algae which are characteristically found throughout the year in this zone are *Enteromorpha* spp., *Chaetomorpha aerea, Chaetomorpha brachygona, Scenedesmus* spp., *Pediastrum* spp., *Coscinodiscus* spp., *Nitzschia* spp., *Micrcystis* spp., *Xenococcus* spp., *Oscillatoria* spp. and *Lyngbya* spp. The red alga *Compsopogon* sp. is absent during summer and the rare blue green algae *Johannesbaptistia pellucida* is present only during monsoon and post- monsoon, thus revealing their preference for low saline environment. The species diversity is maximum during winter and minimum during post monsoon. Nowadays, the diversion of upstream freshwater through the river Padma of Bangladesh, *vis-a-vis*, less amount of upstream freshwater discharge in Sundarbans mangals may have direct correlation with declining of fish species from Sundarbans area of India.

In Pie charts-IIIa, IIIb, IIIc and IIId the seasonal availability of different groups of algae during different seasons in zone I have been depicted. During summer the Chlorophyta contributes to 30%, the Chrysophyta 22%, the Rhodophyta 4% and the Cyanophyta 44% of the total algal population. During monsoon, the Chlorophyta contributes to 48%, the Chrysophyta 13%, the Rhodophyta 9% and the Cyanophyta 30% of the total algal population. During post - monsoon, the Chlorophyta contributes to 44% the Chrysophyta 17%, the Rhodophyta 11% and Cyanophyta 28% of the total algal population. During winter the Chlorophyta contributes to 30% the Chrysophyta 26%, the Rhodophyta 9% and the Cyanophyta 35% of the total algal population.

In Table 22, the above mentioned details are tabulated for zone II, *i.e.*, central part of Indian Sundarbans. The pH ranges between 7.5 to 8.4 with highest value observed during post monsoon. The average water temperature ranges between 17.5 to 28°C. The average dissolved oxygen varies between 8.2 (monsoon) and 10.8 (post-monsoon), the salinity ranges from 12-30 ppt with highest value during summer and lowest during monsoon, the alkalinity ranges between 83.2 to 186 ppm with highest during summer and lowest during monsoon, the average phosphate varies between 0.21 to 1.17 with highest during summer and lowest during monsoon. The average nitrate varies between 0.01 to 0.09ppm, with highest during post - monsoon and monsoon and lowest during winter. The algal species which are constantly available throughout the year in zone II are *Ulva* spp., *Chaetomorpha* spp., *Lola* spp., *Rhizoclonium* spp.,

Boodleopsis sp., *Vaucheria* spp., *Melosira* sp., *Coscinodiscus* spp., *Achnanthes* spp., *Nitzschia* spp., *Merismopedia* spp., *Dermocarpa* spp., *Xenococcus*, spp., *Oscillatoria* spp., *Lyngbya* spp., *Catenella* spp., *Polysiphonia* spp., *Pterosiphonia* sp., and *Bostrychia* spp. The highest species diversity is observed during winter and lowest during monsoon. In summer too a high algal diversity is observed but the blue-green algae dominate in this species composition.

In Pie-Charts-IVa, IVb, IVc and IVd, the seasonal availability of different groups of algal during different seasons in zone-II have been depicted. During summer the Chlorophyta contributes to 17%, the Chrysophyta 26%, the Phaeophyta 2%, the Rhodophyta 17% and Cyanophyta 22% of the total algal population. During monsoon, the Chlorophyta contributes to 31%, the Chrysophyta 22%, the Phaeophyta 4%, the Rhodophyta 17% and the Cyanophyta 22% of the total algal population. During post-monsoon, the Chlorophyta contributes to 31%, the Chrysophyta 26%, the Phaeophyta 4%, the Rhodophyta 17% and the Cyanophyta 22% of the total algal population. During winter the Chlorophyta contributes to 20%, the Chrysophyta 30%, the Phaeophyta 2%, the Rhodophyta 15% and the Cyanophyta 30% of the total algal population.

In Table 23, the detailed physico-chemical report and algal species encountered in zone III, *i.e.*, the southern parts of Indian Sundarbans have been tabulated. The pH ranges between 7.7 and 8.4, with highest during winter and lowest during monsoon. The average annual water temperature varies between 19 to 31°C. The dissolved oxygen varies between 7.8 to 10.6 ppm, with highest during post monsoon and lowest during summer. The salinity ranges between 20 to 37 ppt, with highest during summer and lowest during monsoon.The average alkalinity ranges between 51 to 160 ppm with highest during summer and lowest during monsoon. The average phosphate varies between 0.2 to 0.7 ppm with highest during winter and lowest during monsoon, as well as, post-monsoon. The average nitrate varies between 0.01 and 1.15 ppm, with the highest during summer and lowest during post monsoon.The maximum speices diversity is observed during winter and lowest during monsoon. The species which characteristically occur throughout the year in this zone are *Chaetomorpha* spp., *Lola* spp., *Rhizoclonium* spp., *Coscinodiscus* spp., *Synedra* spp., *Nitzschia* spp., *Chamaesiphon* sp.,

Xenococcus spp., *Oscillatoria* spp., *Lyngbya* spp., *Catenella* spp., *Polysiphonia* spp., *Pterosiphonia* sp. and *Bostrychia* spp.

In Pie-Charts-Va, Vb, Vc and Vd, the seasonal availability of different groups of alga during different seasons in zone-II have been depicted. During summer the Chlorophyta contributes to 11%, the Chrysophyta 40%, the Phaeophyta 4%, the Rhodophyta 15% and Cyanophyta 30% of the total algal population. During monsoon, the Chlorophyta contributes to 33%, the Chrysophyta 27%, the Phaeophyta 6%, the Rhodophyta 22% and the Cyanophyta 22% of the total algal population. During post-monsoon, the Chlorophyta contributes to 28%, the Chrysophyta 19%, the Phaeophyta 5%, the Rhodophyta 19% and the Cyanophyta 29% of the total algal population. During winter the Chlorophyta contributes to 12%, the Chrysophyta 46%, the Phaeophyta 2%, the Rhodophyta 20% and the Cyanophyta 20% of the total algal population.

Thus, it is observed, that gradually as we move from the northern parts of Indian Sundarbans to the central parts and down to the southern parts the algal population is gradually replaced from the Chlorophyta dominated population to Chrysophyta and Rhodophyta dominated population.

7.3 Algal Productivity and Species Diversity Index

Mangrove ecosystems are reputed to be highly productive, their productivity being a function of their photosynthetic ability (Snedekar, 1980); expressed by the amount of other life forms which thrive and prefer this ecosystem for their total life time. The high productivity of this eco-system is also expressed by the amount of litter production which on decomposition and release of nutrients enrich this ecosystem and sustain other groups of living organisms, as well as. The major photosynthetic organism include the mangroves, the macro and micro algal components (benthic or periphytic) and the phytoplankton in the water system.

7.3.1 Shannon-Weaver Function : Species Diversity Index

In the Tables 24, 25, 26 and 27 the plankton distribution/plankton load, species diversity index, species richness, chlorophyll content, light intensity values and primary productivity of water samples from estuarine wetlants or fisheries in three different zones and different seasons on the stretch of an estuarine river-Bidyadhari have been highlighted.

Table 24 : SDI, Species Richness, Chlorophyll a Content and Primary Productivity of Water Samples From Estuarine Wetlands During Summer.

Sample No.	Shannon Weaver Index (SDI) $\Sigma pi \log^2 Pi$			Species Richness		Gross Primary Production (mgC/M³/hr)	Net Primary Production (mgC/M³/hr)	Total Chlorophyll content (mg/l)	Light Intensity (X 100 lux)
	Total	Phyto-plankton	Zoo-plankton	Phyto-plankton	Zoo-plankton				
F1	2.14	1.85	0.99	6	5	259.3	75.5	0.06	1560
F2	2.40	1.96	0.86	6	4	218.6	83.3	0.04	1568
F3	1.97	1.88	1.23	5	4	208.3	83.3	0.03	1742
F4	1.61	2.16	1.54	7	3	245.2	125	0.08	2184
F5	2.63	2.31	1.73	9	5	266.6	104.2	0.11	1634
F6	2.65	2.52	1.56	10	3	364.16	125.0	0.19	1890

F1 & F2 : Sampling sites in the low saline stretch.

F3 & F4 : Sampling sites in the medium saline stretch.

F5 & F6 : Sampling sites in high saline stretch.

Table 25 : SDI, Species Richness, Chlorophyll a Content and Primary Productivity of Water Samples From Estuarine Wetlands During Monsoon

Sample No.	Shannon Weaver Index (SDI) $\Sigma pi \log^2 Pi$ Total	Phyto-plankton	Zoo-plankton	Species Richness Phyto-plankton	Zoo-plankton	Gross Primary Production (mgC/M³/hr)	Net Primary Production (mgC/M³/hr)	Total Chlorophyll content (mg/l)	Light Intensity (X 100 lux)
F1	2.65	1.53	1.82	5	6	229.2	41.6	0.02	645
F2	2.63	1.30	1.91	4	5	166.6	83.3	0.01	657
F3	2.48	1.50	1.60x	3	4	291.6	98.6	0.04	869
F4	2.42	1.5	1.88	3	5	208.3	98.6	0.01	815
F5	1.65	1.48	0.73	3	3	125.0	53.3	0.02	676
F6	2.38	1.80	0.80	7	3	354.2	92.5	0.06	630

F1 & F2 : Sampling sites in the low saline stretch.

F3 & F4 : Sampling sites in the medium saline stretch.

F5 & F6 : Sampling sites in high saline stretch.

Table 26 : SDI, Species Richness, Chlorphyll a Content and Primary (Producivity of Water Samples From Est ıarine Wetlands During Post-Monsoon

Sample No.	Shannon Weaver Index (SDI) $\Sigma pi \log Pi$ Total	Phyto-plankton	Zoo-plankton	Species Richness Phyto-plankton	Zoo-plankton	Gross Primary Production (mgC/M³/hr)	Net Primary Production (mgC/M³/hr)	Total Chlorophyll content (mg/l)	Light Intensity (X 100 lux)
F1	2.53	1.61	1.82	4	6	98.6	56.3	0.03	815
F2	2.14	2.16	0.99	5	4	125	69.3	0.04	809
F3	1.91	1.5	0.99	3	3	83.3	76.2	0.01	823
F4	1.8	1.88	0.98	4	4	166.6	80.8	0.05	861
F5	2.63	2.31	2.14	5	6	125	64.3	0.03	698
F6	2.38	2.63	1.52	7	3	232.3	64.3	0.07	726

Fl & F2 : Sampling sites in the low saline stretch.

F3 & F4 : Sampling sites in the medium saline stretch.

F5 & F6 : Sampling sites in the high saline stretch.

Table 27 : SDI, Species Richness, Chlorophyll a Content and Primary Productivity of Water Samples From Estuarine Wetlands During Winter

Sample No.	Shannon Weaver Index (SDI) $\Sigma pi \log^2 Pi$ Total	Phyto-plankton	Zoo-plankton	Species Richness Phyto-plankton	Zoo-plankton	Gross Primary Production (mgC/M³/hr)	Net Primary Production (mgC/M³/hr)	Total Chlorophyll content (mg/l)	Light Intensity (X 100 lux)
F1	1.53	1.27	1.82	5	9	104.2	20.8	0.04	653
F2	1.76	2.16	0.63	6	7	104.2	62.5	0.18	719
F3	1.76	2.20	0.99	6	4	61.6	40.8	0.02	904
F4	2.12	1.85	1.31	4	2	125	40.8	0.11	984
F5	1.61	2.12	1.46	5	4	98.6	56.4	0.19	996
F6	2.47	1.96	0.99	6	2	83.3	41.6	0.04	1072

F1 & F2 : Sampling sites in the low saline stretch.

F3 & F4 : Sampling sites in the medium saline stretch.

F5 & F6 : Sampling sites in the high saline stretch.

In the estuarine wetlands of Stretch I, *i.e.,* in the upper stretches of the estuarine river, the Shannon Weaver Diversity Index is found to vary between 1.53 and 2.65 for the total plankton, between 1.30 and 2.56 for the phytoplankton and between 0.63 and 2.23 for the zooplankton. Total plankton diversity is found to be highest during monsoon and lowest during summer. The phytoplankton diversity is found to be highest in winter and lowest during monsoon and zooplankton diversity is found to be highest in winter and lowest during summer. Species richness is found to be highest during monsoon with the phytoplankton dominating during summer and monsoon periods and the zooplankton dominate during winter. Gross Primary Production and Net Primary Production is highest during summer and least during summer.

In the estuarine wetlands of Stretch II, *i.e.,* in the middle stretches of the estuarine river, the Shannon Weaver Diversity Index is found to vary between 1.61 and 2.48 for total plankton, between 1.50 and 2.20 for the phytoplankton and between 0.63 and 1.88 for the zooplankton. The total diversity is found to be highest during monsoon and lowest during summer. The phytoplankton diversity is found to be highest during winter and lowest during monsoon and the zooplankton diversity is found to be lowest during winter highest during monsoon. Species richness is highest during summer and lowest during monsoon with the phytoplankton dominating during summer and winter and zooplankton dominating during monsoon. The overall Gross Primary Production and Net Primary Production is highest during summer and lowest during winter.

In the estuarine wetlands of Stretch III, *i.e.,* the lower stretch of the estuarine river, the Shannon Weaver Diversity Index is found to vary between 1.61 and 2.65 for the total plankton, between 1.48 and 2.52 for the phytoplankton and between 0.73 and 1.73 for the zooplankton. The highest plankton diversity is observed during summer and lowest during winter. The phytoplankton diversity is found to highest during summer and lowest during monsoon. The zooplankton diversity is found to highest during summer and lowest during monsoon. The species diversity is found to be highest during summer and lowest during monsoon with the phytoplankton dominating during all the seasons. The Gross Primary Production and Net Primary Production is found to be highest during summer and lowest during winter.

Graph–IIIa, IIIb and IIIc, the effect of Chlorophyll a content on the GPP and NPP in 3 different zones (stretches of the same estuarines river with salinity gradient, *viz.,* Zone-I-low saline, Zone-II-mid saline and Zone-III-high saline regimes) during different seasons, *viz.,* summer, monsoon, post-monsoon and winter has been evaluated. During the summer months there is a steady increase in the chlorophyll a content from stretch I to stretch III. GPP as well as, NPP are both found to increase with the increase in chlorophyll a content and notably the highest production in all the three zones is observed during summer months when temperature and light condition as well are conducive for better production. It is observed that during monsoon, there is a steady increase in the chlorophyll a content with simultaneous increase in the GPP, though the NPP has a slightly lower value in stretch III, specially in one set of wetland, which was found to have more of zooplankton during the sampling time. During the winter months, the average chlorophyll a content is found to decrease in stretch I with least NPP and GPP as well.

Graphs IIa, IIb & IIc, depict the effect of phosphate and nitrate on the NPP in the three stretches. In stretch I, during the summer the phosphate and nitrate are highest with high NPP value as well, in winter though the phosphate and nitrate values are quite low there is not much effect on the NPP. In stretch II, high value of phosphate and nitrate are observed during summer along with the highest NPP value. In stretch III, the highest phosphate is observed during winter and highest nitrate value during summer and a simultaneous high NPP during summer. Graph–Ia, Ib and Ic, represents the effect of salinity and alkalinity on the NPP. In stretch I maximum salinity as well as alkalinity is observed during summer and minimum during monsoon.Maximum NPP is also observed during summer. In stretch II the maximum salinity and alkalinity is observed during summer and minimum during monsoon and the maximum NPP is observed during summer and minimum during winter. In stretch III, the maximum salinity and alkalinity is observed during summer with highest values of NPP. A significant trend is that productivity (NPP or GPP) irrespective of their zones is highest during the summer months and minimum during the winter months. It is generally believed that temperature plays an important role in the aquatic ecosystems, though it is not the only limiting factor (Chako & Krishnamurthy 1954: Nasar, 1977).

7.4 Zooplankton

Though not very extensively studied, the zooplankters definitely have a very important role in maintaining the sustainability of the estuarine and brackish water ecosystems of Sundarbans. The zooplankters feed on the algae and the vast algal groups of this ecosystem serve as the primary food material for not only the zooplankters but also the higher faunal groups, specially the fin-fish and shell-fish species at least in their early juvenile stages; the zooplankters occupy the position of the primary consumers in the trophic levels of the complicated food chains operating in this versatile ecosystem. In this way the zooplankters play quite important and significant role which requires much critical studies covering their annual life cycles and dominant phases, which was beyond the scope of this present study. As such, only a few species of zoolankton encountered in the estuarine and brackish water fisheries were collected with the help of plankton nets and identified.

The zooplankters collected and identified from different habitats of Indian Sundarbans during different seasons and different zones have been highlighted in Table 28.

Table 28: List of Zooplankters Collected & Identified from the Indian Sundarbans.

Rotifers		*Cladocerons*	
1. *Brachionus* spp.	very common	1. *Daphnia* sp.	common
2. *Keratella* spp.	common	2. *Moina* sp.	occasional
3. *Filinia* sp.	common	3. *Diaphanosoma* sp.	common
4. *Lecane* sp.	rare	**COPEPODS**	
5. *Polyarthra* sp.	rare	1. *Cyclops* sp.	very common
6. *Asplanchna* sp.	occassional	2. *Diaptomus* sp.	common
7. *Synchaeta* sp.	occasional	3. *Nauplius* larvae	very common

7.5 Productivity vs. Abiotic Variables

Productivity in the aquatic systems mainly depend on two dominant factors, *viz.*, (a) solar energy and (b) nutritents. The Net Primary Productivity of the estuarine wetlands of Sundarbans was thus correlated with Light Intensity, Salinity, Alkalinity, Phosphate, Nitrate and Phospate: Nitrate.

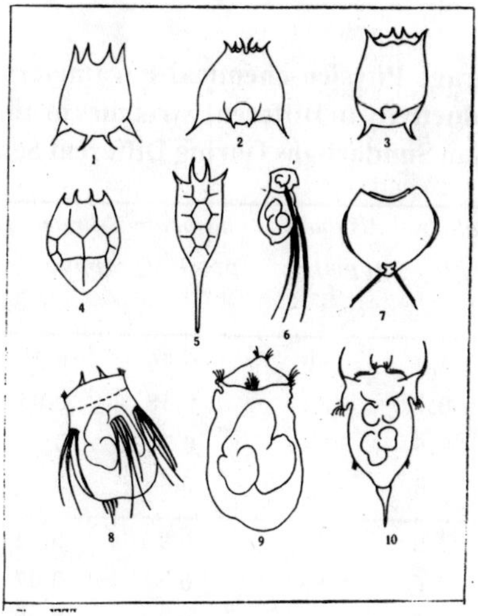

Plate XXVI

Figs.1-3. *Brachionus* spp.; Figs. 4-5. *Keratella* spp.; Fig.6. *Filinia* sp.; Fig. 7. *Lecane* sp.; Fig. 8. *Polyarthra* sp.; 9. *Asplanchna* sp.; Fig. 10 *Synchaeta* sp.

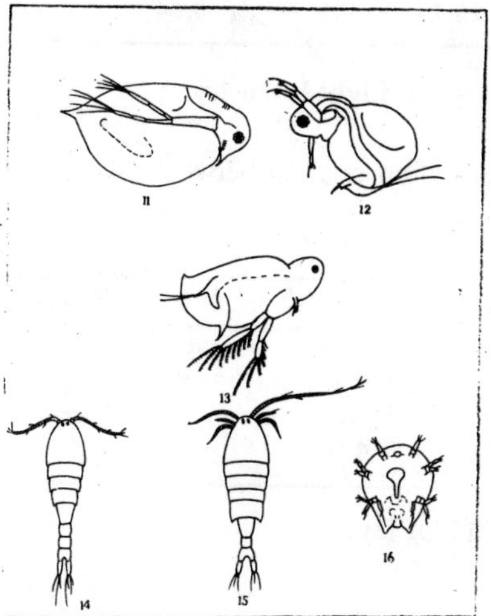

Plate XXVII

Figs.11. *Daphnia* spp.; Fig. 12. *Moina* spp.; Fig.13. *Diaphanosoma* sp.; Fig. 14. *Cyclops* sp.; Fig.15. *Diaptomus* sp.; Fig. 16 *Nauplius* larvae.

Table 29 : Average Physico-chemical Parameters and Net Primary Productivity in Different Stretches of an Estuarine River of Indian Sundarbans During Different Seasons.

	Seasons	*Salinity (ppt)*	*Alkalinity (ppm)*	*Phosphate (ppm)*	*Nitrate (ppm)*	*Light Intensity (X 100 lux)*	*N.P.P. mgC/ m^3/h)*
Upper Stretch	Summer	8.0	150	0.76	0.28	1564	79.4
	Monsoon	0.7	72	0.38	0.05	651	62.45
	P.Monsoon	2.6	77.8	0.36	0.05	812	62.8
	Winter	6.8	96	0.56	0.18	686	41.65
Middle Stretch	Summer	15.0	156	0.96	0.12	1963	104.14
	Monsoon	5.6	88	0.34	0.07	842	98.6
	P.Monsoon	8.8	96	0.40	0.09	712	78.5
	Winter	10.0	108	0.38	0.04	944	40.8
Lower Stretch	Summer	28.0	140	0.52	0.09	1762	114.6
	Monsoon	12.0	80	0.36	0.04	653	72.9
	P.Monsoon	16.4	82	0.42	0.04	712	64.3
	Winter	26.7	110	0.64	0.06	1034	49.0

7.5.1 : Productivity *vs.* Light Intensity

Co-efficient of correlation between Net Primary Productivity and Light Intensity.

	Upper Stretch	*Middle Stretch*	*Lower Stretch*
Summer	+ 0.06	-0.593	-0.002
Monsoon	+ 0.62	+ 0.006	+ 0.86**
Post Monsoon	+ 0.59	- 0.005	- 0.09
Winter	+ 0.56	+ 0.009	+ 0.66

** Significant at 0.01 level.

The co-efficient of correlation between the Net Primary Productivity and light intensity of estuarine wetlands during different seasons at different stretches of the same estuarine river of Indian Sundarbans reveals a positive correlation in most of the cases (66.6%), with one value being significant at 0.01 level. This means that though increased light intensity has a positive effect on the Net Primiary Productivity during most of the time the productivity becomes light saturated; since during monsoon, such a situation may not arise the significant value occurs during this season.

Moore, *et.al.* (1972, 1973) have observed a typical hyperbolic photosynthetic response to increasing light flux densities with photosynthesis becoming light saturated at one third to one half full sunlight. It can be drawn from this observation that in the water near shaded forest areas or under canopy areas with mangrove cover productivity must be significantly high than in the open water system where light is saturated.

7.5.2 : Productivity *vs.* Salinity

Co-efficient of correlation between Net Primary Productivity and Salinity

	Upper Stretch	*Middle Stretch*	*Lower Stretch*
Summer	– 0.77*	–0.149	–0.008
Monsoon	– 0.62	– 0.713*	– 0.217
Post Monsoon	– 0.76*	+ 0.16	– 0.778*
Winter	– 0.08	– 0.116	– 0.749*

* Significant at 0.05 level.

The co-efficient of correlation between the Net Primary Productivity and light intensity of esturarine wetlands during different seasons at different stretches of the same estuarine river of Indian Sundarbans reveals negative correlation in most of the cases (91.6%) with five values significant at 0.05 level. Thus with increase in salinity, the Net Primary Productivity shows a declining trend. This observation has been supported earlier by Babulal (1990), where he studied the productive potentials at different centres of the Hooghly estuary and found that on an average freshwater zone is more productive than saline zone. Furthermore, Clough (1985) has observed that rate of growth and Net Primary Productivity of mangroves could be expected to fall by at least 50% with an increase in salinity from 20$\%_{oo}$ to 35$\%_{oo}$, with a further significant decrease in primary productivity at higher salinity. Also Ball and

Farquhar (1984 a, b) found that photosynthesis decreases with increased salinity.

7.5.3 : Productivity *vs.* Alkalinity

Co-efficient of correlation between Net Primary Productivity and Alkalinity

	Upper Stretch	*Middle Stretch*	*Lower Stretch*
Summer	+ 0.552	+ 0.374	+ 0.654
Monsoon	+ 0.748*	+ 0.989**	- 0.183
Post Monsoon	- 0.558	+ 0.874**	+ 0.758*
Winter	+ 0.749-	+ 0.868**	+ 0.817*

* Significant at 0.05 level.
** Significant at 0.01 level.

The co-efficient of correlation between the Net Primary Productivity and alkalinity of some estuarine wetlands during different seasons at different stretches of the same estuarine rive of Indian Sundarbans reveals positive correlation in most of the cases (83.3%) with 3 values being significant at 0.01 level and 3 values significant at 0.05 level. This shows that increase in alkalinity leads to simultaneous increase in productivity. Chattopadhyay (1999) has noted that the brackish water fish pond soils generally exhibit neutral to slightly alkaline pH values which are conducive for good aquaculture. Furthermore, Chakroborti and Das (1999) have observed that higher alkalinity is an indication of higher productivity potential of the Sundarbans brackish water fisheries.

7.5.4 : Productivity *vs.* Phosphate

Co-efficient of correlation between Net Primary Productivity and Phosphate.

	Upper Stretch	*Middle Stretch*	*Lower Stretch*
Summer	+ 0.96*	+ 0.89**	+ 0.56
Monsoon	+ 0.77*	– 0.58	– 0.10
Post Monsoon	+ 0.69	+ 0.74*	– 0.62
Winter	+ 0.87**	+ 0.73*	+ 0.71*

* Significant at 0.05 level
** Significant at 0.01 level.

The co-efficient of correlation between the Net Primary Productivity and phosphate of some selected estuarine wetlands during different seasons at different stretches of the same estuarine river of Indian Sundarbans reveals a positive correlation in most of the cases (75%) with 2 values significant at 0.01 level and 5 alues significant at 0.05 level, indicating a high possibility of increased production with increase in phosphate.

7.5.5 Productivity *vs.* Nitrate

Co-efficient of correlation between Net Primary Productivity and Nitrate

	Upper Stretch	*Middle Stretch*	*Lower Stretch*
Summer	+ 0.699	− 0.586	+ 0.792*
Monsoon	+ 0.667	+ 0.053	+ 0.384
Post Monsoon	+ 0.904**	+ 0.496	− 0.490
Winter	+ 0.728*	+ 0.033	+ 0.634

* Significant at 0.05 level.
** Significant at 0.01 level.

The co-efficient of correlation between the Net Primary Productivity and phosphate of some selected estuarine wetlands during diferent seasons at different stretches of the same estuarine river of Indian Sundarbans reveals a positive correlation in most of the cases (83.3%). Not many significant results have been obtained in this correlation with only one result being significant at 0.01 level and two results significant at 0.05 level. Kuzmiczewa (1965) has reported that the ratio P : N limits the efficacy of nitrogenous fertilization of water bodies and also formulated that phosphorous deficit clearly had an adverse effect on the full utilization of five times as great nitrogen concentration.

7.5.6 Productivity *vs.* Phosphate : Nitrate

Co-efficient of correlation between Net Primary Productivity and Phosphate : Nitrate

	Upper Stretch	*Middle Stretch*	*Lower Stretch*
Summer	−0.216	+ 0.296	+ 0.848**
Monsoon	+ 0.486	− 0.002	+ 0.902**
Post Monsoon	+ 0.746*	+ 0.885**	+ 0.446
Winter	+ 0.793*	+ 0.945**	+ 0.459

* Significant at 0.05 level.
** Significant at 0.01 level.

The co-efficient of correlation between the Net Primary Productivity and ratio of phosphate and nitrate of some selected estuarine wetlands during different seasons at different stretches of the same estuarine river of Indian Sundarbans reveals a positive correlation with increase in phosphate : nitrate ratio in 91.6% cases, with 4 results significant at 0.01 level and 2 results significant at 0.05 level. Wolny (1967) has reported that for increased production of water bodies the ratio of phosphorus to nitrogen could be kept wider apart in well mineralized waters and even more where pond bottom is alkaline.

7.6 Algal Communities in Indian Sundarbans

Mangrove areas develop with deposition of alluvium, *vis-a-vis,* accretion of soil in the the seaward river bed areas, with low oxygen content and abundant hydrogen sulphide, fine grained, often semi-fluid and ill-consolidated (Mac Nae, 1968). Biological processes involving bacteria, blue green algae, diatoms and green algae particularly nitrifying and sulphur reducing organisms are necessary before this alluvium can be utilized by plants, *i.e.,* mangroves (Schuster, 1952). The blue-green and green algae show substratum preferences and the blue-green algae are responsible for paving way for the green algae. The blue-green algae are noted to prefer a soft, hydrophilic, biologically active mud rich in organic matter, whereas, the green algae prefer a more consolidated type of soil rich in phosphates and nitrates which have previously been released by the activity of the blue-green algae and bacterial flora in the soil. The soil thus inhabited by the green algae subsequently gets inhabited by the pioneer mangrove grass *Porteresia coarctata* followed by the pioneer mangrove species of *Avicennia* spp. or *Sonneratia* spp. (Naskar and Mandal, 1999). Thus, the algae are important in stabilising the soil through aggregation of soil particles, by contributing to soil nutrients, crusts of Cyanophyceae help in the retention of silt and clay and aid in the rearrangement of soil particles leading to increase in organic carbon and nitrogen contents of the soil, improve water infiltration, decrease erosion (Kumar, 1999). These processes pave way for initializing the long succes sion pattern followed by the mangroves in this ecosystem (Naskar and Mandal, 1999).

The algal communities found in Sundarbans indicate some specific associations or assemblages of species that recur under similar ecological conditions in different places. These assemblages may have a single species as the dominant species or may be an assemblage of a few sub dominant species. The most typical community or encountered in the estuarine habitats are :

(i) *Catenella-Bostrychia-Caloglossa* association

This type of association is very commonly seen as dense mats on the bark of *Excoecaria agallocha, Aegialitis rotundifolia,* prop and stilt roots of *Rhizophora* and *Bruguiera*, pneumatophores of *Xylocarpus* and sometimes *Sonneratia* in the sea ward fringes, regularly inundated areas in the northern, central and southern parts of Indian Sundarbans. This association is mainly dominated by *Catenella* spp. and at times a few individual plants of *Polysiphonia/ Pterosiphonia* may also occur alongwith blue-green algae like *Hydrocoleum, Schizothrix* and *Scytonema.*

(ii) *Polysiphonia-Catenella-Bostrychia* association

This association is also as common as the earlier mentioned association and is found as thick mat on the stilt roots of *Rhizophora*, knee roots of *Bruguiera,* pneumatophores of *Xylocarpus* and barks of *Aegialitis rotundifolia, Ceriops, Aegiceras corniculatum, Bruguiera* spp. The dominant algae in this community is *Polysiphonia* and the other co-occurring species are *Pterosiphonia, Microcoleus, Schizothrix* and *Hydrocoleum and* is found in the central and southern parts of Indian Sundarbans.

(iii) *Pterosiphonia-Polysiphonia-Catenella* association

This association is not very commonly encountered and was mainly found in Jharkhali in the central part of Indian Sundarbans, growing on the bark of *Aegialitis rotundifolia. Pterosiphonia* forms the dominant species with few *Caloglossa, Bostrychia* individuals. An epiphyte occurs in this association, *viz., Achnanthes* spp. on the fronds of *Pterosiphonia.*

(iv) *Rhizoclonium-Cladophora-Chaetomorpha* association

This is a very common association encountered in slightly upper littoral zones on the barks and aerial roots of various mangrove species, forming dense green mats on the exposed parts infrequently inundated. *Rhizoclonium* is the dominant species with a few *Cladophora* as an infrequent co-species. *Chaetomorpha* is very low in this association. This association also is encountered in the sea ward fringes of the mangals in the central and lower parts of Sundarbans.

(v) *Gelidiella-Dictyota-Bostrychia* association

This is a very uncommon association and was encountered only in the Dhanchi Island area of Sundarbans growing as dense mats, as forest floor vegetation in upper littoral areas on highly consolidated soil. Both *Gelidiella* and *Dictyota* are equally dominant or can be referred to as co-dominant species. *Catenella* occurs as co-species with a few individuals of *Cladophorella* and *Gelidium*.

(vi) *Boodleopsis-Vaucheria-Cladophorella* association

This is a very common association and occurs on the forest floors in the upper littoral zone as dense greenish brown or yellowish brown mats in the central and southern parts of Sundarbans. A few blue green algae, *viz., Aphanothece* and *Microcoleus* are also present along with this association where both *Boodleopsis* and *Vaucheria* are equally dominant.

(vii) *Enteromorpha-Chaetomorpha-Lola* association

This is a typical algal complex encountered in almost all the brackishwater fisheries in the upper and central parts of Sundarbans with low to mid salinity ranges (0.4-18 ppt). This association changes its species dominace with change in salinity regimes and probably sewage influx. Naskar & Santra (1986) have reported *Enteromorpha tubulosa* from the sewage mixed brackishwater fisheries from Kulti, Minakhan of Sundarbans. In the low saline areas and mid saline areas with more of sewage influx *Enteromorpha* predominantly *E. intestinalis* is found to dominate; in the mid saline areas with less of sewage *Chaetomorpha* and ocassionally *Spirogyra* is found to dominate and in the high saline areas it is the *Lola* spp. which dominates in this association.

(viii) *Ulva-Enteromorpha* association

This type of association is very common on the brick embankments along the river banks in the upper and central parts of Sundarbans areas. *Ulva* dominates in this association with a few *Entermorpha compressa* growing along with it. Also present in this association are *Oscillatoria, Lyngbya* and a few other blue-green algae.

(ix) Associations of the blue-green algae

The blue-green algae form different complexes in different habitats or may be present along with other green, red or brown algae. In the brackishwater environments and in stagnant water areas in the forest floors of upper littoral areas a complex of *Oscillatoria, Lyngbya, Phormidium* is very common. In the mid estuarine regions like Ghusighata, a typical algal growth is seen on the mud banks at low tide and this is mainly an association of *Aphanocapsa, Gloeocapsa* and *Lyngbya* with quite a number of other blue green algae, *viz., Spirulina, Nostoc, Anabaena* growing as co-species. The blue-green algal scums in ponds, other stagnant water bodies and on bricks and other hard substratum usually consist of *Oscillatoria, Lyngbya, Phormidium, Nostoc, Anabaena* and many others. *Dermocarpa, Xenococcus* and *Stichosiphon* occur as epiphytes on other algae, *viz., Lyngbya, Cladophora, Chaetomorpha* and *Lola* respectively.

(x) Phytoplankton as communities

The phytoplankton community structure is found to vary typically with change in chemical environment (salinity, alkalinity and nutrient level) and lentic or lotic condition of the water in which they occur. In the upper stretches with relatively low salinity the dominant phytoplankter include *Lyngbya, Oscillatoria, Microcystis, Coscinodiscus, Synedra, Navicula, Nitzschia, Pediastrum, Scenedesmus, Closterium* and *Volvox.* In the mid-saline areas the dominant phytoplankter include *Oscillatoria, Lyngbya, Coscinodiscus, Merismopedia, Microcystis, Fragilaria, Asterionella, Synedra, Navicula, Pleurosigma, Nitzschia* and *Scenedesmus.* In the high saline areas the dominant phytoplankter include *Oscillatoria, Lyngbya, Coscinodiscus, Melosira, Biddulphia, Chaetoceros, Bacteriastrum, Rhizosolenia, Fragilaria, Asterionella, Synedra* and *Nitzschia.*

These different algal communities, as well as, other algal forms which occur along with these communities add to the nutrient level of the soil and water phases by trapping solar energy through photosynthesis and other biological processes when alive and when dead by adding organic matter, thus rejuvenating the ecosystem as a whole.

The observations made on distribution of phytoplankton reveals the fact that diatoms constitute the most important and dominant algae in both freshwater and marine habitats, whereas the desmids and chlorococcalean green algae are equally important in many freshwater habitats; planktonic desmids normally occur only in calcium-poor waters; diatoms are often abundant in calcareous water with mono : divalent ratio of less than 1.5 and addition of potassium phosphate changes the dominance of algal populations from blue-green to desmids and the blue-green or cyanophyceae blooms dominate the eutropic lakes but at times chlorococcales may also dominate, though inspite of all these observation, it can be said that distribution is governed not by one but a combination of factors (Kumar, 1999).

Chapter 8

Applied Phycology

Over the past two decades the amount of research dedicated to mangroves and the mangrove ecosystem of the Sundarbans as a whole, has increased manifold as a result of growing interest in their role in natue and their value to mankind. Still, the number of research studies focussing on the phycological aspects are very low and sporadic leading to many major gaps in understanding of the ecosystem in all its totality. The algal flora of Sundarbans has a significant role in stabilizing, rejuvenating and sustaining this unique but fragile and vulnerable ecosystem other than its other economic potentialities, it is interesting to explore various aspects of phycology in Sundarbans including its impact on fisheries, biochemistry, soil reclamation potentials, economic uses and other.

8.1 Primary metabolites of Algae

A few macrophytic algae were identified from the Indian Sundarbans to be preferred fish food organisms of shell fish and fin fish species. These algae were investigated for their primary metabolites, *viz.,* proteins, carbohydrates, lipids, minerals and fatty acid composition.

In Table 30, the data regarding the bio-chemical tests, *viz.,* protein, carbohydrate, lipid and mineral cotents of eleven common macrophytic algal species have been highlighted. Tables 31 (a, b & c) deal with the data regarding the fatty acid composition and total lipid content of the major macrophytic algal species, *viz., Enteromorpha* spp., *Lola* spp., *Chaetomorpha* spp., *Rhizoclonium* spp., *Ulva* spp., *Spirogyra* spp. *Catenella* spp., *Polysiphonia* spp., *Pterosiphonia* sp., *Gelidiella* sp. and *Dictyota* sp.

Table 30 : Bio-Chemistry of Selected Common Macrophytic Algae from Sundarbans.

Sl. No.	Name of the macrophytic algae	Bio-Chemical parameter			
		Protein (% on dry basis)	Carbohydrate (% on dry basis)	Lipid (% on dry basis)	Mineral (% on dry basis)
1.	*Enteropmorpha* spp.	2.3%	43.2%	14.6%	28.0%
2.	*Chaetomorpha* spp.	2.9%	21.4%	7.9%	14.8%
3.	*Lola* spp.	2.0%	32.4%	5.4%	16.2%
4.	*Ulva* spp.	3.6%	48.4%	5.6%	29.5%
5.	*Rhizoclonium* spp.	2.0%	34.0%	5.0%	26.3%
6.	*Spirogyra* spp.	1.3%	58.0%	11.4%	13.0%
7.	*Catenella* spp.	0.5%	26.8%	10.7%	18.1%
8.	*Polysiphonia* spp.	1.0%	23.0%	5.7%	17.9%
9.	*Pterosiphonia* sp.	0.5%	28.4%	4.8%	19.8%
10.	*Gelidiella* sp.	3.3%	56.2%	3.2%	22.5%
11.	*Dictyota* sp.	2.5%	39.0%	12.0%	21.3%

In Table 30 the results of the bio-chemical tests have been highlighted. The species selected for these bio-chemical tests are the major macrophytic algal species which are observed to be part of the gut content of various estuarine and brackishwater shell-fish, fin-fish and molluscan species from Indian Sundarbans. The tests were conducted to estimate the protein, carbohydrate, lipid and mineral content of the selected algal species. The highest protein content was observed in *Ulva* spp. (3.64% on dry basis) followed by *Gelidiella* sp. (3.25%, on dry basis), the lowest protein content was observed in *Catenella* spp. (0.52%, on dry basis). The protein content of other algae are *Entermorpha* spp. (2.36%), *Chaetomorpha* spp. (2.92%), *Lola* spp. (1.98%), *Rhizoclonium* spp. (2.0%), *Spirogyra* spp. (1.32%), *Polysiphonia* spp. (1.0%), *Pterosiphonia* spp. (0.54%) and *Dictyota* spp. (2.46%) (on dry basis).

The carbohydrate content was found to be highest in *Spirogyra* spp. (58%, on dry basis) followed by *Gelidiella* sp. (56.2%, on dry basis) and the lowest was observed in *Chaetomorpha* spp. (21.4%, on dry basis). The carbohydrate content of the other algae are *Enteromorpha* spp. (43.2%), *Lola* spp. (32.4%) *Rhizoclonium* spp. (34%), *Ulva* spp. (48.4%)

Polysiphonia spp. (23%), *Pterosiphonia* sp. (28.4%), *Catenella* spp. (26.8%) and *Dictyota* sp. (39%) (on dry basis). The highest mineral content was observed in *Ulva* sp (29.5%) followed by *Enteromorpha* spp. (28%) and the lowest was observed in *Spirogyra* spp. (12.96%). The mineral content of the other algal forms are *Chaetomorpha* spp. (14.8%), *Lola* spp. (16.2%), *Rhizoclonium* spp. (26.3%), *Polysiphonia* spp. (17.9%), *Pterosiphonia* sp. (19.8%), *Catenella* spp. (18.1%), *Gelidiella* sp. and *Dictyota* sp. (21.3% on dry basis).

In Table 31 (a, b & c), the data regarding the fatty acid composition and total lipid content of the major macrophtic algal species, *viz. Enteromorpha* spp., *Lola* spp., *Chaetomorpha* spp., *Rhizoclonium* spp., *Ulva* spp., *Spirogyra* spp., *Catenella* spp., *Polysiphonia* spp., *Pterosiphonia* sp., *Gelidiella* sp. and *Dictyota* sp. have been tabulated. The results show the total lipid contents as follows : *Enteromorpha* spp. (14.44% by wt.), *Lola* spp. (5.40% by wt.), *Chaetomorpha* spp. (7.85% by wt.), *Rhizoclonium* spp. (5.02% by wt.), *Ulva* spp. (5.62% by wt.), *Spirogyra* spp. (11.4 2% by wt.), *Catenella* spp. (10. 66% by wt.), *Polysiphonia* spp. (5.65% by wt.), *Pterosiphonia* sp. (5.00% by wt.), *Gelidiella* sp. (3.2% by wt.) and *Dictyota* (12.0% by wt).

The fatty acid composition as revealed from the above mentioned tables indicate that in *Enteromorpha* spp., *Lola* spp., *Chaetomorpha* spp., *Ulva* spp., *Spirogyra* spp. *Rhizoclonium* spp. which are all green algae, *i.e.,* Chlorophyta members, the major acid is palmitic acid ($C_{16:0}$) contributing 47.1%, 31.7%, 23.8%, 30.3%, 25.7% and 21.3% by wt. respectively. Though in *Ulva* spp. and *Rhizoclonium* spp. the major fatty acid is oleic acid ($C_{18:1}$) contributing 30.8% and 28.9% respectively. Interestingly, however *Catenella* spp., *Pterosiphonia* sp., and *Polysiphonia* spp. all *Rhodophyta* members, *i.e.,* red algal forms, contain eicosapentaenoic acid ($C_{20:5}$) as the major fatty acid, *Catenella* spp. contains 34.5%, *Polysiphonia* spp. contains 49.7% and *Pterosiphonia* sp. contains as high as 56.3% of eicosapentaenoic acid. In *Gelidiella* sp. which is again a Rhodophyta member the major fatty acid is oleic acid contributing 40.6% by wt. Similarly, in *Dictyota* sp. which is a Phaeophyta member, *i.e.,* brown algae, the major fatty acid is again oleic acid contributing 33.9% by wt. (GLC-I-IX).

In the lipid of *Enteromorpha* spp. the major fatty acids are palmitic (47.1%, w/w), oleic (13.0% w/w), linoleic (9.3%, w/w) and linolenic (11.0%

Table 31(a) : Fatty Acid Composition and Total Lipids of the Macrophytic Algal Speices

	Enteromorpha	*Lola*	*Chaetomorpha*	*Ulva*
Lipid content % by weight (on dry basis)	14.55%	5.40%	7.85%	5.62%
$C_{10:0}$	-	-	1.4%	-
$C_{12:0}$	2.9%	0.7%	3.4%	0.3%
$C_{14:0}$	5.6%	11.2%	8.4%	0.5%
$C_{14:1}$	1.2%	0.3%	--	1.5%
$C_{15:0}$			0.1%	
$C_{15:1}$	--	--	1.7%	--
$C_{16:0}$	47.1%	31.7%	23.8%	30.3%
$C_{16:1}$	2.1%	11.6%	2.3%	5.5%
$C_{16:3}$	--	--	--	--
$C_{16:4}$	--	--	--	--
$C_{17:0}$	--	0.5%	0.2%	--
$C_{18:0}$	1.2%	2.3%	1.3%	1.2%
$C_{18:1}$	13.0%	10.4%	8.9%	30.8%
$C_{18:2}$	9.3%	9.4%	4.5%	9.3%
$C_{18:3\,(\alpha)}$	11.0%	4.0%	11.1%	16.1%
$C_{18:3\,(\gamma)}$	--	--	--	--
$C_{18:4}$	0.4%	3.2%	10.1%	--
$C_{20:0}$	1.2%	--	--	--
$(n\text{-}6)C_{20:3}$	--	--	--	0.9%
$(n\text{-}3)C_{20:3}$	--	--	--	--
$C_{20:4}$	--	--	1.7%	1.3%
$C_{20:5}$	--	9.8%	17.3%	--
$C_{22:0}$	--	1.6%	--	--
$C_{22:1}$	--	--	--	1.1%
$C_{22:6}$	--	--	3.5%	--
$C_{24:0}$	--	--	--	--
Unidentified fatty acid	5.0%	3.2%	0.7%	1.2%
Polyunsaturated fatty acid	20.7%	26.5%	48.2%	27.6%
Monounsaturated fatty acid	16.3%	22.3%	12.9%	38.9%
Saturated fatty acid	58.0%	48.0%	38.2%	32.3%

Table 31(b) : Fatty Acid Composition and Total Lipids of the Macrophytic Algal Speices

	Polysiphonia	*Pterosiphonia*	*Catenella*	*Gelidiella*
Lipid content % by weight (on dry basis)	5.65%	5.0%	10.66%	3.2%
$C_{10:0}$	--	--	--	--
$C_{12:0}$	--	0.7%	0.1%	--
$C_{14:0}$	0.9%	2.5%	0.9%	2.5%
$C_{14:1}$	--	--	0.2%	--
$C_{15:0}$	--	--	--	--
$C_{15:1}$	--	--	--	--
$C_{16:0}$	16.5%	15.4%	10.4%	6.5%
$C_{16:1}$	--	--	0.2%	--
$C_{16:3}$	--	1.0%	3.2%	--
$C_{16:4}$	--	--	2.2%	--
$C_{17:0}$	--	--	0.3%	--
$C_{18:0}$	--	1.5%	1.8%	12.5%
$C_{18:1}$	10.6%	4.8%	5.2%	40.6%
$C_{18:2}$	--	2.2%	6.9%	10.8%
$C_{18:3 (\alpha)}$	2.7%	3.0%	4.7%	0.8%
$C_{18:3 (\gamma)}$	--	--	--	--
$C_{18:4}$	2.4%	12.4%	4.9%	--
$C_{20:0}$	--	--	--	--
$(n\text{-}6)C_{20:3}$	--	--	--	--
$(n\text{-}3)C_{20:3}$	3.7%	--	--	--
$C_{20:4}$	--	--	--	16.8%
$C_{20:5}$	49.7%	56.3%	34.5%	--
$C_{22:0}$	--	--	1.9%	--
$C_{22:1}$	--	--	--	--
$C_{22:6}$	4.4%	--	21.3%	9.5%
$C_{24:0}$	9.1%	--	--	--
Unidentified fatty acid	--	0.2%	1.4%	--
Polyunsaturated fatty acid	62.9%	74.9%	77.6%	37.9%
Monounsaturated fatty acid	19.7%	4.8%	5.6%	40.6%
Saturated fatty acid	17.4%	20.1%	15.4%	21.5%

Table 31(c): Fatty Acid Composition and Total Lipids of the Macrophytic Algal Speices

	Spirogyra	*Rhizoclonium*	*Dictyota*
Lipid content % by weight (on dry basis)	11.42%	5.02%	12.0%
$C_{10:0}$	1.2%	--	--
$C_{12:0}$	0.1%	1.7%	--
$C_{14:0}$	5.7%	9.7%	0.6%
$C_{14:1}$	--	3.5%	--
$C_{15:0}$	--	--	--
$C_{15:1}$	3.4%	0.5%	--
$C_{16:0}$	25.7%	21.3%	19.3%
$C_{16:1}$	4.7%	1.9%	--
$C_{16:3}$	--	--	--
$C_{16:4}$	--	--	--
$C_{17:0}$	1.4%	--	--
$C_{18:0}$	0.9%	2.7%	1.3%
$C_{18:1}$	10.8%	28.9%	33.9%
$C_{18:2}$	10.7%	8.6%	18.1%
$C_{18:3(\alpha)}$	17.7%	7.1%	--
$C_{18:3(\gamma)}$	0.8%	--	--
$C_{18:4}$	--	--	--
$C_{20:0}$	--	--	--
$(n-6)C_{20:3}$	0.3%	--	--
$(n-3)C_{20:3}$	0.8%	45%	--
$C_{20:4}$	--	--	11.2%
$C_{20:5}$	10.6%	9.6%	--
$C_{22:0}$	--	--	--
$C_{22:1}$	0.5%	--	--
$C_{22:6}$	4.3%	--	15.6%
$C_{24:0}$	--	--	--
Unidentified fatty acid	0.3%	--	--
Polyunsaturated fatty acid	45.3%	29.8%	44.9%
Monounsaturated fatty acid	19.4%	34.8%	33.9%
Saturated fatty acid	35.0%	35.4%	21.2%

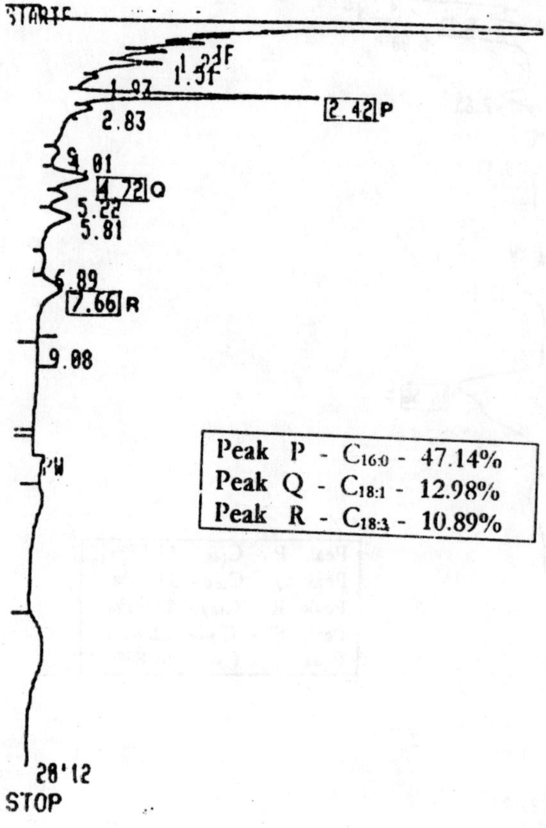

GLC-I. Gas Liquid Chromatogram of Total Lipids of *Enteromorpha* spp.

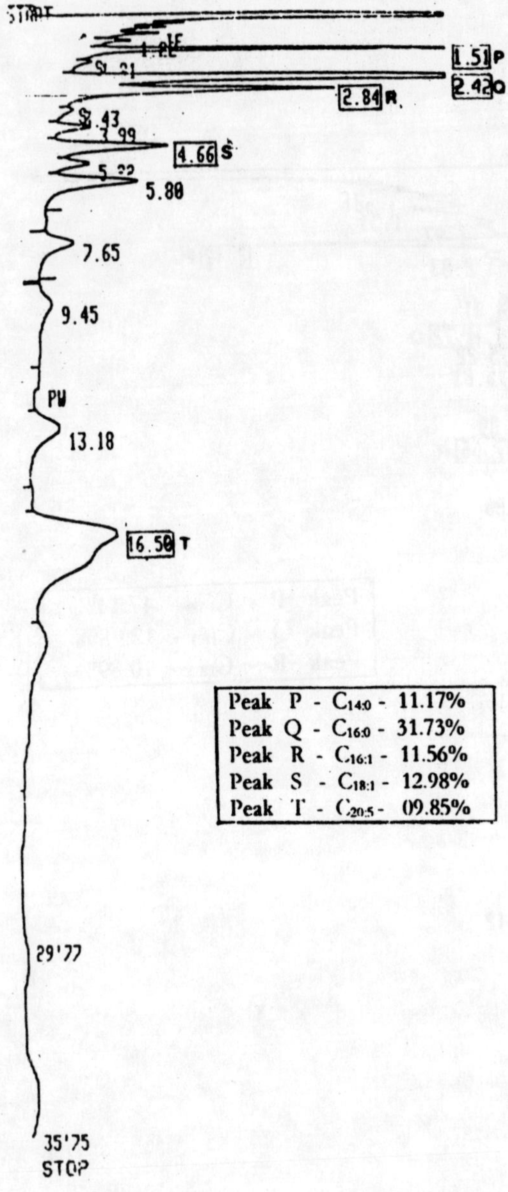

Peak				
Peak	P	$C_{14:0}$	-	11.17%
Peak	Q	$C_{16:0}$	-	31.73%
Peak	R	$C_{16:1}$	-	11.56%
Peak	S	$C_{18:1}$	-	12.98%
Peak	T	$C_{20:5}$	-	09.85%

GLC-II. Gas Liquid Chromatogram of Total Lipids of *Lola* spp.

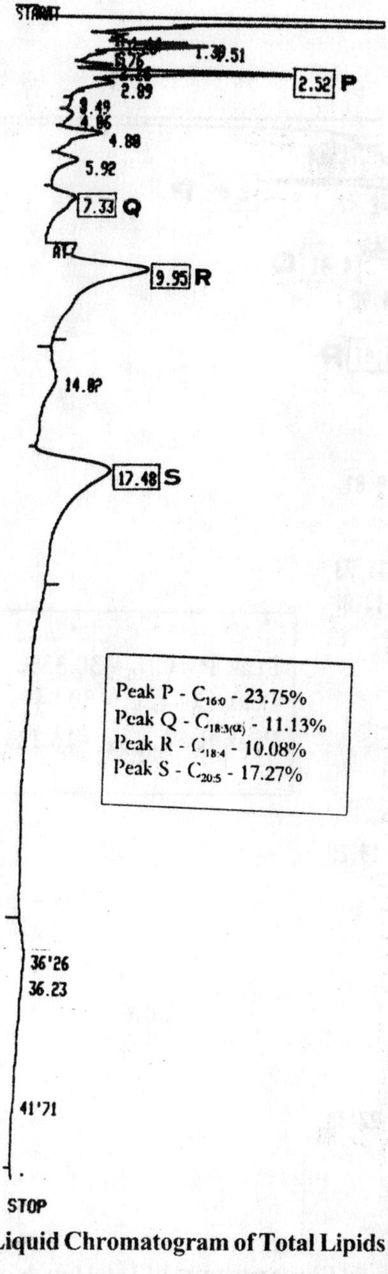

GLC - III. Gas Liquid Chromatogram of Total Lipids of *Chaetomorpha* spp.

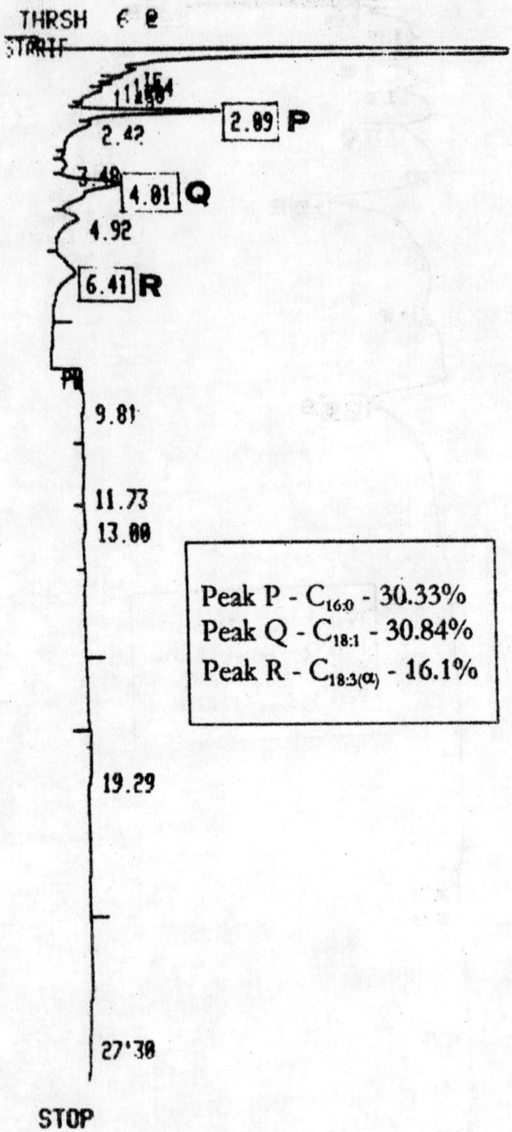

Peak P - $C_{16:0}$ - 30.33%
Peak Q - $C_{18:1}$ - 30.84%
Peak R - $C_{18:3(\alpha)}$ - 16.1%

GLC - IV. Gas Liquid Chromatogram of Total Lipids of *Ulva* spp.

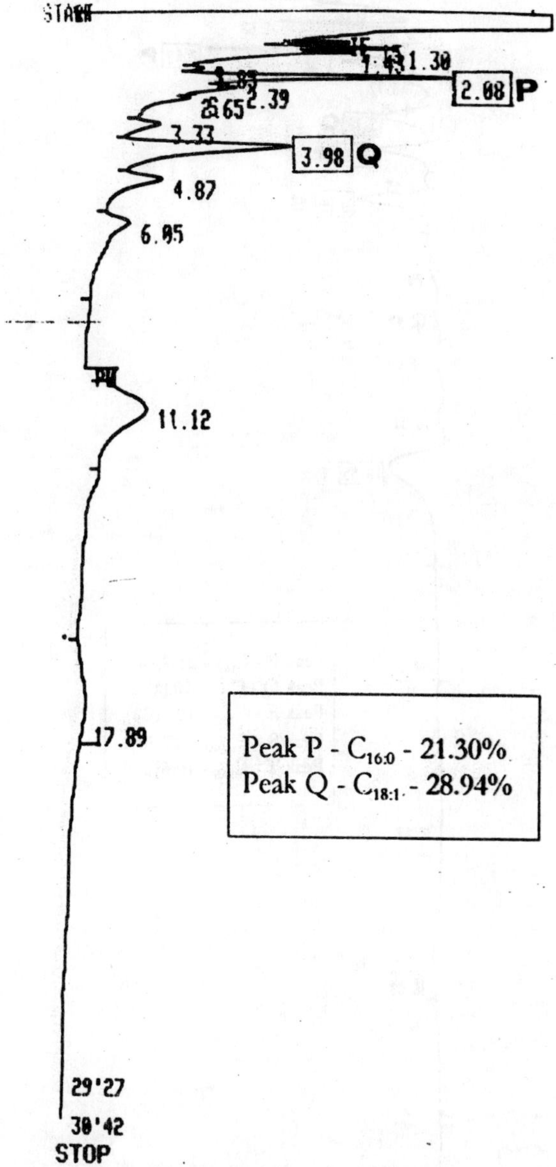

Peak P - $C_{16:0}$ - 21.30%
Peak Q - $C_{18:1}$ - 28.94%

GLC-V. Gas Liquid Chromatogram of Total Lipids of *Rhizoclonium* spp.

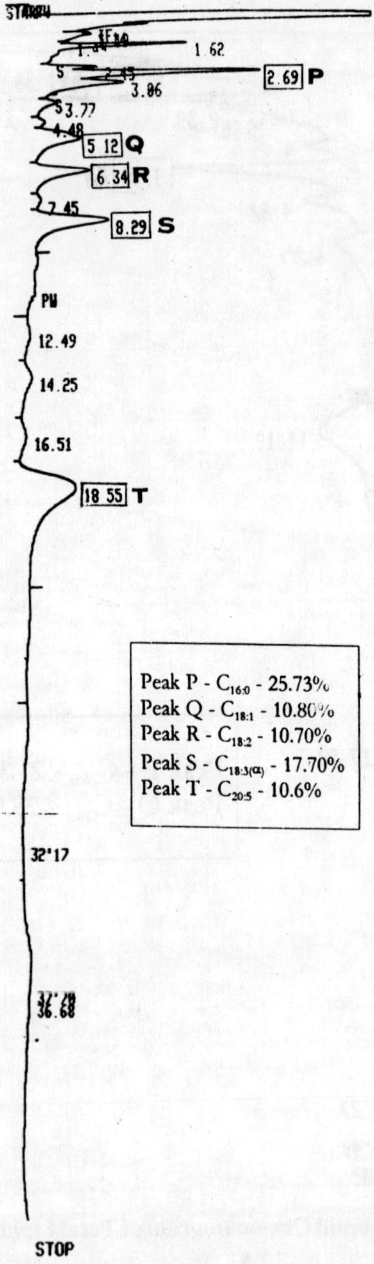

GLC - VI. Gas Liquid Chromatogram of Total Lipids of *Spirogyra* spp.

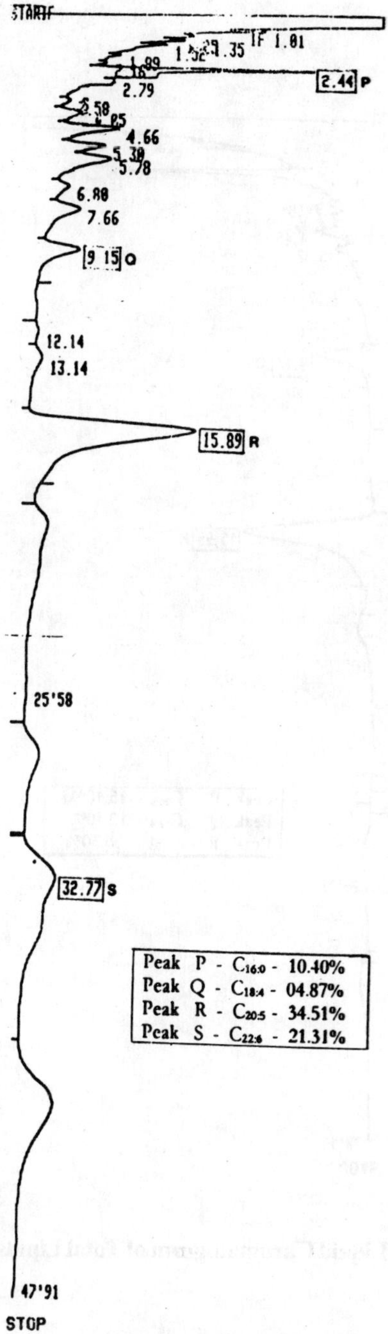

GLC - VII. Gas Liquid Chromatogram of Total Lipids of *Catenella* spp.

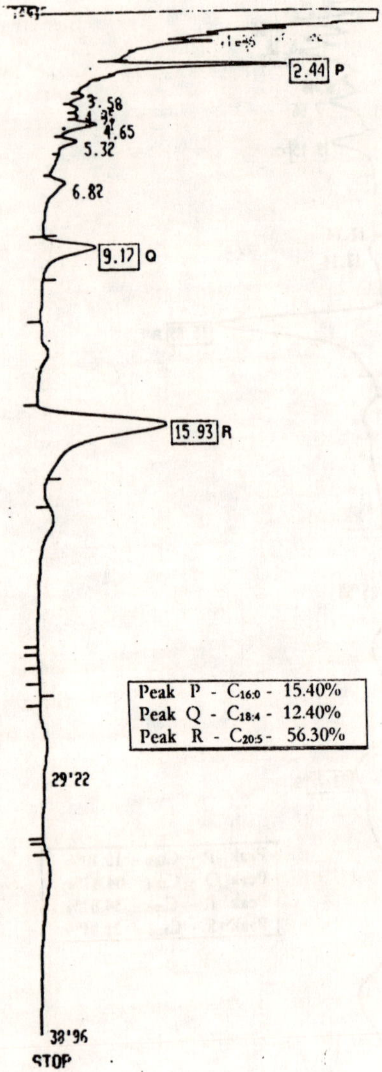

GLC - VIII. Gas Liquid Chromatogram of Total Lipids of *Pterosiphonia* spp.

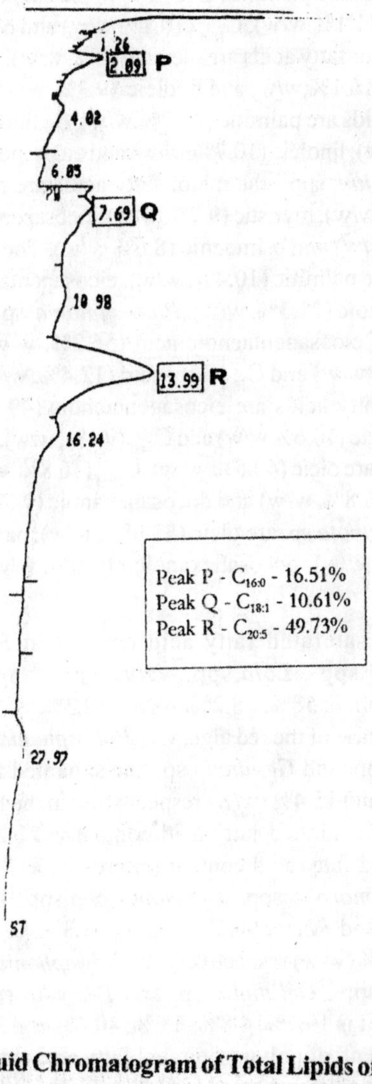

GLC - IX. Gas Liquid Chromatogram of Total Lipids of *Polysiphonia* spp.

w/w). In case of *Lola* spp. the major fatty acids are palmitic (31.7%, w/w), palmitoleic (11.6%, w/w), myristic (11.2%, w/w), oleic (10.4%, w/w), linoleic (9.4%, w/w) and eicosapentaenoic (9.8%, w/w). In *Chaetomorpha* spp. the major fatty acids are palmitic (23.8%, w/w), *eicosapentaenoic* (17.3%,w/w), α linolenic (11.1%,w/w), $C_{18:4}$ (10.1% w/w) and oleic (8.9%, w/w). In *Ulva* spp. the major fatty acids are oleic (30.8%, w/w), palmitic (30.3%, w/w), α linolenic (16.1%,w/w) and linoleic (9.3%, w/w). In *Spirogyra* spp. the major fatty acids are palmitic (25.7%,w/w), α linolenic (17.7%, w/w), oleic (10.1%, w/w), linoleic (10.7%, w/w) and eicosapentaenoic (10.6%, w/w). In *Rhizoclonium* spp. the major fatty acids are oleic (28.9%, w/w), palmitic (21.3%, w/w), myristic (9.7%, w/w), eicosapentaenoic (9.6%, w/w) linoleic (8.6%, w/w) and α linoenic (8.6% w/w). The major fatty acids in *Catenella* spp. are palmitic (10.4%, w/w), eicosapentaenoic (34.5%, w/w) and docosahexanoic (21.3%, w/w). *Pterosiphonia* sp. contains unusually high amounts of eicosapentaenoic acid (56.3%, w/w); it also contains palmitic (15.4%, w/w) and $C_{18:4}$ fatty acid (12.4%, w/w). In *Polysiphonia* spp. the major fatty acids are eicosapentaenoic (49.7%, w/w), palmitic (16.5%, w/w), oleic (10.6% w/w) and $C_{24:0}$ (9.1%, w/w). In *Gelidiella* sp. the major fatty acids are oleic (40.6%, w/w), $C_{20:4}$ (16.8%, w/w), stearic (12.5%, w/w), linoleic (10.8%, w/w) and docosahexanoic (9.5%, w/w). The major fatty acids in *Dictyota* sp. are oleic (33.9% , w/w), palmitic (19.3%,w/w), linoleic (18.1%, w/w), docosahexanoic (15.6%, w/w) and $C_{20:4}$ (11.2%, w/w).

The total saturated fatty acid content in *Enteromorpha* spp. *Chaetomorpha* spp., *Lola*,spp., *Ulva* spp., *Spirogyra* spp. and *Rhizoclonium* spp. is 58%, 38.2%, 48%, 32.3%, 35% and 35.4% (w/w) respectively. In case of the red alge, *viz., Polysiphonia* spp., *Pterosiphonia* sp., *Catenella* spp. and *Gelidiella* sp. the saturated fatty acid content is 17.4%, 20.1 %, and 15.4% (w/w) respectively; in the Phaeophyta member *Dictyota* sp. the saturated fatty acid content is 21.2% (w/w). The total monounsaturated fatty acid content (expressed as % by weight on dry basis) in *Enteromorpha* spp., *Chaetomorpha* spp., *Lola* spp., *Ulva* spp. *Spirogyra* spp. and *Rhizoclonium* spp. is 16.3%, 12.9%, 22.3%, 38.9%, 19.4% and 34.8% (w/w) respectively. In *Polysiphonia* spp., *Pterosiphonia* sp., *Catenella* spp., *Gelidiella* sp. and *Dictyota* sp. monounsaturated fatty acid content is 19.7%, 4.8%, 5.6%, 40.6% and 33.9% (w/w) respectively. The content, of polyunsaturated fatty acids important for pharmaceutical applications (expressed as % by weight on dry basis) is 20.7% in

Enteromorpha spp., 26.5% in *Lola* spp., 48.2% in *Chaetomorpha* spp., 27.6% in *Ulva* spp., 45.3% in *Spirogyra* spp., 29.8% in *Rhizoclonium* spp., 77.6% in *Catenella* spp., 74.9% in *Pterosiphonia* sp., 62.9% in *Polysiphonia* spp., 37.9% in *Gelidiella* sp. and 44.9% in *Dictyota* sp.

8.2 Impact on Fisheries Development

This large and varied algal groups and communities of Sundarbans form the chief foundation on which the aquatic food webs depend. Starting from the zooplankton to large herbivorous fin-fish and shell-fish species are dependent on these algae as their source of food. The plankton are grazed by the crustaceans, rotifers and the protozoans; the crustaceans and rotifers engulf the algal cells, whereas the protozoans enter the cells and ingest them or attach themselves to the alga and use their contents (Golterman, 1975). Whatever may be the mode of predation it is an established fact that the herbivorous fish species (both shell-fish and fin-fish), amphipods and filter feeders feed on these algae with their own preferences for green, blue-green, brown or red algae. As observed by Mac Nae (1968), *Tilapia mossambirca* at all stages of development prefer the green algae whereas the *Chanos chanos* at least in its youth seem to prefer the softer blue-green algae, as such, the dense growth of algae are sometimes seen to be grazed by *Tilapia* and keep the ecosystem under control. Porter (1973) conducted grazing experiments and found that *Anabaena affinis* and *A. flos-aquae* were not ingested by crustaceans. Tang and Hwang (1967) evaluated the relative suitability of four major groups of brackish water algae as feed for *Chanos chanos* (reported earlier from Sundarbans but not found during the present study, as it has become very rare) and found the nutritive ratio of fresh *Chaetomorpha*, phytoflagellates, filamentous blue - green algae and diatoms as 1 : 33.44, 1: 2.47, 1 : 1.18 and 1 : 1.54 respectively; the detritus produced by *Chaetomorpha* has a nutritive ratio of 1 : 1.66. Tang and Hwang (1967) also reported that fresh *Chaetomorpha* is undersirable because, the *Chanos chanos* of harvestable sizes are unable to consume *Chaetomorpha* due to its coarseness and fibrous nature and it also interferes with the manipulation of fish population but its detritus can serve as fish food; filamentous blue - greens and benthic diatoms growing on pond bottoms are the most desirable food for all age groups of *Chanos chanos*, a luxuriant growth of these algae dispense out undesirable algae and possible other harmful organisms in the ponds and create better environment conditions for the fish.

Many algae are directly used as fodder for live stock, *viz.*, cattle, pigs, sheep, fish and poultry (Kumar, 1999). The algal mixture obtained as by product of sewage treatment process can be fed to fishes, poultry and cattle to improve their health and productivity. For this the National Botanical Research Institute at Lucknow has used *Spirulina* grown on waste water for this purpose and reported that even if one- fifth of the conventional animal feeds are substituted by microalgae, a remarkable improvement in the health and milk yield of cattle is observed (Kumar, 1999).

The algae other than serving as natural food for the innumerable faunal components thriving in the Sundarbans mangals enrich them with proteins, carbohydrates minerals and important essential fatty acids, some of which are precursor of prostaglandins, prostacyclins, thromboxines and leukotrienes (collectively called eicosanoids) which are ultimately harvested for human consumption. The studies have revealed the fact that the algae are not only rich in proteins and carbohydrates but also contain certain pharmaceutically valuable fatty acids as these are health care liquid components, *viz.* eicosapentaenoic acid and docosahexanoic acid which have immense value in prevention and treatment of heart and circulatory diseases, inflammation, asthma, arthritis, psoriasis, atherosclerosis, thrombosis, vascularspasm, diabetes, cancer, ageing and viral infections, Yamachi, *et. al* (1997). Dyerberg (1986); Simopoulos (1991); Ziboh, *et. al.* (1986); Voorhees (1986); De, *et. al.* (1998); Gunstone, *et. al.* (1994); Kendrick & Ratledge (1992); Suzuki (1998), Cohen (1990); Cohen & Cohen (1991). The synthesis of these fatty acids is found to be low in the shell fish and fin fish species from which they are commercially exploited (Watenbe, *et. al.,* 1980; Yongmanitchai & Ward, 1992), but are found in high quantities in the microorganisms, *viz.,* algae in which they primarily occur (Cartens, *et. al.* 1996 : Seto *et. al.* 1984; Owen, *et. al.*,1972; Yamada, *et. al.* 1980; Ackman, 1982). Marine fish oil is the traditional source of polyunsturated fatty acids. The conventional source of polyunsaturated fatty acids is the oil of sardine, white seer hilsa *(Hilsa hilsa)* etc. Howerer, this traditional source inherently possesses some problems like its lack of accessibility throughout the world due to which it cannot meet the demand for PUFA and its disagreeable odour. The most useful source of PUFA is therefore the microbial source, which may be of two types, *viz.,* fungal source and algae source. PUFA from cultures of fungi and algae are cholestrol-free (Cohen & Cohen, 1991), contamination-free and of good taste (Medina, *et al.* 1995). The conventional and commercial source of ϒ–linolenic acid is evening primrose *(Oenothera biennis L.)* oil which contains nearly 8% ϒ-linolenic acid. Borage *(Borago officinalis)* oil also contains as high as 21% ϒ-linolenic acid (Kleiman, 1988). Algae, specially *Spirulina platensis* is an alternative source for the ϒ– linolenic acid and it possesses, some inherent advantage like lesser bulk volume requirement than the above mentioned commercial sources (Manabe, *et. al.* 1994 ; Hirano *et. al.* 1990).

Apart from this, the algal thalli mainly of *Tiffaniella, Pterosiphonia* and *Murrayellopsis* are used by certain fish species to build nests upon which eggs are laid; the fish usually isolate the vegetative thalli of these algae for this purpose (Foster, 1972).

8.3. Food and Feeding Habitats of shell-fish and fin-fish

In Table 32, the food and feeding habits of some common estuarine and brackishwater shell-fish and fin - fish species have been highlighted. Photo Plate XXXVI & XXXVII.

For the the study, 6 common fin fish (estuarine and brackish water), *viz., Scatophagus argus, Liza parsia, Liza tade, Mugil cephalus, Rhinomugil corsula* and *Valamugil cunnesius* were selected. Of these 6 fin-fish species other than *Scatophagus argus* which is a scat the rest are mullets. *Scatophagus argus* and *Rhinomugil corsula* were found to be solely herbivores and the others, *viz., Liza parsia, Liza tade, Mugil cephalus,* and *Valamugil cunnesius* were found to be primarily herbivores but also fed upon zooplankters and at times also detrivorous depending on the availability and the nature of food in the environment. *Scatophagus argus* reveals a preference for the red alga *Catenella* spp., filamentous green algae and diatoms. *Rhinomugil corsula* shows preference for green algae, diatoms and other green algae, diatoms and other green plant parts in the water. *Liza parsia, Mugil cephalus* and *Valamugil cunnesius* show preference for green algae, diatoms, copepods polychaetes and at times also feed upon debris and detritus. *Liza tade* shows preference for blue-green algae, green algae, diatoms and also feeds upon decaying matter. The shell - fish species selected for the study include *Penaeus monodon, Macrobrachium rosenbergii, Scylla serrata, Uca* sp. Both the shrimp species, *viz., Penaeus monodon* and *Macrobrachium rosenbergii* are omnivores and their gut content analysis reveal varied food materials ranging from zooplankters, green filamentous algal forms and also decaying

Table-32 : Food and Feeding Habit of Some Common Estuarine and Brackish Water Shell-Fish and Fin-Fish Species.

Species	*Food (from gut content analysis)*	*Feeding Habit*
FIN-FISH SPECIES		
Scatophagus argus	Preference for *Catenella* spp. (red algae), filamentous algae & diatoms etc.	Herbivore
Liza parsia	Preference for green algae, diatoms, copepods, polychaetes, debris and detritus.	Primarily Herbivore but also feeds on zooplankton/ Detrivore.
Liza tade	Preference for BGA, green algae, diatoms and decaying matter.	Primarily Herbivore but also Detrivore.
Mugil cephalus	Preference for green algae, diatoms, copepods, polychaetes, debris and detritus.	Primarily Herbivore but also feeds on zooplankton/ Detrivore.
Rhinomugil corsula	Preference for green algae, diatoms and other higher plant materials in the water.	Herbivore
Valamugil cunnesius	Preference for green algae, diatoms, copepods, polychaetes, debris and detritus.	Primarily Herbivore but also feeds on zooplankton/ Detrivore.
SHELL-FISH SPECIES		
Penaeus monodon	Preference for animal matter in the form of Zooplankton, also decomposing green filamentous algae and other decaying matter.	Omnivore
Macrobrachium rosenbergii	Preference for animal matter in the form of Zooplankton, also decomposing green filamentous algae and other decaying matter.	Omnivore
Scylla serrta	Preference for attached/periphytic green filamentous algae and decaying matter in soil.	Primarily Herbivore but also Detrivore.
Uca sp.	Preference for both algal, as well as, animal matter.	Omnivore
Telescopia sp.	Preference for attached/periphytic green filamentous algae and decaying matter in soil and water	Primarily Herbivore but also Detrivore.

material or detritus. The crabs *viz., Sylla serrata, Uca* sp. and *Telescopium* sp. are found to be primarily herbivores but are also found to feed upon decaying matter in the mud. Their feeding habit and also gut content analysis reveal their preference for attached or periphytic algal forms and decaying matter in mud and water, which they commonly seen to graze upon in the mangals of Sundarbans. The molluscan species *Telescopium telescopium* is also primarily a herbivore and grazes the attached or periphytic forms of algae and also feeds upon decaying matter in the soil.

Another important function of the algae in the ecosystem is their ability to act as waste traps and these algae can be harvested to remove the toxic compounds, *viz.,* mercury, cadmium, copper, uranium, lead and pesticides from the water bodies (Rai, *et. al.*1981). This harvested algae may be put into use as raw material for bio-gas production ; *Sargassum tenerrimum* has been found to be quite suitable for use in bio-gas plants at Bhavnagar (Sarma, *et. al.,* 1984).

The other aspect of algal growth is that while moderate growth of algae in water bodies is beneficial in oxygenating the water and preventing purefaction of organic matter, excessive growth often creates serious nuisance. These are in the form of blooms and as aquatic weeds choking the water bodies as seen with *Entermorpha* and *Spirogyra* in the brackish water fisheries in the areas like Haroa, Minakhan, Sarberia and Malancha in the Indian Sundarbans. This excessive growth often results in severe depletion of oxygen and may lead to mass mortality. Tang and Hwang (1967) have reported that Phytoflagellate blooms are harmful in ponds where, culture of *Chanos chanos* is done, as it results in interference with schooling and grazing activities of juvenile fingerlings as a result of obstruction of light penetration, oxygen depletion and decomposition of dead organisms. Fish mortality may also occur due to release of some toxic substances by certain algae. Certain other manifestation of blooms translate into nuisances, these are: (i) water quality deterioration; (ii) chronic or intermittent health hazards; and (iii) loss of aesthetic and recreational value of the affected water (Kumar. 1999). *Lyngbya majuscula* is known to cause dermatitis, which manifests itself as a skin rash followed by blistering, *Fragilaria striatula* and *Cladophora* spp. also cause dermatitis and itching in fishermen (Hashimoto, 1979). Species of *Chlorella, Oscillatoria, Anabaena* are suspected to be potent allergens causing asthama in Venezuela and U. S. A. (Stein, and Borden, 1984). Some algae are also producers of toxins which are ultimately ingested by humans, *viz.*, certain dinoflagellates *Pyrodimum* and *Protogonyaulax* produce toxins that cause fish and shell-fish poisoning and when these fish are consumed by birds or mammals they cause neurological and muscular disorder, (Hashimoto, 1979). The toxic blue-green alga *Microcystics aeruginosa* is grazed by fish and become concentrated through bio-accumulation and bio-magnification, when these fish are consumed by human they can develop neuromuscular disorders; species of *Anabaena, Oscillatoria* and *Schizothrix* produce lipopolysaccharide endotoxins which probably cause gastrointestinal disorders (Carmichael, 1981). The dinoflagellates produce toxins that cause paralytic shellfish poisoning (PSP) inducing derangement of nervous system or respiratory paralysis and often death may result within 12 hours of illness preceded by symptoms such as nausea, vomitting, weakness and muscular paralysis (Kumar, 1999). The algae of course have both positive and negative impacts in polluted water. Other than creating blooms in such water bodies certain algae flourish in water polluted with sewage or organic wastes, playing important role in 'self purification' of such a water body (Kumar, 1999).

One use that the immense algal diversity and population can be put into is in the reclamation of wastelands. The occurrence of alkaline and saline soils is widespread in many parts of India and particularly in Sundarbans, where saline blanks are common due to the highly saline environment and widespread deforestation ; in the forest zones the mangroves are responsible for taking up the saline water through their physiological processes, and also release the excess soil through leaf surface, lenticels and galls, but with widespread felling of mangroves the saline water comes to the surface of the soil through capillary action and after evaporation of the water the salt is left behind to gradually form a salt crust on the surface called saline blanks (Naskar and Mandal, 1999). Such areas are suitable for algal growth and cultivation (Kumar, 1999). In case of members of the order Charales the high calcium encrustration of the plants play an important role when used as a manure, particularly in preventing acidity of the soil and to ameliorate dense clayey soils and in view of their abundance it may be worthwhile to try them as manures, as they appear to furnish a cheap source of calcium in acidic conditions (Pal, *et. al*. 1962). Zanevald (1940) summarized the economic potentialities of the algae belonging to Charales, *viz.*, fisheries, polishes, mud baths, therapeutic values, sugar clarification and insect control. The Charophytes are also known food for water fowls and Gordon (1943) observed that these plants might not be a direct source of food but always supported abundant epiphyte and small herbivorous animal growth, which in turn are used as food by the other fish in nature and also in the aquarium.

As such, with effective research and application of various roles played by algae including economic and ecological potentialities the rural economy can be enhanced generating employment opportunity and enhancing their production and potentialities as food, medicine and bio-fertilizers.

References

Ackman, R. G. (1982) In : *Nutritional evaluation of long chain fatty acids in fish oil,* eds. S. M. Barlow & M. E. Stanley, Academic Press, London, 25 - 28.

Adhya, T. K. & S. C. Santra (1977) *Draparnaldiopsis indica* Bharadwaja– a new record from West Bengal, *Sci. & Cult.* 43 : 399.

Agarkar, D. S. (1969) Contribution to the desmids of Gwalior, Madhya Pradesh (India). *Phykos* 8(1&2) : 1-10.

Agarkar, D. S. (1971) Contribution to the desmids of Gwalior, Madhya Pradesh (India) – II. *Phykos* 8 (1&2) : 54 – 69.

Agharkar, S. P. & B. C. Kundu (1937) Charophyta of Bengal. *J. Dept. Sci. Calcutta Univ. n. s.* 1(1) : 1 - 23.

Akhter, N. & A. K. M. N. Islam (1992) Addition to the lists of desmids flora of Bangladesh, *Dhaka Univ. Stud. Part E,* 7(1) : 95 - 99.

Alben, W. E. & E. E. Cupp (1935) Plankton diatoms of the Java Sea. *Annales der Jardin Botanique de Buitenzorg,* XI.IX(2).

Allen, W. E. (1936) Occurrence of marine plankton diatoms in a ten year series of daily catches in Southern California, *Amer. J. Bot.* 23 : 60- 63.

Al-Saadi, H. A., S. E. Antoine & A. K. M. N. Islam (1981) Limnological investigations in Al-Hammara marsh area in southern Iraq, *Nova Hedwigia,* 35 : 157 - 166.

Anonymous, (1995) Hugli-Matla Estuary, West Bengal, *Estuarine Ecosystem,* ZSI Series, Part II, pp. 1 - 542.

Anonymous, (1999) **Report of the National Fellow Scheme,** for the project entitled *"Impact of Mangrove Ecosystem of the Sundarbans, West Bengal in estuarine fisheries development with special reference to* *identification and estimation of different mangroves and algae"*, submitted by Dr. K. R. Naskar to the **Indian Council of Agricultural Research.**

APHA (1992) **Standard Methods for the Examination of Water and Waste Water,** 17th Ed. American Public Health Association.

Aziz, A. & A. K. M. N. Islam (1979) Marine dinoflagellates from the Bay of Bengal, Bagladesh. *J. Bangladesh Acad. Sci.* 3(1-2) : 41 - 49.

Aziz, A. & A. K. M. N. Islam (1986) Lagoon algae of St. Martin's Island, Bangladesh. *Dhaka Univ. Stud., Part E,* 1(1) : 45 - 52.

Aziz, A., J. Alam & A. K. M. N. Islam (1991) Studies on the members of Oedogoniales epiphytic on deepwater rice plants near Sonargaon, *Bangladesh. Dhaka Univ. Stud.,* Part E, 6(2) : 117 - 121.

Aziz, K. M. S. & A. K. M. N. Islam (1962) A new species of *Fritschiella* from East Pakistan, *Trans Amer. Micros. Soc,* 81(2) : 185 - 189. (USA).

Bainbridge, R. (1953), Studies on the interrelationships of zoo - and phyto- plankton, *J. Mar. Biol. Ass.,* U. K., 32 : 385 - 443.

Ball, M. C. & G. D. Farquhar (1984a) Photosynthetic and stomatal responses of two mangrove species, *Aegiceras corniculatum & Avicennia marina,* to long term salinity and humidity conditions, *Plant Physiology,* 74 : 1-6.

Ball, M. C. & G. D. Farquhar (1984b) Photosynthetic and stomatal responses of the grey mangrove, *Avicennia marina* to transient salinity. *Plant Physiology,* 74 : 7-11.

Banerji, J. C. (1935) On algae found on soil samples from alluvial paddy field of Faridpur, Bengal, *Sci. & Cult.,* 1 : 298 - 299.

Banerji, J. C. (1936) Studies on the Myxophycèae of lower Bengal, I, *J. Indian Bot. Soc.,* 15 : 285-302.

Banerji, J. C. (1938) Studies on the Myxophyceae of lower Bengal, II, *J. Dept. Sci. Calcutta Univ.,* 1 : 95-109.

Banerjee, Manju, Mausumi Basu & P. K. Sen (1986) Study of the bioassemblage from the recent deltaic surface and subsurface sediments in the Sundarbans as an aid to understand the depositional environment of the ancient sediments. In : *Proc. Spl. Ind. Geo. Con., Poona,* 1-8.

Barton, E. S. (1903) List of Marine Algae collected at the Maldives and Laccadive Islands by J. S. Gardiner, Esq. M. A., *J. Univ. Soc. London Bot.*, 35(247) : 475 - 482.

Be. A. W. H., J. M. Forns & O. A. Roels (1971) Plankton abundance in the North Atlantic Ocean. In : *Fertility of the Sea* (J. D. Costlow, Jr. ed.), 1: 17-50. Gordon and Breach Sci. Publ. New York.

Begum, A. & A. K. M. N. Islam (1992), New records of desmids for Bangladesh. *Dhaka Univ. Stud,* Part E, 7(1) : 91-93.

Begum, Z. T. & A. K. M. N. Islam (1982) Preliminary studies on the effects of blue-green algae in rice yield, *Dacca Univ. Stud.,* B30(1) : 145 - 147.

Begum, Z. T., A. K. M. N. Islam & S. B. Chowdhury (1987) Studies on the chemical composition of blue - green algae. *Bangl. J. Bot.,* 16(2) : 219-220.

Begum, Z. T., R. Akhtar, A. K. M. N. Islam & A. Aziz (1994) Taxonomy of Stigonematalean algae in culture. In : *Algal Biotechnology in the Asia-Pacific region* (Phang, *et.al.,* eds.), pp. 257-262.

Bharadwaja, Y. (1934) The taxonomy of *Scytonema* and *Tolypothrix* including some new records and new species from India and Ceylon. Rev. Algo., 7 : 149-178.

Bharati, S. G. & K. M. Pai (1972) Some desmids from Kodaikanal Lake, S. India, *Phykos,* 11(1&2) : 27-36.

Bhuiya, Z. H., A. K. M. N. Islam, M. A. Harsham, Z. T. Begum & M.M. Rehman (1984) Effect of blue - green algae as biofertilizers on rice, *Bangladesh J. Agric.,* 9(2) : 47 - 51.

Bienati, N. L. & R. A. Comes (1971), Seasonal variation in the physico - chemical composition of sea water in Paradise Harbour-West Antartica; In : *Fertility of the Sea* (J. D. Costlow, Jr. ed.) 1 : 51–70. Gordon and Breach Sci. Publ. New York.

Biswas, K. (1925) Road slimes of Calcutta, *J. Dept. Sci. Univ. Calcutta,* 7 : 1- 10.

Biswas, K. (1926) Flora of Salt lakes, Calcutta, *J. Dept. Sci. Univ. Calcutta,* 8 : 1 - 47.

Biswas, K. (1930), Contributions to the knowledge of the freshwater algae of Manipur, Assam, *J. Bombay Nat. Hist. Soc.,* 34 :189 - 192.

Biswas, K. (1932a) Notes on the organisms in the filtered water of Calcutta, *J. & Proc. Asiatic Soc. Bengal,* 21 (4) : 533 - 539.

Biswas, K. (1932b) The role of aerophilous algae in producing colour - effects on the bark of *Oreodoxa regia* of the Oreodoxa avenue of the Royal Botanic garden, Calcutta. *Hedwigia,* 72 : 31 - 41.

Biswas, K. (1934), Observations on the algal collections from the Khasia and Jaintia Hills, Assam, India, *Hegwigia,* 74 : 1 - 28.

Biswas, K. (1936). A new nannandrous *Oedocladium* from India, *J. Roy. Bot. garden,* Calcutta. 10 : 341 - 345.

Biswas, K. (1937) Common diatoms of the *Loktak* Lake - Manipur, Assam, *J. Roy. Asiatic Soc. Bengal,* n. s. 11 (2).

Biswas, K. (1942) The role of common algal communities of the river Hooghly on the drinking water, Calcutta, *150th Ann. Vol. Roy. Bot. Garden* : 189 - 206.

Bligh, E. G. & W. J. Dyer (1959) A rapid method of total lipid extraction and purification, *Can. J. Biochem. Physiol,* 37 : 911 - 917.

Boergesen, F. (1913) The marine algae of the Danish West India, I, Chlorophyceae. *Dansk, Bo. Arkiv.* 1(4) : 1 - 158 + 2, 126 figs. Map.

Boergesen, F. (1914a), The marine algae of the Danish West India, II, Phaeophyceae, *Ibid.,* 2(2) : 1 - 66 + 2, figs. 1 - 44.

Boergesen, F. (1915) The marine algae of the Danish West India. III, Rhodophyceae, a, *Ibid.,* 3 : 1 - 80, figs. 1 - 86.

Boergesen, F. (1916) The marine algae of the Danish West India. b, *Ibid.,* 3: 81 - 144, figs. 87 - 148.

Boergesen, F. (1917) The marine algae of the Danish West India. C, *Ibid.,* 3: 145 - 240, figs. 149 - 230.

Boergesen, F. (1918) The marine algae of the Danish West India. d, *Ibid.,* 3: 241-304, figs. 231-307.

Boergesen, F. (1919) The marine algae of the Danish West India. e, *Ibid.,* 3: 305-368, figs. 308-360.

Boergesen, F. (1920) The marine algae of the Danish West India. f, *Ibid.,* 3: 369-504, figs. 361-435.

Boergesen, F. (1933) Some Indian Rhodophyceae especially from the shores of the Presidency of Bombay III, *Kew Bull. Misc. Infn.*, 3 : 113 - 42.

Boergesen, F. (1935) A list of marine algae from Bombay. *Det. Kgl. Dasnsk Videnskaab, Selskab. Biol. Medd.*, 12 : 1-64.

Boergesen, F. (1936) Some marine algae from Ceylon. *Ceylon J. Sci., Sect. A. Bot.* 5(2) : 57-96.

Bold, H. C. & Wynne, M. J. (1978) **Introduction to the algae**, Prentice Hall, Englewood Cliffs, New Jersey.

Boney, A. D. (1975) **Phytoplankton**, *Studies in Biology*. No. 52. Edward Arnold, London.

Borge, O. (1899) Ueber tropische und subtropische Süsswasser - Chlorophyceae. Bihang till *K. Svenska Vet.-Akad. Handle.*, 24(12) : 33, pl. 2.

Braun, Al. (1854) In : De Bary, *Abhandle. D. Senck bergischen Gesellsch*, p. 94.

Brook, A. J. (1959) Notes on desmids of the genus *Staurastrum*, II. *St. leptodermum, St. longispinum, St. brasiliense, St. setigerum, St. clevei* and *St. tohopekaligense* var. *trifurcatum*, Freshwater Fish. Lab. Pitlochry, Scotland. 91-96.

Brunnthaler, J. (1913) Die Algengatting *Radiofilum* Schmidle und ihre systematische stellung, Oesterr - Bot. Zeitschr. 63 : 1-8. 3 figs.

Camacho, A. S. (1984) Some aspects of the management and utilization of Phillipines mangroves in relation to fisheries development. Paper presented during the 19th Convention of the Biology Teachers Association of the Phillipines (BIOTA) Iloito City, Phillipines.

Carmichael, W. W. (ed.) (1981) *The Water Environment : Algal toxins and health*, Plenum Press, New York.

Cartens, M., E. Molina Grima, A. Robles Medina, A. Grimenez Grimenez & J. Ibanez Gonzalez (1996) Eicosapentaenoic acid (20:5 n-3) from the marine microalgae *Phaeodactylum tricornutum*. *J. Am. Oil Chem. Soc.*, 73(8): 1025 - 1031.

Catling, H. D., M. R. Martinez & Z. Islam (1981) Survey of algae associated with deepwater rice in Bangladesh. *Crypt. Algol.* 11(2) : 109-121.

Chacko, P. I. (1950) Marine plankton from water around the Krusadai Island. In : *Proc. Ind. Acad. Sci.* 31B : 162-174.

Chacko, P. I. & J. B. Krishnamurthy (1954) On the plankton of three freshwater fish ponds in Madras city, In : *Ind. Symp. Mar. Freshwater Plankton.* Indo-Pacific Fish Coun., UNESCO.

Chadha, Asha & D. C. Pandey (1978) Addition to the algal flora of Allahabad II (Chlorophyta, Chlorococcales). *Phykos* 17 (1&2) : 73-80.

Chakraborti, R. K. & S. R. Das (1999), Chemistry of the brackishwater impoundments in the Indian Sundarbans. In : *Sundarbans Mangals*, pp. 150 - 155.

Chapman, V. J. (1957) The societies of algae of salt marsh and mangroves/ les Societes 'des algues des marais sales et des mangroves. Coll. Intern. CNRS Dinard, 1957, *Ecologie des algues marines,* 81 : 153 - 163.

Chapman, V. J. (1962) *The Algae,* Macmillan & Co. Ltd. pp. 472.

Chapman, V. J. (1970) Mangrove phytosociology, *Trop. Ecol.,* 2 : 1-19.

Chatterjee, P. (1975) Some additions to the Charophytes of West Bengal. *Bull. Bot. Soc. Bengal,* 29 : 105-109.

Chatterjee, P. (1976a) Cytotaxonomical studies of West Bengal Charophyta: Karyotype analysis in *Chara braunii. Hydrobiologia,* 49 : 171-174.

Chatterjee, P. (1976b) *Chara socotrensis f. nuda :* a new record for West Bengal, *Sci. & Cult.*, 42 : 51 - 52.

Chatterjee, P. (1976c), Cytotaxonomic studies on *Chara socotrensis* Nordst. in Kuhn f. *nuda* (Pal) RDW from West Bengal, India. *Cytologia,* 41 : 659 - 663.

Chatterjee, P. (1979a) *Chara fibrosa* var. *fibrosa f. longicorollata* : a new record for India and its cytology. *Curr. Sci.,* 48(2) : 545-547

Chatterjee, P. (1979b) Karyological investigation of *Chara wallichi* Br. from West Bengal. *Cell & Chromosome News Letter,* 2 (1) : 21-23.

Chatterjee, P. (1980) *Aphanochaete repens* A. Br., a member of Chlorophyceae new to West Bengal. *Sci. & Culture,* 46 : 93-94.

Chattopadhyay, G. N. (1999), Chemistry of coastal wetlands of India with relation to aquaculture, In : *Sundarbans Mangals,* pp. 218-232.

Chattopadhyay, P., A. Roychoudhury, R. Pal & A. Lahiri Majumdar (1996), Evaluation of algal biomass quality cultivated at rural level by low cost method, *Ind. J. Exp. Biol.,* 34 : 226-231.

Chaudhuri, A. B. & A. Choudhury (1994) *Mangroves of the Sundarbans.* vol. 1 : India, IUCN, Bangkok, Thailand, pp. 1-247.

Choudhury, M. C. (1969) On forms of *Chara zeylanica* Kl. ex Willd. in India and in the south eastern United States of America. *Phykos.,* 8 (1&2) : 18-27.

Chowdhury, S. C. & A. Khair (1982) The phytoplankton member of Kaptai lake, Chittagong Hill – Tracts. I. Desmidiaceae. *Chittagong Univ. Stud.,* pt. II. 6 : 129-136.

Cleve - Euler, A. (1953) Die Diatomeen von Schweden und Finnland, K. S. vet. Akad. Handle, *Fjarde Ser.* II, Bd. 4(1).

Cleve - Euler, A. (1955) Die Diatomeen von Schweden und Finnland. *K. Svenska Vetensk Akad. Handle. Ser.,* 4. IV. 5,4,1 - 232.

Cleve, P. T. (1904) Plankton table for the North Sea, Bulletin, Part D, May, 1904, Conseil Permanent pour L' Exploration de la Mer, *Bulletin des Resultats aquit pendant les courses periodiques,* 216.

Cleve, P. T. & J. D. Moller (1882) **"Diatoms"** Part 6, no. 277-324, pp. 1-6. Uppsala, Sweden.

Clough, B. F. (ed.) (1982) *Mangrove ecosystems in Australia : Structure, Function and Management,* Australian National University Press, Canberra, 302 pp.

Clough, B. F. (1986) Factors regulating mangrove ecosystem primary productivity. Workshop on Mangrove Ecosystems Dynamics. UNDP/ UNESCO Regional Project RAS/ 72/ 002, pp. 79-85.

Cohen, Zui & Simon Cohen (1991) Preparation of Eicosapentaenoic acid (EPA) concentrate from *Porphyridium cruentum. J. Am. Oil Chem. Soc,* 68(1) : 16-19.

Cohen, Zui (1990) The production potential of Eicosapentaenoic and Arachidonic acids by the red alga *Porphyridium cruentum. J. Oil Chem. Soc.,* 67(12) : 916-920.

Cole, C. A. (1979) *Text book of Limnology.* The C.V.Mosby Company, St. Louis, 2nd Ed.

Collins, F. S. (1901) The algae of Jamaica, *Proc. Amer. Acad. Arts and Sci.,* 37 : 229-270.

Collins, F. S. & A. B. Hervey (1917) The algae of Bermuda. *Proc. Amer. Acad. Arts and Sci.,* 53 : 1-195.

Cooksey, K.E. (1984) Role of diatoms in the mangrove habitat. In : S.C. Snedaker and J. E. Snedaker (eds.) *The Mangrove Ecosystem : Research Methods.* UNESCO.

Costlow, Jr. J. D. (eds.) (1971) *Fertility of the Sea.* vol. I. Gordon & Breach Sci Publ.

Cox, J. D. (1901) Algae; In : *Plant Life of Alabama,* (Mohr, C. ed.), *Contri. U. S. Nat. Herb.* 6 : 1-921.

Croasdale, H. T. & A. M. Scott (1976) New or otherwise interesting desmids from Northern Australia, *Nova Hedwigia,* 27 : 501–596.

Cupp, E. E. (1943) Marine plankton diatoms of the West Coast of North America. *Bull. 'Scripps Institute of Oceanography',* University of California Press, Berkeley & Los Angeles. 1-237.

Darley, W. M. (1977) Biochemical compositions. In : *Biology of Diatoms* (ed. D. Werner), pp. 198-223.

Das, A. K. & N. C. Nandi (1999) Fauna of the Indian Sundarbans mangals and their role in the ecosystem. In : *Sundarbans Mangals* (eds. D. N. Guha Bakshi, P. Sanyal and K. R. Naskar, pp. 417 - 427.

Das, C. R. (1961) A new report on *Spirogyra rugulosa* Iwanoff from Senchal lake, Darjeeling, India. *Bull. Bot. Surv. India,* 3 & 4 : 389.

Das, C. R. (1962) A new species of *Zygnemopsis* (Skuja) Transeau 1934 from the Calcutta 1961 National Agricultural fair. *Curr. Sci.,* 31 : 255.

Das, C. R. (1963) The Compsopogonales in India. *Proc. Nat. Inst. Sci. India,* (B)29 : 239-243.

Davis, C. C. (1955) The Marine and Fresh Water lankton. Mich. St. Univ. Press, U.S.A. pp. 1-562.

Dawson, E. Y. (1952) Marine red algae of Pacific Mexico. Part I. Bangiales to Corallinaceae Subf. Coralinoideae, *Allan Hancock. Pacific Exped*, 17 : 1-239.

De, B. K., S. Chaudhury & D. K. Bhattacharya (1998), Effects of nitrogen sources on V-linolenic acid accumulation in *Spirulina platensis. J. Am. Oil Chem. Soc.*, 75(12):1-4.

De, D. K. (1999) The river Ganga-environment and fishery. In : *Ecology, Fisheries and Fish Stock Assessment in Indian Rivers,* CIFRI, Barrackpore, pp. 129-140.

De, Toni, J. B. (1895) Sylloge Algarum 3 (Fucoidae) : 1-638. *Omnium Hucsque Cognitarum* Pata vii.

De, Toni, J. B. (1897-1903) Sylloge Algarum, vol. 4. Florideae. Vol. 6, Florideae . pata vii.

Desikachary, T. V. (1959) *Cyanophyta.* ICAR Publication, New Delhi.

Dickie, G. (1874) On the marine algae of Barbados. *Joun. Linn. Soc. Bot.,* 14: 146-152, pl. 11.

Dixit, S. C. (1940) Algae of the Malwar Harbour. *J. Univ. Bombay,* 8(5) : 1-19.

Dokulil, M. (1975) Horizontal-und Vertikal-gradienten in einem Flachsee (Neusiedlersee, Österreich). *Verh. Ökol. Wien,* 177-187.

Drebes, G. (1977) Sexuality. In : *Biology of Diatoms* (ed. D. Werner), pp. 250-283.

Duke, E. L. & B. E. F. Reimann (1977) The ultrastructure of the diatom cell. In : *Biology of Diatoms* (ed. D. Werner), pp. 65-109.

Dyerberg, J. (1986) Linoleate - derived polyunsaturated fatty acids and prevention of Artherosclerosis, *Nutr. Rev.,* 44 : 125 - 134.

Eaton, J. W. & B. Moss (1966) The estimation of numbers and pigment content in epipelic algal populations. *Limnol. Oceanog,* 11 : 584-595.

Eppey, R. W. (1977) The growth and culture of Diatoms. In : *Biology of Diatoms* (ed. D. Werner), pp. 24-64.

Evensen, D. L. & G. R. Hasle (1975) The morphology of some *Chaetoceros* (Bacillariophyceae) species as seen in the electron microscope. *Beihft zur Nova Hedwigia,* 53 : 153-174.

Feldmann, J. & G. Lami (1934) Observations sur quelques Gelidiaceae, *Rev. gen. Bot.,* 46 : 528 - 49.

Feldmann, J. & R. Lami (1937) Sur la vegetation marine de la Guadelloupe, *Comptes Rendus Acad. Sci. (Paris),* 203(17) : 883 - 885.

Fjerdingstad, E. (1964) Pollution of streams estimated by benthal phyto-micro-organisms I. A saprobic system based on communities of organisms and ecological factors. *Int. Rev. ges. Hydrobiol.,* 49(1) : 63-131.

Foersten, K. (1972) Desmideen aus dem Sudosten der vereinigten von Amerika, *Nova Hedwigia,* 23 : 515- 644.

Foged, N. (1971) Diatoms found in a bottom sediment sample from a small deep lake on the northern slope, Alaska. *Nova Hedwigia,* XXI(2-4) : 923-1034.

Fogg, G. E. (1969) The physiology of an algal nuisance. *Proc. Roy. Soc. (London),* B173, pp. 175-189.

Foster, M. S. (1972) The algal turf community in the nest of ocean gold fish, Hypsypops rubicunda. In: Proceedings of the VII International Seaweed Symposium (ed. Nissizawa, K.) University of Tokyo Press, Tokyo, 55-60.

French, D. P., M. J. Furnas & T. J. Smayda (1983) Diel changes in nitrite concentration in the chlorophyll maximum in the Gulf of Mexico, Deep Sea Res., 30 : 707-722.

Frenguelli, J. (1960) Diatomeas y silicoflageledos recogidasin Tierra Adalia duranta les Expediaones Polares Francesas de Paul - Emile Victor (1950 - 52). *Rev. Algologique,* V(1) : 3-48.

Fritsch, F. E. (1935) *The Structure and Reproduction of the Algae.* Vols. I & II, pp. 939, Cambridge University Press, Cambridge.

Fritsch, F. E. (1949) The genus Anabaena with special reference to the species recorded from India and the adjacent Asiatic mainland. *J. Indian bot. Soc.,* 28 : 135-161.

Frost, B. W. (1981) Grazing. In : *The Physiological Ecology of Phytoplankton* (ed. Morris), pp. 465-491, Blackwell, Oxford.

Golterman, H. L. (1975). *Physiological Limnology,* Elsevier, Amsterdam.

Gons, H. J. (1982) Structural and functional characteristics of epiphyton and epipelon in relation to their distribution in Lake Vechten. *Hydrobiol,* 95 : 79-114.

Gopalakrishnan, P. (1971) *Rhodymenia dissecta* Boergs. from Gujarat. *Phykos* 10 (1&2) : 154-155.

Gopalakrishnan, P. (1974) On the occurrence of *Hormophysa triquetra* (L.) Kutzing and *Rhodomenia palmata* Grev. on the west coast of India. *Phykos,* 13(1) : 6-9.

Gopalakrishnan, P. (1974) *Turbinaria indica* sp. nov. A new marine alga from the Gulf of Kutch, *Phykos,* 13(1) : 10 - 15.

Gopalan, C. (1971) *A manual of laboratory techniques*, National Institute of Nutrition. Hyderabad, India, pp. 434.

Gordon, M. (1943) Charales, plant benefactors of fishes. *Aquarium* 12(7): 117 - 120.

Gréville, R. K. (1833) Algae, vol. 1 & 2. In: *A. de Saint Hilaire, Voyage dans le district des diamens et sur le littoral du Brésil.*, pp. 1-456.

Grieve, S. (1909) Note on some sea - weeds from the Island of Dominica, British West Indies. *Trans. Bot. Soc. Edinburgh,* 24(1) : 7 - 12.

Griffith, W. (1844) The cryptogamous plants of Dr. Roxburgh forming the fourth and last part of the Flora Indica. *Calcutta J. Nat. Hist.,* 4 : 463-520.

Griffith, W. (1849) Notulae ad plantas Asiaticas. *pt.* 2 : 275-284.

Guillerd, R. R. L. & P. Kilham (1977) The ecology of marine planktonic diatom. In : *Biology of Diatoms* (ed. D. Werner), pp. 372-469.

Guilten, O., B. R. E. D. Mendiola & R. I. D. Rondan (1971) Primary productivity and phytoplankton in the coastal Peruvian waters. In : *Fertility of the Sea* (J. D. Costlow, Jr. ed.) 1 : 157-185.

Gunstone, F. D., J. L. Harwood & F. B. Padley (eds.) (1994) *The Lipid Handbook.* Chapman and Hall, London, pp. 173-175.

Gupta, D. (1965) Some new records of blue - green algae from West Bengal. *Bull. Bot. Soc. Bengal,* 19 : 1-2.

Gupta, D. (1975) Some new records of blue-green algae from West Bengal, II, *Bull. Bot. Soc. Bengal,* 29 : 29-31.

Gupta, Meera & D. C. Pandey (1974) On an interesting species of *Cylindrospermum* Kuetzing (Cyanophyceae). *Phykos* 13(1) : 16- 21.

Hadi, A. M., Rafia Islam, A. K. Y. Haroon & Azahar, A. Al - Saboochi (1993) Marine benthic algae from coastal water of Iraq. *J. Asiatic Soc. Bangl. (Sci.)* 19 : 123-144.

Hamel, G. & A. Hamel - Joukov (1937) *Algue des Antilles Fracaises.* Fasc. 1-3, Paris.

Hammer, V.T., K. F. Walker & W. D. Williams (1973) Derivation of daily phytoplankton production estimates from short term experiments in some shallow, eutrophic Australian saline lakes, *Austr. J. Mar. Freshwater* Res., 24 : 259-266.

Hanson, J, M. & W. C. Leggett (1982) Emperical prediction of fish biomass and yield. *Can. J. Fish Aq. Sci.,* 39 : 257-63.

Hargraves, P. E. (1972) Studies on the marine plankton diatoms. I. *Chaetoceros diadema* (Ehr.) Gran. : Life cycle, structural morphology and regional distribution. *Phycolog.* 11(3&4) : 247-258.

Harper, M. A. (1977) Movements. In : *Biology of Diatoms* (ed. D. Werner) pp. 224 – 249.

Harris, G. P. (1973) Diel and annual cycles of net plankton photosynthesis in Lake Ontario. *J. Fish. Res. Bd. Can.,* 30 : 1779-1787.

Harris, G. P. (1986) *Phytoplankton Ecology : Structure, Function & Fluctuation.* Chapman and Hall Ltd. New York , pp. 384.

Hashimoto, Y. (1979) *Marine Toxins and Other Bioactive Marine Metabolites,* Japan Scientific Societies Press, Tokyo.

Hasle, G. R. & E. E. Syvertsen (1997) Marine Diatoms. In : *Identifying Marine Phytoplankton* (ed. C. R. Tomas) Academic Press, pp. 5-385.

Hasle, G. R. (1977) Use of Electron Microscopy in morphological and taxonomical diatom studies. In : *Biology of Diatoms* (D. Werner, ed.) Blackwell Scientific Publications. pp. 18-23.

Hasle, G. R. H. & B. R. E. D. Mendiola (1967) The fine structure of some *Thalassionema & Thalassiothrix* species. *Phycolg.* 6(2+3) : 107-125.

Hauck, F. (1888) Meersalgen von Puerto - Rico. *Engler's Bot. Jahrb.* 9 : 457-470.

Hayes, P. R. & E. H. Anthony (1964) Production capacity of North American Lakes as related to the quantity and trophic level of fish, the lake dimensions and the water chemistry. *Trans.Am. Fish. Soc.,* 93 : 53-57.

Heckey, R. E. & H. J. Kling (1981) The phytoplankton and protozooplankton of the euphotic zone of Lake Tanganyike : species composition, biomass, chlorophyll content and spation-temporal distribution, *Limnol. Oceanogr.* 26 : 548-564.

Hellebust, J. A. & J. Lewin (1977) Heterotrophic nutritions. In : *Biology of Diatoms* (ed. D. Werner). pp. 169-197.

Hellebust, J. A. & J. S. Craige (eds.) (1978) *Handbook of Phycological Methods,* Physiological & Biochemical Methods, Cambridge University Press, Cambridge.

Hemsley, W. B. (1884) Algae. vol. i, ii, pp. 104 - 128. In : *Report on the botany of the Bermudas and various other islands of the Atlantic and southern oceans. Rep. Sci. Res. Exploring Voyage of H. M. S. "Challenger",* 1873-76. **Botany** I, 2 : 1-135.

Hendey, N. I. (1954) A preliminary check list of British marine diatom. *J. Mar. Biol. Ass. U.K.* 33 : 537-560.

Hendey, N. I. (1964) An introductory account of the smaller algae of the British coastal water, Part V : Bacillariophyceae (Diatoms) *Fisheries Investigations Ser. IV.* 317 pp., HMSO, London.

Hendey, N. I. (1971) Some marine diatoms from the Galapagos islands, *Nova Hedwigia,* 22(1+2) : 371- 422.

Heret, M. (1960) General principle of pond construction and management, productivity and stocking. *ETAP, FAO lecture at the IIIrd International Training Centre,* Bogar, Indonesia. 1955. 2.

Hirano, M. (1955-1960) Flora Desmidiarum Japonicarum. *Contr. Biol. Lab. Kyoto Univ.* Nos. 1,2,4,5,7,9,11 : 1 - 474, pls. 54.

Hoyt, W. D. (1920) Marine algae of Beaufort, North Carolina. *Bull. Bur. Fisheries* (U. S.) 36: 367-556, 3 maps, 47 figs, pls. 84-109.

Howe, M. A. (1918) Algae, pp. 489-540, In: *Flora of Bermuda,* Britton, N. L. (ed.), ix + 585 pp., illus. New York.

Howe, M. A. (1920) Algae, pp. 553-618. In : The Bahama Flora, Britton, N. L. & C. F. Millspough (eds.), viii + 695 pp, New York.

Howe, M. A. (1928) Notes on some marine algae from Brazil and Barbados. *Journ. Washington Acad. Sci.,* 18(7) : 186-194.

Hustedt, F. (1930) Die Kieselalgen Deutschlands, Osterreichs und der Schweiz. 7(1) : 1-920.

Hustedt, F. (1959) Die Kieselalgen Deutschlands, Osterreichs und der Schweiz. Dr. L. Rabenhorsts Kryptogamen Flora von Deutschland, Osterreichs und der Schweiz, 7(2) : 1-845.

Islam, A. K. M. N. (1960) Some subaerial green algae from East Pakistan, *Trans. Amer. Micros. Soc.,* 79(4) : 471 - 479.

Islam, A. K. M. N. (1961) The genus *Cloniophora. Revue Algologique. n.s.t.* 6(1) : 7-32.

Islam, A. K. M. N. (1963) A revision of the genus *Stigeoclonium.* Beiheft zur. *Nova Hedwigia.,* 10 : 1-165 + 47 plates. J. Cramer Publisher, Weinheim

Islam, A. K. M. N. (1964) The genus *Cladophorella* newly found in East Pakistan. *Revue Algologique.* 7(4) : 275-289.

Islam, A. K. M. N. (1965) Taxonomic study of the species of *Dichotomosiphon* and *Vaucheria* found in East Pakistan. *Pak. Acad. Sci.,* 2(1) : 47-56 + 9 plates.

Islam, A. K. M. N. (1966) Occurrence of the genus *Sirocladium* in East Pakistan. *Pak. Jour. Biol. & Agric. Sci.,* 8(2) : 264 - 270.

Islam, A. K. M. N. (1969a) *Kirchneriellasaccus lunatus* gen. et *sp. nov.* a new member of the order Chlorococcales. *Rev. Alg. n.s.t.* 9(4) : 348-358.

Islam, A. K. M. N. (1969b) Some rare planktonic green algae found in East Pakistan. *Jour. Pak. Bot. Soc.,* 1 : 19-32.

Islam, A. K. M N. (1969c) A preliminary report on the phytoplankton and other algae of Chittagong Hill-Tracts. *J. Asiatic Soc. Pak.,* 14(3) : 343-363.

Islam, A. K. M. N. (1970a) *Contribution to the knowledge of desmids of East Pakistan.* Part I. *Nova Hedwigia.* 20 : 903-983.

Islam, A. K. M. N. (1970b) Preliminary ecological report on the marine algal flora of the St. Martin's Island. *East Pakistan. Jour. Asiatic Soc. Pak.,* 15(3) : 273-282.

Islam, A. K. M. N. (1970c) The genus *Tetrasporidium* in East Pakistan. *Rev. Alg.* 1 : 85-89.

Islam, A. K. M. N. (1972a) New and rare species of some green algae from Bangladesh. *Nova Hedwigia.* 22 : 655 - 663 + 14 pls.

Islam, A. K. M. N. (1972b) Subaerial algae of Bangladesh. *Bangladesh Jour. Bot.* 1(1&2) : 13-64.

Islam, A. K. M. N. (1972c) The genus Bulbochaete in Bangladesh. *Bangladesh Jour. Bot.* 1(1&2) : 1-12 + 14 pls.

Islam, A. K. M. N. (1973) The algal flora of Sundarbans mangrove forest, *Bangladesh. Bangladesh J. Bot.* 2(2) : 11-36.

Islam, A. K. M. N. (1973a) A new species of *Boodleopsis* (Chlorophyta) from Bangladesh. *Bangladesh J. Bot.* 2(1) : 53-67.

Islam, A. K. M. N. (1973b) Freshwater algae of Bangladesh I. Chlorophyceae, Xanthophyceae and Chrysophyceae. *Dacca Univ. Stud.* B21(1) : 89-84.

Islam, A. K. M. N. (1973c) Freshwater algae of Bangladesh. III. Cyanophyceae. *Dacca Univ. Stud.,* B21(2) : 133-139.

Islam, A. K. M. N. (1974a) A preliminary list of benthic marine algae from the Bay of Bengal, Bangladesh. *Bangladesh J. Bot.,* 3(1) : 83-91.

Islam, A. K. M. N. (1974b) Freshwater algae of Bangladesh. IV. *Aphanochaete, Coleochaete and Chaetosphaeridium. Bangladesh J. Bot.,* 3(1) ; 35-43.

Islam, A. K. M. N. (1974c) Freshwater algae of Bangladesh. VII. Flagellates. Volvocales. *Bangladesh J. Bot.,* 3(2): 7-15.

Islam, A. K. M. N. (1974d) Preliminary studies on the food of some fish. *Dacca Univ. Stud.,* B22(1) : 47-51.

Islam, A. K. M. N. (1975a) Contribution to the study of desmids of Bangladesh. II. *Dacca Univ. Stud.,* B23(2) : 31-37.

Islam, A. K. M. N. (1975b) Further addition to the genus *Coleochaete* from Bangladesh. J. *Asiatic Soc. Bangladesh,* 1(1) : 71-72.

Islam, A. K. M. N. (1976) Contribution to the study of benthic marine algae of Bangladesh. *Bibliotheca Phycologica.* 19 : 1-253 + 7 pls.

Islam, A. K. M. N. (1976a) Freshwater algae of Bangladesh. VI. Genus *Oedognium. Nova Hedwigia,* 27 (3&4) : 919-925 + 4 pls.

Islam, A. K. M. N. (1976c) Addition to the species of *Vaucheria* in Bangladesh, *Dacca Univ. Stud.,* B24(1) : 59-62.

Islam, A. K. M. N. (1977) Studies on the members of *Zygnemaceae* from Bangladesh. I. *Mougeotia, Zygnema* and *Sirogonium. Dacca Univ. Stud.,* B25(1) : 7-22.

Islam, A. K. M. N. (1978) A new species of *Vaucheria* from Bangladesh. *Bangladesh J. Bot.,* 7(1) : 13-16.

Islam, A. K. M. N. (1979a) Addition to the list of Oedogoniaceae from Bangladesh. *Dacca Univ. Stud.,* B27(1) : 47-52.

Islam, A. K. M. N. (1979b) Genus *Chloroclonium* in Bangladesh. *Bangladesh J. Bot.* 8(162) : 113-115.

Islam, A. K. M. N. (1979c) *Vaucheria longicaulis* Hoppaush from Iraq with a note on the species. *Bull. Torrey Bot. Club.* 106 : 167-173.

Islam, A. K. M. N. (1980a) Revision of the members of Oedogoniaceae-some proposals. In : *Taxonomy of algae* (ed. T.V. Desikachary). University of Madras, 533-536.

Islam, A. K. M. N. (1980b) Study of *Triplastrum* found in Bangladesh with a note on its species. *Bangladesh J. Bot.,* 9(1) : 1-12.

Islam, A. K. M. N. (1981) Study of algal flora of Tibet and Bangladesh. In: *Proc. Symp. on Qinghai-Xizang (Tibet) Plateau.* Geological and ecological studies on Qinghai-Xizang Plateau 2 : 1141-1143. Science Press, Beijing.

Islam, A. K. M. N. (1982) Marsh algae from southern Iraq. *Intl. Rev. ges. Hydrobiol.,* 67(2) : 245-260.

Islam, A. K. M. N. (1983) Gall formation in *Vaucheria* spp. by parasitic rotatorian members. *Bangl. J. Bot.,* 12(1) : 87 - 89.

Islam, A. K. M. N. (1984a) Studies on the genus *Vaucheria* (Xanthophyceae) in Iraq. *Intl. Rev. ges. Hydrobiol,* 69 : 877-902.

Islam, A. K. M. N. (1984b) Studies on the members of Zygnemaceae from Bangladesh. II. *Spirogyra. Bang. J. Bot.* 13(2) : 194 -213.

Islam, A. K. M. N. (1985a) Occurrence of diatom - bloom in the coastal area of Bangladesh. *Bangladesh J. Bot.* 14(2) : 185-187.

Islam, A. K. M. N. (1985b) Occurrence of *Johanneshaptistia* (Cyanophyceae) in Bangladesh. *Bangladesh J. Bot.,* 14(1) : 73-75.

Islam, A. K. M. N. (1985c) Some new and rare algae from Iraq. *Intl. Rev. ges. Hydrobiol,* 70 : 755-766.

Islam, A. K. M. N. (1986) New records of algae from Bangladesh. I. *Dactylococcopsis* and *Onychonema. Bangladesh J. Bot.,* 15(1) 109-110.

Islam, A. K. M. N. (1987) Studies on the Characeae of Iraq. I. Genus *Nitella. Bangl. J. Bot.* 16(2) : 229-231.

Islam, A. K. M. N. (1991) Phycology. In : *Two Centuries of Plant Sciences in Bangladesh and adjoining regions.* (Ed. A. K. M. N. Islam) pp. 97-153. *Asiatic Society of Bangladesh.* Dhaka.

Islam, A. K. M. N. (1992) Freshwater red algae of Bangladesh. *J. Asiatic Soc. Bangladesh (Sci.)* 18(1) : 91-93.

Islam, A. K. M. N. & P. Sarma (1963) Two new species of terrestrial *Oedogonium* from East Pakistan. *Trans. Amer. Micros. Soc.,* 82(1) : 74-77 (USA).

Islam, A. K. M. N. & A. N. M. Ahia (1964) Contribution to the knowledge of Chaetophoraceae of Dacca district. *Pak. Jour. Biol. & Agric. Sci.,* 7(1) : 103-110.

Islam, A. K. M. N. & P. Sarma (1964a) Contribution to the knowledge of Oedogoniales of Dacca district, East Pakistan. *Pak. Jour. Biol. & Agric. Sci.,* 7(1) : 132-135.

Islam, A. K. M. N. & P. Sarma (1964b) Preliminary survey of the epizoic species of *Oedogonium* on the shells of freshwater molluscs in East Pakistan. *Rev. Algologique.* 7(2) : 178-186.

Islam, A. K. M. N. & P. Sarma (1965) New and rare species and varieties of the Oedogoniales from Dacca district. *East Pakistan. Pak. Jour. Biol. & Agric. Sci.*, 8(1) : 169-188.

Islam, A. K. M. N. & M. Khatun (1966) Preliminary studies on the phytoplankton of polluted waters. *Sci. Res.* 3(2) : 94-109.

Islam, A. K. M. N. & L. Nahar (1967) Preliminary studies on the phytoplankton of polluted waters. II. Blue-Green algae. *Sci. Res.* 4 (2&3) : 141-149.

Islam, A. K. M. N. & D. Sarma (1968) The Characeae of East Pakistan. I. *Chara* and *Lychnothamnus*. *Jour. Asiatic Soc. Pak.* 13(3) : 357-376.

Islam, A. K. M. N. & M. A. Uddin (1969a) A preliminary report on the phytoplankton and other algae of Chittagong Hill - Tracts. *Jour. Asiatic Soc. Pak.* 14(3) : 3453 - 363 + 13 pls.

Islam, A. K. M. N. & M. A. Uddin (1969b) Studies on the blue-green algae of Dacca district. *Dacca University Studies.* Part B, 17 : 85-102.

Islam, A. K. M. N. & Z. T. Begum (1970) Studies on the phytoplankton of Dacca district. *Jour. Asiatic Soc. Pak.*, 15(3) : 227-271 + 8 pls.

Islam, A. K. M. N. & M. A. Uddin (1973) Freshwater algae of Bangladesh. II. Cyanophyceae. *Dacca Univ. Stud.*, B21(2) : 127-132.

Islam, A. K. M. N. & A. M. S. Zaman (1974) Freshwater algae of Bangladesh. VIII. Ulotrichales. *Dacca Univ. Stud.*, B22(2) : 83-98.

Islam, A. K. M. N., A. K. Y. Haroon & K. M. Zaman (1974) Limnological studies of the river Buriganga, I. Physical and Chemical aspects. *Dacca Univ. Stud.*, B22(2) : 99-111.

Islam, A. K. M. N. & A. Aziz (1975) Study of marine phytoplankton from the north-eastern Bay of Bengal, Bangladesh. *Bangladesh J. Botany.* 4(1-2) : 1-32.

Islam, A. K. M. N. & A. K. Y. Haroon (1975a) Limnological studies of the river Buriganga. II. Biological aspect. *Dacca Univ. Stud.*, B21(1) : 25-44.

Islam, A. K. M. N. & J. K. Saha (1975b) Limnological studies of the Ramna lake at Dacca. *Dacca Univ. Stud.*, B23(2) : 39 - 46.

Islam, A. K. M. N. & K. M. Zaman (1975) Limnological studies of the river Buriganga III. Biological aspect. *J. Asiatic Soc. Bangladesh (Sc.)* 1 : 45-65.

Islam, A. K. M. N. & F. Mendes (1976) Limnological studies of a jheel in Sher-e-Bangla Nagar, Dacca. *Dacca Univ. Stud.* B24 (2) : 63-67.

Islam, A. K. M. N. & F. Mendes (1976) Proteins from *Azolla pinnata* R. Br. (aquatic fern) and *Lyngbya majuscula* Harvey ex Gomont (blue - green algae). *Bangladesh Jour. Bot.* 5(1&2) : 89-92.

Islam, A. K. M. N. & A. Sobhan (1976) Ecology and periodicity of the members of Zygnemaceae in Dacca district. *J. Asiatic Soc. Bangladesh.* B1(2) : 113-116.

Islam, A. K. M. N. & D. Sarma (1976) The Characeae of Bangladesh. II. The genus *Nitella. J. Asiatic Soc. Bangl.* 2(1) : 43-61.

Islam, A. K. M. N. & P. Sarma (1976) Freshwater algae of Bangladesh. V Genus *Oedogonium. Nova Hedwigia.* 27(1) : 425-454.

Islam, A. K. M. N., B. C. Chowdhury & S. Begum (1976) The genus *Botrydium* in Bangladesh. *Dacca Univ. Stud.,* B24(1) : 63-67.

Islam, A. K. M. N. & S. N. Paul (1977) Limnological studies on *Wolffia arrhiza* (L.) Wiom. *J. Asiatic Soc. Bangladesh (sci.).* 3(1) : 111-123.

Islam, A. K. M. N. & F. Mendes (1977) Studies on the proteins in blue - green algae . II. Extractable proteins. *Dacca Univ. Stud.,* B25(2) : 51-54.

Islam, A. K. M. N. & F. Mendes (1977) Studies on the proteins in the blue-green algae of Bangladesh. I. Preliminary investigations. *Dacca Univ. Stud.* B25(1) : 23-28.

Islam, A. K. M. N. & A. Aziz (1977) Studies on the phytoplankton of the Karnaphuli river estuary. *J. Bangladesh Acad. Sci.,* 1(2) : 141-154.

Islam, A. K. M. N. & M. A. Uddin (1977) Blue-green algae from Dacca, Bangladesh. I. Chroococcaceae and Pleurocapsaceae. *J. Asiatic Soc. Bangladesh (Sci.)* 2(2) : 75-81.

Islam, A. K M. N. & N. Paul (1978) Hydrobiological study of the haor Hakaluki in Sylhet. *J. Asiatic Soc. Bangladesh (Sci.)* 4(1) : 83-91.

Islam, A. K. M. N. & M. A. Uddin (1978) Blue-green algae from Dacca, Bangladesh III. Nostocaceae, Scytonemaceae and Stigonemataceae. *Dacca Univ. Stud.* B26(1) : 85-93.

Islam, A. K. M. N. & A. K. Y. Haroon (1978) New reports of some members of Chaetophoraceae from Bangladesh. *Nova Hedwigia.* 29(3/4) : 537.

Islam, A. K. M. N. & A. Khair (1978) Addition to the list of marine algae of St. Martin's Island. I. Genus *Codium* Stackhouse. *J. Asiatic Soc. Bangladesh (Sci.).* 4(1) : 123-126.

Islam, A. K. M. N. & A. Khair (1978) Report of some phytoplankton from lake Kaptai, Chittagong hill tracts. *Dacca Univ. Stud,.* B26(2) : 53-61.

Islam, A. K. M. N. & M. A. Uddin (1978) Blue-green algae from Dacca, Bangladesh. II. Oscillatoriaceae. *Dacca Univ. Stud.* B26 (1) : 73 - 84.

Islam, A. K. M. N. & S. K. Tariquat Hossain (1978) Algal flora of the ablution tanks of mosques in Dacca city. *J. Asiatic Soc. Bangladesh (Sci.)* 4(1) : 103 - 113.

Islam, A. K. M. N. & M. Hossain (1979) Preliminary studies on the algal flora of Bagerhat, Khulna. *J. Asiatic Soc. Bangladesh (Sci.)* 5(1) : 37-45.

Islam, A. K. M. N. & A. Aziz (1979) Algal flora of Moheshkhali Island, Bangladesh. *Dacca Univ. Stud.,* B27(2) : 105 - 122.

Islam, A. K. M. N. & A. R. Chowdhury (1979) Hydrobiological studies of Dhanmondi lake, Dacca II. Phytoplankton. *J. Asiatic Soc. Bangladesh (Sci.)* 5(2) : 47-57.

Islam, A. K. M. N., M. Rahman & A. R. Choudhury (1979) Hydrobiological studies of Dhanmondi lake, Dacca. I. Macrophytes and benthic flora. *J. Asiatic Soc. Bangladesh (Sci.)* 59(1) : 59-75.

Islam, A. K. M. N. & A. Aziz (1980) Marine diatoms from the Bay of Bengal. *Bangladesh J. Bot.* 9(1) : 29-35.

Islam, A. K. M. N. & A. Aziz (1980) Marine diatoms from the Bay of Bengal. *Bangladesh J. Bot.* 9(1) : 29-35.

Islam, A. K. M. N. & A. K. Y. Haroon (1980) Desmids of Bangladesh. *Intl. Rev. ges. Hydrobiol.* 65(4) : 543-594.

Islam, A. K. M. N. & R. J. M. Shamshul Alam (1980) Members of Cladophoraceae of Dacca district. *Dacca Univ. Stud.*, B28(1) : 61-70.

Islam, A. K. M. N., Anatunnesa & A. K. Y. Haroon (1980) Hydrobiological studies in and around Naogaon, Rajshahi. *Dacca Univ. Stud.*, B28(2): 31-47.

Islam, A. K. M. N. & K. Moniruzzaman (1981) Contribution to the study of Euglenophyta. I. Genus *Trachelomonas. Intl. Rev. Ges. Hydrobiol.*, 66(1) : 109-125.

Islam, A. K. M. N. & Z. T. Begum (1981) Addition to the list of blue-green algae of Bangladesh. II. *Bangladesh J. Bot.*, 10(1) : 1-15.

Islam, A. K. M. N. & H. A. Hameed (1982) Addition to the list of marine algae of St. Martin's Island, Bangladesh. II. Brown, red and blue - green algae. *Nova Hedwigia.* 36 : 643-657.

Islam, A. K. M. N. & A. K. Y. Haroon (1983) Studies on Chaetophoraceae from southern Iraq. *Intl. Rev. ges. Hydrobiol.*, 68 : 297-315.

Islam, A. K. M. N. & A. K. Y. Haroon (1985) Desmids from Iraq. *Intl. Rev. ges. Hydrobiol.*, 70(6) : 877-889.

Islam, A. K. M. N. & H. A. Hameed (1985) Check-list of algae with anote on the limnological and oceanographic studies in Iraq (1942-1982). *Asiatic Soc. Bangladesh.* pp 63.

Islam, A. K. M. N. & M. G. Morshed (1985) Occurrence of diatoms-bloom in the coastal area of Bangladesh. *Bang. J. Bot.* 14(2) : 185-187.

Islam, A. K. M. N. & M. A. Mannan (1986) Algal flora of some brackish water shrimp culture ponds at Satkhira. *Dhaka Univ. Stud.* E1(1) : 7-18.

Islam, A. K. M. N. & A. Aziz (1987) Addition to the list of marine algae of St. Martin's Island, Bangladesh III. Red Algae. *Nova Hedwigia,* 45(1-2) : 211-221.

Islam, A. K. M. N. & A. Aziz (1987) New record of algae from Bangladesh. II. Genus *Radiococcus* Schmidle (Chlorophyta). *Bangl. J. Bot.* 16(1): 89-92.

Islam, A. K. M. N. & Z. T. Begum (1987) New record of algae from Bangladesh. III. Genus *Pseudobohlinia* (Chlorococcales). *Bangladesh J. Bot.* 16(1) : 103-106.

Islam, A. K. M. N., R. A. M. Hadi & A. Aziz (1988) Studies on the Characeae of Iraq. II *Chara and Nitellopsis. Bangl. J. Bot.* 17 : 57-64.

Islam, A. K. M. N., A. Begum & N. Akhtar (1991) Some observations on the phytogeography of desmids. *J. Asiatic Soc. Bangladesh (Sci.).* 17(2) : 171-178.

Islam, A. K. M. N., M. Khondker & S. Haque (1991) Euglenoid algae of four polluted ponds in and around Dhaka city. *Bangl. J. Bot.* 20(1) : 7-15.

Islam, A. K. M. N. & M. R. Khan (1992) A new combination of a *Coleochaete* taxon (*C. reptans* (Duringer) Islam & Khan comb. nov.) based on Bangladesh materials. *Bangladesh J. Bot.* 21(2) : 287-290.

Islam, A. K. M. N., Ashrafi Begum & Nasima Akhter (1992) Study of the desmids (Chlorophyta) from Cox's Bazar, Bangladesh. *Bangladesh J. Bot.* 21 (1) : 43-51.

Islam, A. K. M. N., M. Khondker, A. Begum & N. Akhter (1992) Hydrobiological studies in two habitats at Dhaka. *J. Asiatic Soc. Bangl. (Sci.)* 17(2) : 171-178.

Islam, A. K. M.N., A. Begum & N. Akhtar (1992) Study on the desmids (Chlorophyta) from Cox's Bazar, Bangladesh. *Bangl. J. Bot.,* 21(1) : 43-51.

Islam, A. K. M. N. & M. Khondker (1993) Some unicellular flagellate algae of Bangladesh. *J. Asiatic Soc. Bangladesh (Sci.).* 19 : 75-79.

Islam, A. K. M. N. & M. S. Islam (1993) Hydrophytes, eutrophication and diseases. In : *Hypertrophical and polluted freshwater ecosystems : Ecological bases for water resource management* (Tilzer, M. M. & M. Khonker eds.) pp. 173 - 178. *Proc. Intl. Symp. Limnol.,* Dept. of Botany, Dhaka University Press.

Islam, A. K. M. N. & M. Khonker (1994) New records of algae from Bangladesh IV. *Heteromastrix* and *Gonyostomum. Bangladesh J. Bot.* 23 (2) : 199-203.

Iyengar, M. O. P. (1971) Contribution to our knowledge of South Indian algae IV. *Phykos* 10(1&2) : 141-151.

Jagtap, T. G. (1985) Studies on littoral flora of Andaman Islands. In : *Marine Plants.* Paper presented at the All India Symposium on marine

Plants, their Biology, Chemistry and Utilization. Dona Paula –Goa, Oct. 30- Nov. 1, 1983, Krishnamurthy, V. & A. G. Untawale (eds.). pp. 43-50.

Jha, B.C. (1986) *Comparitive periphytic studies of certain freshwater bodies like Getalsud reservoir and Subarnarekha river of the tribal areas of Ranchi, Bihar.* Ph.D. Thesis, Ranchi University, Bihar.

Jhingran, A. G. & Babulal (1990) Primary production; energy fixation and fishery potential of Hooghly-Matla estuarine system. In : *Contr. to the Fisheries of Inland Open Water Systems in India.*, pp. 130-134.

Jhingran, V. G., A. V. Natarajan, S. M. Banerjee & A. David (1988) *Methodology on Reservoir Fisheries Investigations in India.* CICFRI. Bulletin No. 12.

Joly, A. B. (1954) The genus *Bostrychia* Montagne, 1838, in northern Brazil. Taxonomic and ecological data, *Bot. Fac. Filos., liências e Letras da Univ. da São Paulo,* 173 (Bot. 17) : 53-74, 4 pls.

Joly, A. B. (1957) *Contribuiçãao as conhecimento da flore ficológica manniha da Ba la de Santos e Arredores.* Thesis. 196 pp. Map, 19 pls. Sâo Paulo.

Jones, J. R. & M. V. Hoyer (1982) Sportfish harvest predicted by summer chlorophyll a concentration in mid western lakes and reservoirs. *Trans. Am. Fish. Soc.* III : 176-179.

Jones, W. (1798) A catalogue of Indian plants comprehending their Sanskrit and as many of their Linnaen generic names. *Asiatic Res.*, 4: 234 - 323.

Jorgensen, E. G. (1977) Photosynthesis. In : *Biology of Diatoms* (ed. D. Werner) pp. 150 - 168.

Kannan, L. & K. Krishnamurthy (1985) Nutrients and their impact on phytoplankton. In : *Marine Plants.* Paper presented at the *All India Symposium on marine Plants, their Biology, Chemistry and Utilization.* Dona Paula - Goa, Oct. 30 – Nov. 1, 1983, Krishnamurthy, V. & A. G. Untawale (eds.). pp. 73-78.

Kant, Shashi & V. K. Anand (1978) New records of *Scenedesmus* spp. from Jammu (J & K). Phykos 17(1&2) : 81- 85.

Kargupta, A. N. & P. Sarma (1987) Rare species of *Bulbochaete* (Chlorophyceae : Oedogoniales) from Birbhum and Murshidabad districts of West Bengal, India. *Hydrobiologia,* 153 : 13-21.

Kargupta, A. N. & P. Sarma (1991) New records of *Coleochaete* species from West Bengal, India. *Bangladesh J. Bot.,* 20.

Kargupta, A. N., P. Sarma & V. Jha (1986) New record of Chaetophoralean alga *Leptosiropsis* from eastern India. *Envir. Ecology,* 4(1) : 179 - 180.

Karim, M. A. & A. K. M. N. Islam (1976) Eutrophication of the surface water of some lakes studies at Dhaka (Bangladesh). *Proc. Inter. Symp. on Eutrophication and Rehabilitation of Surface Waters,* **Eutrosym III.** 200-208.

Kendrick, A. & C. Ratledge (1992) Lipids of selected moulds grown for production of n-3 & n-6 polyunsaturated fatty acids. *Lipids.* 27 : 15 - 20.

Khan, M. (1978) On *Thorea* Bory (Nemalionales, Rhodophyta), a freshwater red alga new to India. *Phykos* 17(1&2) : 55-58.

Khan, M. R. & A. K. M. N. Islam (1988) Two new records of green algae from Bangladesh. *Bangl. J. Bot.* 17(2) : 167-171.

Khan, M. R. & A. K. M. N. Islam (1993) Ecology of the Chaetophoralean algae of Bangladesh. *J. Asiatic Soc. Bangladesh (Sci.)* 19 : 145-153.

Khan, M. R. & A. K. M. N. Islam (1993) New records of green algae from Bangladesh. I. *Aphanochaete* and *Chaetosphaeridium. J. Asiatic Soc. Bangl. (Sci.)* 19(1) : 49-54.

Khan, M.A., K.P.Srivastava, R.K.Dwivedi, D.N.Singh, R.K.Tyagi & S.N.Mehrotra (1990) Significance of ecological parameters in fisheries management of a newly impounded reservoir–Bachhra Reservoir. In : *Contribution to the Fisheries of Inland Open Water Systems in India* (eds. A.G. Jhingran, V.K. Unnithan & A. Ghosh). IFSI, pp. 100-108.

Khondker, M. & M. Dokulil (1987), Contribution to the systematics of epipelic algae in a shallow lake, Neusiedlersee (Austria). *Bangladesh J. Bot.* 16(2) : 181 – 186.

Khondker, M., A. K. M. N. Inslam, Z. N. T. Begum & S. Haque (1990), Limnological studies of four polluted ponds in and around Dhaka city with reference to indicator species. *Bang. J. Bot.* 19(1) : 51-63.

Khondker, M., A. K. M. N. Islam & R. Islam (1988) Studies on the primary productivity of Dhanmondi lake. *Dhaka Univ. Stud.* Part. E 3(1) : 15 - 21.

Khonker, M. & M. Dokulil (1988) Seasonality, bio-mass and primary productivity of epipelic algae in a shallow lake (Neusiedlersee, Austria). *Acta Hydrochim et hydrobiol.*

Kjellman, F. R (1890) *Handbk. Skandinaviens Hafsalgflora, Fucoidea;* 1-03. Stockholm : O. M. Lamm.

Kleiman, R (1988) *Proceedings World Conference on Biotechnology for the Fat and Oil Industry,* edited by T. H. Applewhite. American Oil Chemist's Society, p. 73.

Knoechel, R. & J. Kalff (1978) An *in situ* study of the productivity and population dynamics of fine freshwater plankton diatom species. *Limnol. Oceangr.,* 23 : 195-218.

Krieger, W. & J. Gerloff (1969) Die Gatting *Cosmarium.* J. Cramer, Weinheim, pp. 410, pls. 71.

Kumar, H. D. (1999) Introductory Phycology. Affiliated East - West Press Private Ltd., New Delhi, pp. 651.

Kundu, B. C. (1929) Studies of the Charophytes of Bengal. *Proc. India Sci. Congr.,* p. 248.

Kundu, B. C. (1934) Charophytes of Bengal, II. Charophyte notes from the district of Rajshahi, Bengal. *Proc. Indian Sci. Congr.,* : 293-294.

Kundu, B. C. (1935) Charophyte notes from the district of Dinajpur. *Proc. India Sci. Congr.,* 247.

Kundu, B. C. (1937) A new *Nitella* from Rajshahi, Bengal. *J. Indian Bot. Soc.,* 16 (4) : 223 - 226.

Kundu, B. C. (1941) Two new *Nitellas. J. Bombay Nat. Hist. Soc.,* 12 : 843-846.

Kundu, B. C. (1959) A note on the occurrence of *Nitella transluscens* Ag. in Rajshahi, East Pakistan. *Sci. & Cult.,* 25 : 385-386.

Kutzing, F. T. (1859) *Tabulae Phycologicae, oder abbildungen der Tange.* 9:viii+42, 100 pls., Nordhausen.

Kuzmickzeva, V. I. (1965) Pervichnoe deistvie mineralnykh udobrenii v vyrostnykh prudakh. *Trudy vses.nauchno-issled. Inst. prud. kyb. Khoz.,* 13 : 131-138.

Lakshminarayana, S. (1976) A note on *Sirocladium vandalurense* Randh. from Andhra Pradesh. *Phykos* 17(1&2) : 49-50.

Lal, Babu (1990) Impact of Farakka Barrage on the hydrological changes and productivity potential of Hooghly estuary. *J. Inland Fish. Soc. India,* 22(1&2) : 38-42.

Laloraya, V. K. & A. K. Mitra (1971) Some new *Anabaena* from paddy fields of India. *Phykos* 10(1&2) : 118-126.

Lamourox, J. V. F. (1809) Observations sur la physiologie des Algues Marinas, et description de cing nouveaux generes de cette famille. Nouv. *Bull. Sci. Soc. Philom.* 1 : 330-33.

Lanjouw, J. (1956) *International Code of Botanical Nomenclature,* (ed.) Utrecht, Netherlands.

Lewin, J. & R. E. Norris (1970) Surface zone diatoms of the coasts of Washington and New Zealand *(Chaetoceros armatum T. West & Asterionella spp.) Phycologia* 9(2) : 143-150.

Lewin, R. A. (ed.) (1962) *Physiology and Biochemistry of Algae.* Acad. Press. New York & London.

Lewis, Irene M. (1972) Studies on free and combined amino acid composition in some blue-green algae. *Phykos* 11(1&2) : 50-58.

Lowe, Rex L. (1974) Environmental requirements and pollution tolerance of freshwater diatoms. *Env. Monit. Ser.* EPA 670/4 - 74-005. Nat. Env. Res. Centre Res. & Dev. U. S. Env. Prot. Agency, Cincinnati, Ohio. 334 pp.

Lugo, A. E. & S. C. Snedaker (1974) The ecology of mangroves. *Annual Review Ecology and System.* 5 : 39-64.

Macintosh, D. J. (1984) Ecology and productivity of Malaysian mangrove crab population. pp. 354 - 377. In : E. Soepadma, A. N. Rao & D. J. Macintosh (eds.) *Proc. sian Symp. Mangrove Environment Res. Manage.*, Univ. of Malaya, Kualalumpur, Malaysia.

MacNae, W. (1968) A general account of flora and fauna of mangrove swamps and prospects in the Indo-West pacific region, *Adv. Mar. Biol.*, 6 : 73-270.

Maguira, B. & W. E. Neill (1971) Species and individual productivity in phytoplankton communities. *Ecology* 52 : 903-907.

Majumdar, S., S. C. Santra, Anjana Sengupta, K. K. Sengupta & K. Pati (1997) Biofouling of fresh water pearl mussels *(Lamellidens marginalis* Lamark) in West Bengal (India). *J. Inland Fish Soc., India,* 29(1) : 65-70.

Maity, H. & S. C. Santra (1985) Blue-green algal flora of 24 Parganas, West Bengal (India), *Phykos.* 24 : 46-51.

Manabe, E., H. Morio, T. Hiroyuki, I. D. Naoka, S. Koji & M. Tadashi (1994) *Applied Biochemistry and Biotechnology,* 34 35, 274.

Mandal, A. K. & N. C. Nandi (1989) Fauna of Sundarbans mangrove ecosystem, West Bengal, India, Fauna of Conservation areas, *Zoological Survey of India,* Calcutta, 3 : 1-116.

Mandal, A. K. & R. K. Ghosh (1989) *Sundarbans–A socio-biological study,* Bookland, Calcutta, pp. xv + 194.

Mandal, R. N. & K. R. Naskar (1994) Studies on the periphytic algae on the aerial roots of the mangroves of Sundarbans in West Bengal, In : *Environmental Pollution & Impact of Technology on Life,* Malabika Roy (ed.), Recent Researches in Ecology, Environment & Pollution, 9: 91-104.

Mandal, R. N. (1996) Morpho-anatomical studies of mangroves and halophytic algal flora of the Sundarbans in West Bengal (India) with special reference to their ecological adaptations. Ph. D. Thesis. University of Calcutta.

Mann, F. D. & T. D. Steinke (1989) Biological nitorgen fixation (acetylene reduction) associated with blue green algal (cyanobacterial) communities in the Beachwood Mangrove Nature Reserve. The effect of environmental factors on acetylene reduction activity. *S. Afr. J. Bot., S. Afr.* TYDSKR. *Plant Kd. vol.* 55(4) : 438-446.

Mannan, M. A., A. K. M. N. Islam & A. Aziz (1986) Use of blue-green algae as biofertilizers. I. A preliminary study with *Scytonema mirabile* D610 in pot culture. *Dhaka Univ. Stud.* E 1(2) : 157-160.

Manning, W. M., C. Juday & M. Wolf (1939) Photosynthesis of aquatic plants at different depths in Trout Lake, *Wisconsin, Trans. Wisconsis Acad.* 31 : 377-410.

Margalef, Ramon (1963) On certain unifying principles in ecology. *Amer. Nat.* 97 : 357-374.

Martens, G. (1870) Conspectus algarum Brasiliae haectenus detectarum. *Vidensk. Medd. Natur-hist. Foren. Kjøbenh.* iii. (18 - 20) : 297-314.

Mattos, A. De (1952) Notes sôbre algae do litoral Paranaense. *Arquivos do Museu Paranaense,* 9 : 245 - 260, pls. 16-36.

Mazé, H. & A. Schramm (1870-77) *Essai de classification des algues de la Guadeloupe.* xix + 283 + iii pp.

McIntire, C. D. & B. L. Wulff (1969) A laboratory method for the study of marine benthic diatoms. *Limnol. Oceanogr.* 14 : 667-678.

McIntire, C. D. & W. M. Moore (1977) Marine littoral diatoms - ecological considerations. In : *Biology of Diatoms* (ed. D. Werner) pp. 333-371.

Misra, J. N. (1960) The ecology, distribution and seasonal succession of the littoral algae on the west coast of India. *Proc. Symp. Alg. ICAR* : 187-203.

Misra, J. N. (1966) *Phaeophyceae in India.* ICAR Publication, New Delhi.

Misra, K. C. (1989) *Manual of Plant Ecology.* 3rd Edition. Oxford & IBH Publishing Co. Pvt. Ltd. pp. 491.

Möbius, M. (1889) Bearbeitung der von H. Schenck in Brasilien gesammelten Algen. *Hedwigia,* 28 : 309-347, pls. 10, 11.

Möbius, M. (1890) Bearbeitung der von H. Schenck in Brasilien gesammelten Algen. *Hedwigia,* 28 : 309 - 347, pls. 10, 11.

Möbius, M. (1892) Üben einige brasiliannische Algen. *Ber. D.d. Bot. Gesellsch.*, 10 : 17-26, pl. 1.

Möbius, M. (1895) Üben einige brasiliannische Algen. *Hedwigia,* 34 : 173-180, pl. 2.

Montagne, J.F.C. (1850). Cryptogamic Guayanensis, seu plant-arumcellularium in Guayana Gallica annis 1835 - 1849a cl. Leprieur Collect anum enumeratio universalis Note sur la station insolite de quelques florides dams les eaux douces et courantes des ruisseux des Montagnes de la Guayane. *Ann. Sci. Nat., Bot.*, iii, 14 : 283 309.

Montagne, J. F. C. (1863) Algae. pp. 1-102. In: *Ramon de la Sagra, Icones plantarum in flora Lubana descripptanum ex historia physica, politica et naturelli.* 64 pp.

Moore, R. T., P. C. Miller, D. Albright & L. L. Tieszen (1972) Comparitive gas exchange characteristics of three mangrove species during the winter. *Photosynthetica,* 6 : 387-393.

Moore, R. T., P. C. Miller, J. Ehleriiger & W. Lawrence (1973) Seasonal trends in gas exchange characterisitcs of three mangrove species. *Photosynthetica,* 7 : 387-394.

Morgan, J. P. & W. G. McIntire (1959) Quaternary Geology of the Bengal Basin, East Pakistan and Burma, *Bull. Geol. Sic. Am.*, 70 : 319-342.

Murray, G. (1889). Catalogue of the Marine Algae of the West Indian Regime. *Jour. Bot., Brit. & For.*, 27 : 237-242, 257-262, 298-305.

Nandi, N. C. & A. Misra (1987) Bibliography of the Indian Sundarbans with special reference to fauna, *Records of the Zoological Survey of India,* Occasional Paper No. 97, pp. 1 -135.

Nandi, N. C. & S. K. Pramanik (1994) *Crab and crab fisheries of Sundarbans,* Hindusthan Publishing Corporation (India), Delhi, pp. 1-192.

Nandi, N. C., A. K. Das & N. C. Sarkar (1993) Protozoa fauna of Sundarbans mangrove ecosystems, *Rec. Zool. Surv. of India,* 93 (1-2) : 83-101.

Nandy, A. C., M. M. Bagchi, S. K. Majumdar & B. Saha (1990) Observations on the phytoplankton populations of the Hooghly estuary. In : *Contribution to the Fish. of Inland Open Water Systems in India,* pp. 159-164.

Nasar, S. A. K. (1977) Investigation on the seasonal periodicity of zooplankton in a freshwater pond in Bhagalpur, India. *Acta hydrochem et hyrobiol.*, 5(6) : 577 - 584.

Naskar, K. R. (1981) *Floristic survey of the district 24 Parganas in West Bengal (India) with Special reference to the mangrove vegetation of the Sundarbans,* Ph. D. Thesis, University of Calcutta.

Naskar, K. R. (1983) Halophytes and their unique adaptations in the Sundarbans mangroves swamps. *J. Indian Soc. Coastal Agric. Soc.,* 1(2) : 91-105.

Naskar, K. R. (1985a) Population pressure on the tidal mangrove forest of the Sundarbans–Its cause and problems. *Monograph-4,* pp. 159-161. National Atlas and Thematic Mapping Organisation, Calcutta, India.

Naskar, K.R. (1985b) A short history and the present trends of brackish water fish culture in paddy fields of the Kulti-Minakhan areas of the Sundarbans in West Bengal. *J. Indian Soc. Coastal Agric. Res.* 3(2): 115-26.

Naskar, K. R. (1987) Mangrove ecology of the Sundarbans delta in West Bengal and its role in brakish water fisheries. *J. Aqua.,* 1(1) : 1-9.

Naskar, K. R. (1988) Economic potentialities of the tidal mangrove forests of Sundarbans in India. *J. Indian Soc. Coastal Agric. Res.,* 6(2) 149-157.

Naskar, K. R. (1990) *Aquatic and semi-aquatic plants of the lower Ganga delta,* Daya Publishing House, Delhi. pp. 1-402.

Naskar, K. R. (1993) *Plant wealth of the lower Ganga delta–an eco-taxonomical approach* , Daya Publishing House, Delhi, vols.2, pp. 1-810.

Naskar, K. R. (1999) Status of the mangroves in Indian Sundarbans–In the perspectives of India and world mangals. In: *Sundarbans Mangals* (ed. Guha Bakshi, D. N., Sanyal, P. & K. R. Naskar), pp 20-89.

Naskar, K. R. (1999) Mangrove ecosystems in the Indian sub-continent. In : *Sundarbans Mangals* (ed. Guha Bakshi, D.N., Sanyal, P. & K. R. Naskar). pp. 90-108.

Naskar, K. R. (1999) The Sundarbans mangrove forests in India and their ecological stresses. In : *Sundarbans Mangals* (ed. Guha Bakshi, D. N., Sanyal, P. & K. R. Naskar), pp. 235-250.

Naskar K. R. (1999) Ecodynamics and importance of mangroves in sustaining the coatal fisheries and aquatic biodiversity. In : *Sundarbans Mangals* (ed. Guha Bakshi, D. N., Sanyal, P. & K. R. Naskar), pp. 493-525.

Naskar, K. R. (2000) Ecology and environment of Indian Sundarbans in West Bengal, (in Bengal). Paschimbanga, Published by the Department of Information and Culture, Government of West Bengal.

Naskar, K. R. & D. N. Guha Bakshi (1982) Sundarbans–the world famous mangrove forests of the district 24 Parganas in West Bengal (India). *J. Econ. Tax. Bot.,* 3 : 883-18.

Naskar, K.R.& D.N. Gupta Bakshi (1983) A brief review on some less familiar plants of the Sundarbans. *J. Ecom. Tax. Botony,* 4(3):699-711.

Naskar, K. R. & N. M. Chakraborty (1984) Studies on the economic fauna from the Sundarbans in West Bengal. *J. Indian Soc. Coastal Agric. Res.,* 2(1): 17-25.

Naskar, K. R. & S. C. Santra (1985) A note on *Enteromorpha tubulosa* in brackish water mixed sewage fed fisheries from the Sundarbans in West Bengal. *J. Indian Soc. Coastal Agric. Res.,* 5(2) : 471-72.

Naskar, K. R. & D. N. Guha Bakshi (1987) *Mangrove Swamps of the Sundarbans-An Ecological Perspective,* Naya Prokash, Calcutta.

Naskar, K. R., D. N. Guha Bakshi & R. N. Mandal (1999) Mangrove forests of the Indian sub-continent–its reckless exploitation with special reference to Sundarbans. In : *Sundarbans Mangals* (ed. Guha Bakshi, D. N., Sanyal, P & K. R. Naskar). pp. 660-689.

Naskar, K. R. & R. N. Mandal (1999) *Ecology and Biodiversity of Indian Mangroves.* vol. I & II. Daya Publishing House, New Delhi.

Nath, D. & S. D. Tripathi (1997) Primary production of some semi-intensive fish ponds of West Bengal, India. *J. Inland Fish Soc. India,* 29(1): 59-64.

Oberdorfer, E. (1927) Lichtverhältnisse und Algen besiedlung in Bodensee. *Zeitschr. Bot.,* 20 : 465-568.

Odum, H. T. (1956) Primary productivity in flowing waters. *Limnol. Oceanogr.*, 1 : 102-117.

Odum, H. T. (1960) The role of tidal marshes in estuarine production. *N. Y. St. Conserv.*, 60 : 1-4.

Okamura, K. (1907) *Icones of Japanese Algae. Kazamashobo,* Tokyo, 2 : 1-191.

Paerl, H. W. (1988) Nuisance phytoplanktonic blooms in coastal, estuarine and inland waters, *Limnol. Oceangr.*, 33 : 823 - 847.

Pal, B. P., B. C. Kundu, V. S. Sundarilingam & G. S. Venkataraman (1962) *Charophyta.* ICAR Publication. pp. 130.

Pal, T. K. & S. C. Santra (1982) Contributions to the Cyanophyceae of Murshidabad. *Phykos,* 21 : 150-152.

Pal, T. K. & S. C. Santra (1984) New additions to the algal flora of Murshidabad, West Bengal. *Phykos,* 23(1&2) : 139-141.

Pal, U. C. (1990) *Floristic survey of algal flora of Midnapore (West Bengal) with particular reference to their utilization.* Ph. D. Thesis, Univ. of Kalyani pp. 584.

Pal, U. C. & S. C. Santra (1985) Algal flora of Midnapore, West Bengal (India), Cyanphyceae. *Phykos,* 24 : 12-17.

Pal, U. C. & S. C. Santra (1986) Leaf inhabiting blue - green algal flora of West Bengal. *Indian J. Bot.*, 9(2) : 175-176.

Pandey, U. C., G. L. Tiwari & D. C. Pandey (1983) Diatom flora of Allahabad, India II. *Bibliotheca Phycologica,* 66 : 127-139.

Parke, M. (1953) A preliminary check - list of the British marine algae. *J. Mar. Biol. Ass. U. K.*, 32 : 497-520.

Patel, R. J. & M. A. Francis (1969) Some interesting observations on *Compsopogon aeruginosus* (J. Ag.) Kutzing, a species new to India, Phykos, 8 (1&2) : 46-51.

Patel, R. J. (1971) Cytotaxonomical studies of British marine species of *Chaetomorpha*-I, *Chaetomorpha linum* Kutz., and *Chaetomorpha aerea* Kutz. *Phykos,* 10 (11&2) : 127-136.

Patrick, R. (1977) Ecology of freshwater diatoms - diatom communities. In: *Biology of Diatoms* (ed. D. Werner). pp. 284-332.

Philiposé, M. T. (1960) Freshwater phytoplankton of Inland Fisheries. *Proc. Symp. Algology, ICAR,* N.D. 272-291.

Pomeroy, L. R. (1959) Algal productivity in salt marshes of Georgia. *Limnol. Oceanogr.* 4 : 386-397.

Porter, K. G. (1973) Selective grazing and differential digestion of algae by zooplankton, *Nature,* 244: 179-180.

Post, E. (1936) System atische und pflanzengeographische Notizen zur *Bostrychia-Caloglossa*-Assoziation. *Rev. Algol.,* 9 : 1-84, 4 figs.

Prasad, B. N. & R. K. Mehrotra (1976) *Calothrix lucknowense* - a new species from paddy field soil. *Phykos* 15 (1&2): 69-73.

Prasad, B. N. & R. K. Mehrotra (1976) *Nostoc citrisporum* sp. nov. from paddy field soil. *Phykos,* 15(1&2) : 65- 67.

Prasad, B. N., Shipra Dutta & V. Jain (1973) Cytological studies in the genus *Rhizoclonium. Phykos,* 12(1&2) : 106-113.

Prasad, R. R. (1963) Study of primary production and its importance in an integrated fisheries research programme. *Fish Tech.* 1 : 1-4.

Prescott, G. W. (1969) The algae : A review. Hughtton Mifflin Co., Boston.

Qasim, S. Z., P.M.A. Bhattathiri & S. A. H. Abidi (1968), Solar radiation and its penetration in a tropical estuary. *J. exp. mar. Biol. Ecol.,* 2(1) : 87-103.

Questel, A. (1942) Marine Algae, p. 69. In : *The flora of the Island of St. Bartholomew and its origin.* vii + 224 pp. Basse - Terre, Guageloupe.

Rai, L. C., J. P. Gaur & H. D. Kumar (1981) Phycology and heavy metal pollution, *Biol. Rev.,* 56 : 99-151.

Randhawa, M. S. (1959) *Zygnemaceae* ICAR Publication, New Delhi.

Rao, Rama, K., P. C. Thomas & K. Subbaramaiah (1976) Fruiting behaviour of *Gelidiella acerosa* (Forsskal) Feldmann et Hamel, the perennial agarophyte. *Phykos,* 15(1&2) : 35-37.

Rao, S. P. (1970) Systematics of Indian Gelidiales. *Phykos,* 9 (2) : 63-78 + plates.

Rao, S. P. (1974) Studies on *Gelidiella acerosa* (Forsskal) Feldmann et Hamel II. Growth and phenological events in the annual life of alga. *Phykos,* 13(2) : 7-15.

Rao, Sreenivasa P. (1976) Studies on *Gelidiella acerosa* (Forsskal) Feldmann et Hamel III, regeneration of *Gelidiella* at Jaleshwar reef, Veraval. *Phykos,* 15(1&2) : 23-27.

Rao, U. M. (1974) Addition to the algal flora of the Gulf of Mannar and Palk Bay from Mandapam area - II. *Phykos* 13(2) : 56-59.

Reynolds, C. S. & A. E. Walsby (1975) Water blooms, *Biol. Rev.,* 50 : 437-481.

Rines, J. E. B. & P. E. Hargraves (1988) The *Chaetoceros* Ehrenberg (Bacillariophyceae) flora of Narragansett Bay, Rhode Island, U.S.A. *Bibliotheca Phycologica,* 79 : 1-196.

Round, F. E. & C. M. Happy (1965) Persistent vertical migration rhythm of the epipelic diatom association in non-tidal waters. *Brit. Phycol. Bull.,* 2 : 463-471.

Round, F. E. & J. W. Eaton (1966) Persistent vertical migration rhythm in benthic microflora. III. The rhythms of epipelic algae in a fresh - water pond. *J. Ecol.,* 54 : 609-615.

Round, F. E. (1960) Studies on bottom living algae in some lakes of the English lake District. Part IV. The seasonal cycles of Bacillariophyceae. J. Ecol. 48 : 529-547.

Round, F. E. (1965) *The Biology of Algae*. 269 pp. St. Martin's Press, New York.

Round, F. E. (1972) The formation of girdle, intercalary bands and septa in diatoms. *Nova Hedwigia.* 23(2+3) : 449-464.

Round, F. E., R. M. Crawford & D. G. Mann (1990) *The Diatoms, Biology and Morphology of the Genera.* pp 747. Cambridge University Press., Cambridge.

Roxburgh, W. (1832) *Flora Indica.* 3 : 1-375. Serampore. Thacker & Co.

Ryder, R.A. (1965) A method for estimating the fish production of north temperate lakes. *Trans Am. Fish. Soc.,* 94 : 214 - 218.

Sadasivam, S. & A. Manickam (1991) *Biochemical Methods for Agricultural Sciences.* Wiley Eastern Limited & Tamil Nadu Agricultural University.

Saenger, P., M. M. Specht, R. L. Specht & V. J. Chapman (1972) Mangal and coastal salt - marsh communities in Australasia. In : *Ecosystems of the World I. Wet Coastal Ecosystems,* Ed. V. J. Chapman. Elsevier, Amsterdam. 428 pp.

Saenger, P., E. J. Hegere & J. D. S. Davie (1983) Global status of mangrove ecosystem, IUCN and Natural Resources, Switzerland.

Saha, S. B., B. B. Ghosh & V. Gopalakrishnan (1971) Plankton of the Hooghly estuary with special reference to salinity and temperature. In : *Symposium on Indian Ocean and Adjacent Seas,* The Marine Biological Association of India. 12 - 18 Jan. 1971.

Saha, S. B., B. B. Ghosh & V. Gopalakrishnan (1975) Plankton of the Hooghly estuary with special reference to salinity and temperature, *J. Mar. Biol. Asso. India,* 17(1) : 107-120.

Salam, A. A. M. & Y. S. A. Khan (1978) Algal flora of the Karnafully estuary. *Phykos* 17(1&2) : 99-103.

Santra, S. C. (1984) Algal flora of Sikkim Himalaya. I. A contribution to our knowledge of the blue - green algae of north - east Sikkim (India). *J. Econ. Tax. Bot.* 5(5) :1209-1219.

Santra, S. C. & T. K. Adhya (1973) Zygnemataceae of eastern Himalayas. *Spirogyra. Bull. Bot. Surv. India,* 15(3&4) : 281-282.

Santra, S. C. & T. K. Adhya (1976a) Zygnemaceae of eastern Himalays. II. *Spirogyra. Bull. Bot. Soc. Bengal.* 30 : 47-49.

Santra, S. C. & T. K. Adhya (1976b) Vaucheriaceae of eastern Himalayas (India). *Nova Hedwigia* 27 : 655-659.

Santra, S. C. & T. K. Adhya (1977) *Sirocladium himalayensis,* a new species from India. Phykos. 16 (1 &2) : 65-68.

Sarma, P. (1971) The genus *Cladophorella* Fritsch newly found in India. *Proc. 58th India Sci. Congr.,* Pt. III : 387.

Sarma, Y. S. R. K. (1970) Contributions to the karyology of the Ulotrichales. V. *Enteromorpha Link. Phykos.* 9(1) : 29-35.

Sarma, Y. S. R. K., R. B. Chandarana, B. K. Ramavat & S. J. Tarwadi (1984) Utilization of sea weed for liquid fertilizer energy and poultry feed, In: *Proceedings Seminar on Sun, Sea and Shore,* 90-91.

Sarode, P. T. & N. D. Kamat (1983) The diatom flora of Vidarbh, India - I. *Bibliotheca Phycologica,* 66 : 259-319.

Schuster, W. H. (1952) Fish culture in brackish water ponds of Java. I. P. F. C. *Spl. Publ.,* (2).

Schütt, F. (1895) Arten von Chaetoceros und Peragallia. Ein *Beitrag zur Hochseeflora* Berichte *der deutschen botanischen Geoellschaft,* 13: 35-48.

Scott, A. M. & G. W. Prescott (1961) Indonesian Desmids. *Hydrobiologia* 17 : 1-132, pls. 63.

Scott, A. M. & R. Groenblad (1957) New and interesting desmids from the southern United States. *Acta Soc. Sci. Fennicae, n. s.* B 11(8) : 1- 62, pls. 37.

Setchell, W. A. & N. L. Gardner (1925) The marine algae of the Pacific Coast of North America. III. Melanophyceae. *Univ. Calif. Publ. Bot.,* 8 : 383-898, 74 pls.

Seto, Akira, H. L. Wang & C. W. Hesseltine (1984), Culture conditions affect Eicosapentaenoic acid content of Chlorella minutissima. *J. Am. Oil Chem. Soc.,* 61(5) : 892-894.

Shannon, C. E. & W. Weaver (1963) *The Mathematical Theory of Communication,* Univ. III. Press, 117 pp.

Sherman, J. B. & H. K. Phinney (1971) Benthic algal communities of the Metoluis River. J. *Phycol,* 7 : 269-273.

Shetty, H. P. C., S. B. Saha & B. B. Ghosh (1961) Observations on the distribution and fluctuations of the plankton in the Hooghly - Matla estuarine systems with notes on their relation to commercial fish landings. *Indian J. Fish.,* 8(2) : 326-363.

Simopoulos. A. P. (1991) Omega-3 fatty acids in health and disease and growth and development. *Am. J. Cli. Nutr.,* 544:438-463.

Singh, P. K. (1973) Occurrence of green algae Pithophora sp. and Hydrodictyon reticulatum as weed in rice fields of Cuttack. *Phykos,* 12(&2) : 82-85.

Singh, Shivaji & Y. B. K. Chowdhury (1974) Some observations on the cytology of *Halicystic boergesenii* Iyengar and Ramanathan. *Phykos*, 13(1) : 90-93.

Singh, V. P., P. N. Saxena, A. Tewari & M. A. Khan (1970) Studies on the seasonal variation of algal flora of sewage. *Phykos* 9(2) : 57-62.

Sinha, B. D. & N. K. Srivastava (1978) *Comsopogon iyengarii* Krishnamurthy var. nov. a new record for Bihar. *Phykos*, 17(1&2) : 93-96.

Sinha, J. P. & S. Pandey (1972) A new myxophyceae from Chotanagpur, Bihar-*Cylindrospermum gloeotrchioides* sp. nov. *Phykos* 11(1 &2) : 78-79.

Sinha, M., D. K. De & B. C. Jha (1998) The Ganga-Environment and Fishery, CIFRI, Barrackpore, 142 pp.

Sinha, R. K., Ratna Sinha & P. N. Mehrotra (1993) Species diversity indices and related habitat characteristics. *Poll. Res.*, 12(4) : 207-212.

Smith, G. M. (1950) *The Fresh-Water Algae of the United States.* McGraw Hill Book Company, Inc. 1-719.

Smith, G. M. (1958) *Cryptogamic Botany-Algae & Fungi.*, vol. 1. Tata McGraw - Hill Publishing Company Limited, New Delhi, pp. 165-216.

Snedekar, S. C. & J. G. Snedekar (eds.) (1984) *The mangrove ecosystem. Research Methods.* Paris-France UNESCO. 265 pp. 8 volumes.

Sreenivasa Rao, P. & Sudha R. Kale (1969). Marine algae from little known places of Gujarat coast-I : Algae from Gopnath. Phykos 8 (1&2) : 71-82.

Srinivasan, K. S. (1946) Ecology and seasonal succession of the marine algae at Mahabalipuram (Seven Pagodas) near Madras. *J. Indian bot. Soc., Iyengar Commem. vol.* : 267-78.

Staub, R., J. W. Appling, A. M. Hofsteiler & I. J. Hass (1970) The effect of industrial waste of Memphis and Shelby county on primary plankton producers. *Bioscience*, 20 : 905-912.

Stein, J. R. (1973) Handbook of phycological methods. vol. I. Cambridge University Press. 448 pp.

Stein, J. R. & C. A. Borden (1984) Causative and beneficial algae in human disease conditions : a review, *Phycologia*, 23 : 485 - 501.

Stevenson, R. J. & E. F. Stoermer (1981) Differences between benthic algal communities along a depth gradient in Lake Michigan. *J. Phycol.*, 17: 29-36.

Stewart, W. D. P. (ed.) (1974) Algal Physiology and Biochemistry. Blackwell Scientific Publications. Oxford.

Steyaert, J. & E. M. Bailleux (1978) *The structure of Eucampia belaustium* Castr. as revealed by the Stereo Scanning Electron Microscope. *Nova Hedwigia*. XXVI : 195-204.

Subhrahmanyan, R. (1946). A systematic account of the marine plankton diatoms of Madras coasts. *Proc. Ind. Acad. Sci.*, 24B : 85-197.

Subrahmanyan, A. (1973) Two new species of *Camptylonemopsis*. *Phykos*, 12(1&2): 114-116.

Subrahmanyan, R. (1946) A systematic account of the marine plankton diatoms of the Madras *Coast. Proc. Ind. Acad. Sci.*, 24B : 85-197.

Subrahmanyan, R. (1958) Phytoplankton organisms of the Arabian Sea off the west coast of India. *J. Ind. Bot. Soc.*, XXVII(4) : 435-441.

Subrahmanyan, R. (1960) Observations on the effect of monsoon in the production of phytoplankton. *J. Ind. Bot. Soc.*, XXXIX(1) : 78-89.

Subrahmanyan, R. (1965) Studies on the plankton of the east coast of India, Part II. *Proc. Ind. Acad. Sci.*, LXI : 12 - 24.

Subrahmanyan, R. (1968) The Dinophyceae of the Indian Sea I & II. *Mar. Biol. Ass.*, India.

Subramanyam, K. (1977) History of botanical exploration in India. *Bull. Bot. Surv. India 19* (1-4) : 1-4.

Suxena, M. R. & V. Venkateswarlu (1970). Desmids of Andhra Pradesh-V. Mostly from Warangal and its environs. *Phykos* 9(1) : 16-28.

Suxena, M. R. (1983) Algae from Kodaikanal Hill, South India, *Bibliotheca Phycologica* 66, 43-99.

Suzuki, O. (1998) Production of γ -linolenic acid by fungi and its industrialization. In : *Proceedings World Conference on*

Biotechnology for the Fats and Oils Industry, edited by T. H. Applewhite. American Oil Chemist's Society, Champaign, 110-116.

Tahmida Begum, Z. N., A. K. M. N. Islam & Santana Begum Chowdhury (1987) Studies on the chemical composition of three green algae. *Bangladesh J. Bot.*, 16(2) : 219-220.

Tang, Yun - An & Hwang, Ting-Lang (1967) Evaluation of the relative suitability of various groups of algae as food of milk fish in brackishwater ponds. *Proceedings of the World Symposium of Warm - Water Pond Fish Culture*, FAO, Rome, 365-372.

Taylor, W. R. (1928) The marine algae of Florida, with special reference to the dry Tortugas, *Carnegie Inst. Wash., Publ.*, 379; Papers from the Tortugas lab., 25 : v + 219. 3 figs, 37 pls.

Taylor, W. R. (1929) Notes on the algae from the tropical Atlantic Ocean (1). *Amer. Journ. Bot.*, 16 : 621-630, 13 figs., p. 162.

Taylor, W. R. (1931) A synopsis of the marine algae of Brazil. *Rev. Algol.* 5: 279-313.

Taylor, W. R. (1933) Notes on algae from the Tropical Atlantic Ocean, II. *Papers Mich. Acad. Sci., Arts & Lett.* 17 : 395-407, pl. 26.

Taylor, W. R. (1935) Marine algae from the Yucatan Peninsula. *Carnegie Inst. Wash. Publ.*, 461 : 115-124.

Taylor, W. R. (1940) Marine algae of the Smithsonian - Hartford Expedition to the West Indies. *Contr. U. S. Nat. Herb.*, 28 : 549-562, pl. 20.

Taylor, W. R. (1941) Notes on the marine algae of Texas. *Paper Michigan Acad.*, 26 : 69-79.

Taylor, W. R. (1942) *Carribean marine algae of the Allen Hancock Atlantic Exped.*, 2 : 193 pp., 20 pls. Los Angeles.

Taylor, W. R. (1943) Marine algae from Haiti collected by H. H. Barlett in 1941. *Papers Michigan Acad.*, 28 : 143-163 4 pls.

Taylor, W. R. (1954) Distribution of marine algae in the Gulf of Mexico. *Papers Michigan Acad.*, 39 : 85-109.

Taylor, W. R. (1960) *Marine algae of the eastern tropical & sub tropical coasts of the Americas* by William Randolph Taylor, University of Michigan Press, pp. 870.

Taylor, W. R. & C. H. Taylor (1929) The marine algae of the south-eastern penninsular of Hispaniola. *Amer. Journ. Bot.,* 16 : 651-62, 10 figs.

Tiffany, E. N. (1930) The Oedogoniaceae. *Monograph.* Columbus, Ohio. pp. 1 - 253.

Tiwari, G. L. & D. C. Pandey (1972) Observations on *Anabaena arnoldii* ptekarj. *Phykos.* 11(1&2) : 23 - 26.

Tiwari, G. L. & D. C. Pandey (1976) On the morphology and nitrogen fixation in *Nodularia harveyana* (Thwaites) Thuret, a new record from India. *Phykos* 15(1&2) : 59-63.

Tomlinson, P. B. (1986) *The Botany of Mangroves.* Cambridge University Press, Cambridge, London. New York, pp.1-414.

Turner, W. B. (1892) Freshwater algae of East India. *Kg. sv. vet. Acad. Handl.,* 25(5) : 1-187, pls. 23.

UNESCO (1986) *Mangroves of Asia and the Pacific. Status and Management.* Technical Report of the UNDP/ UNESCO Research and Training Pilot Programme on Mangrove Ecosystems in Asia and the Pacific (RAS/79/002).

Vareschi, E. (1977) Biomasse und Fressrate der Zwergflamingoes in Lake Nakuru (Kenia); *Verh. Dtsch. Zool. Ges. p. 41* (Abst.)

Vedajanani, K. & Y. S. R. K. Sarma (1978) Karyological studies on Indian Conjugales – IV : *Desmids. Phykos,* 17 (1&2) : 63-72.

Venkataraman, G. S. (1958) A contribution to the knowledge of the Diatomaceae of Kanyakumari, India – II. *Proc. Nat. Sci. India*–II. *Proc. Nat. Sci. India.* 24B(6) : 307-313.

Venkateswarlu, V. (1985) Taxonomy and ecology of algae in the river Moosi, Hyderabad, India, II. Bacillariophyceae. *Bibliotheca Phycologica,* 66: 1-41.

Verma, B. N. & M. P. Verma (1978) New record of *Dichotomosiphon tuberosus* (A. Br.) Ernst. var. indicum var. nov. from Bihar. *Phykos* 17(1&2) : 59 - 62.

Verma, B. N. & S. K. Verma (1976) A note on the cytology and life history of *Chaetomorpha gracilis* Kutz. *Phykos,* 15(1&2) : 89-94.

Vickers, A. (1905) Liste des algues marines de la Barbade. *Ann. Sci. Nat. Bot.*, ix, 1 : 45-66.

Wallich, G. C. (1860) On the siliceous organisms found in the digestive cavities of the salpae and their relation to the flint nodules of the chalk formation. *Transaction of the Microscopical Society New Series,* 8 : 36-55.

Watanabe, M. & A. K. M. N. Islam (1980) Freshwater algae from lake Akan. 4. *Japanese J. Phycol.*, 28(1) : 37-45.

Werner, D. (1977) Silicate metabolism. In : *Biology of Diatoms* (ed. D. Werner) pp. 65-109.

Werner, D. (1977) *The Biology of Diatoms.* Blackwell Scientific Publications, pp. 1-498.

Wetzel, R. G. (1964) A comparitive study of the primary productivity of higher plants, periphyton and phytoplankton in a large shallow lake. *Int. Rev. ges. Hydrobiol.* 49 : 1-61.

Whittaker, R. H. (1977) Evolution of species diversity in land communities. In: *Evolutionary Biology* (M. K. Hecht, W. C. Streeve & B Wallace eds.), vol. 10, Plenum Press, New York, pp. 1-55.

Wildeman, E. De. (1900) Les algues de la flora de Buitengorg. *Librairie et imprimerie.* E. J. Brill-Leide.

Wildenow, C. L. (1805) Du Genre nomme *Chara. Mem. Acad. roy. Berlin,* 1803 : 79-90.

Wilhum, J. L. & T. C. Dorris (1966) Species diversity in benthic macroinvertebrates in a stream receiving domestic and oil refinery effluents. *Amm. Midl. Nat.,* 76 : 427-449.

Williams, L. G. (1949) Seasonal alternation of marine floras at Cape Lookout, North Carolina, *Amer. Jour. Bot.,* 35(10) : 683 - 695, 20 figs.

Winkler, L. W. (1888) The determination of dissolved oxygen in water. *Berlin Deut. Chem. Ges.* 21 : 2843.

Wolny, P. (1967) Fertilization of warm - water fish ponds in Europe. In : *Proceedings of the World Symposium on Warm-Water Pond Fish Culture.* FAO Publication, Rome, pp. 64-81.

Wood, E. J. F. (1954) Dinoflagellates in the Australian regions. *Aust. J. Mar. Freshwat. Res.*, 5(2) : 171-351.

Yamachi, T., S. Tamaka, T. Yokochi, T. Nakahara & T. Hingashihara (1997) Production of high yield of Docosahexanoic Acid by *Schizochytrium* sp. strain SR 21. *J. Am. Oil Chem. Soc.*, 74(11):1431-1440.

Yamada, Y. (1950) A list of marine algae from Ryukyusho, Formosa. I. *Sci. Pap. Inst. Alg. Res. Pac. Sci.*, Hokkaido Imp. Univ., 3(1) : 31-45.

Yamaji, I. (1968) *The plankton of Japanese coastal waters.* Hoishuki, Osaka, Japan, 1-238.

Yongmanitchai, W. & O. P . Ward (1992) Growth and Eicosap entaenoic acid production by *Phaeodactylum tricornutum* in batch and continuous culture systems. *J. Am. Oil Chem. Soc.* 69(6) : 584-590.

Zanevald, J. S. (1939) *Nitella madagascarensis* nov. spec., with notes on the Charophyta of Madagascar. *Blumea,* 3(2) : 372-387.

Zeller, G. (1876) Algae Brazilienses circa Rio de Janeiro a Dr. Glezion horti publici directore, collectae. In : Warming, E. ed. *Symbolae ad floram Brasiliae centralis cognoscendam,* XXII. *Videnske. Medd. Naturrh. Foren.* København, 1875 : 426 - 432.

Ziboh, V. A., K. A. Cohen, C. N. Ellis, C. Miller, T. A. Hamilton, K. Kragsbelle, C. R. Hydrick & V. J. Voorhees (1986) Effects of dietary supplementation of fish oil on neutrophil and epidermal fatty acids : Modulation of clinical course of psoriatic subjects. *Arch. Dermatol.* 122 : 1277-1282.

Zoto, A., D. D. Dillon & H. E. Schlichting (1973) A rapid method for cleaning diatoms for taxonomical and ecological studies. *Phycologia,* 13 : 69-70.

Algal Species Index

Species (Cyanophyta)

Order	*Family*	*Name of Species*
Chroococcales	Chroococcaceae	*Microcystic litoralis*
		Microcytis benglensis
		Gloeocapsa decorticans
		Gloeocapsa punctata
		Gloeocapsa aeruginosa
		Gloeocapsa rupestris
		Gleocapsa kuetzingiana
		Aphanocapsa littoralis
		Aphnocapsa pulchra
		Aphanothece stagnina
		Merismopedia tenuissima
	Entophysalidaceae	*Johannesbaptistia pellucida*
Chamae-siphonales	Chamaesi-phonaceae	*Chamaesiphon curvatus*
		Dermocarpa leibleiniae
	Dermocarpaceae	*Dermocarpa hemisphaerica*
		Stichosiphon sansibaricus
Pleurocarpales	Hyellaceae	*Xenococcus chaetomorphae*
		Xenococcus cladophorae
Nostocales	Oscillatoriaceae	*Microcoleus chthonoplastes*
		Hydrocoleum lyngbyaceum
		Schizothrix lamyi
		Lyngbya birgei
		Lyngbya hieronymusii
		Lyngbya lutea
		Lyngbya majuscula
		Lyngbya confervoides
		Lyngbya semiplena
		Phormidium fragile
		Phormidium stagnina
		Oscillatoria chlorina
		Oscillatoria limosa
		Oscillatoria subbrevis
		Oscillatoria curviceps

Contd...

Order	*Family*	*Name of Species*
		Oscillatoria princeps
		Oscillatoria nigroviridis
		Oscillatoria tenuis
		Spirulina major
		Spirulina princeps
		Arthrospira platensis
	Nostocaceae	*Raphidiopsis curvata*
	Sub Family–	*Raphidiopsis indica*
	Anabaenae	*Anabaenopsis arnoldii*
		Anabaena doliolum
		Anabaena anomala
		Anabaena gelatinicola
		Nostoc punctiforme
		Nostoc linckia
	Scytonemataceae	*Scytonema hofmanni*
	Rivulariaceae	*Calothrix contarenii*
Stigonematales	Nostochopsidaceae	*Mastigocoleus testarum*

(Chlorophyta)

CHLOROPHYCEAE

Class, Order & Sub-order	*Family*	*Name of Species*
Volvocales	Volvocaceae	*Pandorina morum*
		Volvox sp.
Ulotrichales	Ulotrichaceae	*Uronema confervicola*
Ulvales	Ulvaceae	*Enteromorpha clathrata*
		Enteromorpha prolifera
		Enteromorpha compressa
		Enteromorpha intestinalis
		Ulva lactuca
		Ulva fasciata
		Ulva patengensis
Oedogoniales	Oedogoniaceae	*Oedogonium undulatum*
Cladophorales	Cladophoraceae	*Cladophora echinus*
		Cladophorella sundarbanensis
		Lola capillaries
		Lola implexa
		Lola tortuosa
		Chaetomorpha brachygona
		Chaetomorpha gracilis
		Chaetomorpha aerea
		Rhizoclonium grande
		Rhizoclonium riparim
		Rhizoclonium hookeri

Contd....

Class, Order Sub-order	*Family*	*Name of Species*
Zygenematales	Zygnemataceae	*Spirogyra dubia*
		Spirogyra setiformis
		Spirogyra ternata
Desmidiales	Desmidiaceae	*Closterium acutum*
		Cosmarium regenlli
		Cosmarium paucigranulatum
		Triploceras gracile
		Triplastrum abbreviatum
Chlorococcales	Hydrodictyaceae	*Pediastrum boryanum*
		Pediastrum tetras
		Pediastrum duplex
	Chlorellaceae	*Chlorella vulgaris*
	Radiococcaceae	*Radiococcus sp.*
	Scenedesmaceae	*Scenedesmus bijuga*
		Scenedesmus quadricauda
Siphonales	Codiaceae	*Boodleopsis sundarbanensis*
CHAROPHYCEAE		
CHRYSOPHYTA & PHAEOPHYTA		
CHRYSOPHYTA		
Xanthophyceoe	Heterosiphonales	*Voucheria prescoffi*
		Voucheria sp.
Bacillariophyceae	Melosiraceae	*Melosira moniliformis*
Biddulphiales		*Melosira sol*
Coscinodiscineae		*Stephanopyxis palmeriana*
	Thalassiosiraceae	*Cyclotella glomerata*
	Coscinodisaceae	*Coscinodiscus eccentricus*
		Coscinodiscus gigas
		Coscinodiscus granii
	Hemidiscaceae	*Hemidiscus cuneiformis*
	Leptocylindraceae	*Corethron hystrix*
Rhizosolenilineae	Rhizosoleniaceae	*Rhizosolenia imbricata*
		Rhizosolenia setigera
Biddulphineae	Hemiaulaceae	*Climacodium frauenfedianum*
	Chaetocerotaceae	*Chaetoceros curvisetus*
		Chaetoceros flexuosus
		Chaetoceros laciniosus
		Chaetoceros tenuissimus
		Chaetoceros subsecundus
		Bacteriastrum varians
		Bacteriastrum cosmosum
		Bacteriastrum delicatulum

Contd...

Class, Order Sub-order	Family	Name of Species
	Biddulphiaceae	*Biddulphia mobiliensis*
		Biddulphia sinensis
Fragilariineae	Fragilariaceae	*Asterionella japonica*
		Fragilaria vaucheriae
		Synedra ulna
	Diatomaceae	*Diatoma vulgare*
Bacillariineae	Achnanthaceae	*Achnanthes microcephala*
		Achnanthes minutissima
		Cocconeis placentula
Bacillariineae	Naviculaceae	*Navicula cryptocephala*
		Navicula radiosa
		Stauroneis phoenicenteron
		Pinnularia viridis
		Pleurosigma angulatum
		Gyrosigma acuminatum
		Gomphonema sphareophorum
		Anomoeoneis exilis
	Cymbellaceae	*Cymbella ehrenbergii*
		Amphora veneta
	Bacillariaceae	*Nitzschia obtusa*
		Nitzschia sublinearis
		Nitzschia acicularis
PHAEOPHYTA		
Ectocarpales	Encoeliaceae	*Colpomenia sinuosa*
Dictyotales	Dictyotaceae	*Dictyota ceylanica*

Class, Order Sub-order	Family	Name of Species
RHODOPHYTA		
Bangioideae		
Compsopogonales	Erythrotrichiaceae	*Compsopogon coeruleus*
Floridae		
Gigartinales	Rhabdoniaceae	*Catenella repens*
		Catenella impudica
		Catenellanipae
Ceramiales	Delesseriaceae	*Caloglossa leprieurii*
		Caloglossa adnata
	Rhodomelaceae	*Bostrychia radicans*
		Bostrychia tenella
		Polysiphonia mollis
		Polysiphonia denudata
		Pterosiphonia pinnata
		Herposiphonia dendroidea
	Dasyaceae	*Heterosiphonia sp.*
Gelidiales	Gelidiaceae	*Gelidiella acerosa*
		Gelidium pusillum